SMEs and Open Innovation:

Global Cases and Initiatives

Hakikur Rahman
University of Minho, Portugal

Isabel Ramos
University of Minho, Portugal

BUSINESS SCIENCE
Reference

Managing Director:	Lindsay Johnston
Senior Editorial Director:	Heather Probst
Book Production Manager:	Sean Woznicki
Development Manager:	Joel Gamon
Development Editor:	Michael Killian
Acquisitions Editor:	Erika Gallagher
Typesetter:	Jennifer Romanchak
Print Coordinator:	Jamie Snavely
Cover Design:	Nick Newcomer

Published in the United States of America by
Business Science Reference (an imprint of IGI Global)
701 E. Chocolate Avenue
Hershey PA 17033
Tel: 717-533-8845
Fax: 717-533-8661
E-mail: cust@igi-global.com
Web site: http://www.igi-global.com

Library of Congress Cataloging-in-Publication Data

SMEs and open innovation: global cases and initiatives / Hakikur Rahman and Isabel Ramos, editors.
p. cm.
Includes bibliographical references and index.
Summary: "This book unites knowledge on how SMEs can apply open innovation strategies to development by incorporating academic, entrepreneurial, institutional, research, and empirical cases"--Provided by publisher.
ISBN 978-1-61350-519-9 (hardcover) -- ISBN 978-1-61350-520-5 (ebook) -- ISBN 978-1-61350-521-2 (print & perpetual access) 1. Small business--Case studies. 2. Research, Industrial--Case studies. 3. Diffusion of innovations--Case studies. 4. Technological innovations--Management--Case studies. 5. Strategic alliances (Business)--Case studies. I. Rahman, Hakikur, 1957- II. Ramos, Isabel, 1966-
HD2341.S6236 2012
658.02'2--dc23
2011038370

British Cataloguing in Publication Data
A Cataloguing in Publication record for this book is available from the British Library.

All work contributed to this book is new, previously-unpublished material. The views expressed in this book are those of the authors, but not necessarily of the publisher.

Table of Contents

Foreword

It is widely recognized that small and medium enterprises (SMEs) play an important role in any country's economy. They are important contributors to the national GDP and also relevant creators of jobs. Innovative entrepreneurship is often presented as the main strength of SMEs, which compensates for a lack of resources and of advanced managerial skills. Although innovation might be present at the moment of creation of a SME, its sustained competitiveness demands for continuous pursuit of new products, services, or processes. However, due to their nature, SMEs have difficulties on establishing R&D units to address those innovation issues.

Open innovation is an alternative to organizational innovation approaches based on internal resources and efforts only. It is being widely accepted by large corporations, in many sectors of the economy, which recognize the benefits of opening their innovation processes and initiatives to the participation of external talent or to profit from ideas that have been excluded from internal utilization.

Many accounts on the use of open innovation can nowadays be found in the literature, either academic or commercial. However, such accounts seldom address SMEs. Although they might be viewed as natural candidates to the adoption of open innovation, SMEs seem to be absent from the exploitation of these novel and promising approaches to enterprise competitiveness.

The editors of this book created an opportunity for publication of reports of open innovation use in SMEs. At the same time the book addresses other important aspects related to SMEs and open innovation. One of such aspects is how to enable SMEs to use open innovation or how to access open innovation services. Another aspect is concerned with laying the basis for a research program on the subject.

The good response the call for the book had from researchers and open innovation practitioners from all over the world is a sign that open innovation in SMEs is currently an important topic that is worth fostering and investigation.

João Alvaro Carvalho
University of Minho, Portugal

***João Alvaro Carvalho** is Full Professor at Department of Information Systems, School of Engineering, University of Minho, Portugal. He holds a PhD in Information Systems from the UMIST (University of Manchester Institute of Science and technology), UK (1991). His research and teaching interests include: the foundations of information systems, information systems development, meta-modelling, requirements engineering, knowledge management, and R&D methods. He currently is the Director of the Doctoral Program on Information Systems and Technology at University of Minho.*

Preface

INTRODUCTION

Innovation has become a key success factor for the economic development of and a prerequisite for sustainable development in a country. In a complex and highly competitive global scenario, to compete and survive, enterprises have to innovate and develop commercially viable products and services faster than ever before. Furthermore, to meet these new challenges, enterprises must adopt new approaches to their innovation strategies and processes (Nørager, 2009). These include traditional concepts like R&D, but are increasingly inclined towards the concept of open innovation. Concepts like, technology exploitation or technology exploration, or more specifically, licensing in or licensing out, are becoming issues of common interest in the arena of modern entrepreneurships. Ranging from large multinational enterprises to corporate business houses to small firms, all are striving towards attempts to grab the market in ahead of others for economic gain, or value addition.

Enterprises consider innovations as a major driver to enhance their performance and to strengthen their competitive position in the market. In doing so, many firms have paid most of their management attention to a greater focus on internal efficiencies of the development process, team structures, decision making and cross functional interaction. However, as more and more companies are bringing innovation straight to the heart of their corporate strategies, developing internal innovation capabilities is no longer sufficient to gain and sustain competitive advantage over global competitors. Since innovation strategies look increasingly commoditized, more and more organizations try to further improve their innovation performance through intensifying collaboration across industry networks and partnerships, and opening up their innovation processes in line with the open innovation (Chesbrough, 2003, 2006; EIRMA, 2009).

However, while applying open innovation concepts and strategies to the cluster of enterprises, namely, small and medium enterprises (SMEs), situations do vary. Despite, being a major component in terms of contributions towards national economy or job creation, SMEs in many countries are lagging behind in adopting open innovation strategies for their business development. While searching for impediments along this route, researchers, academia, and practitioners have found almost similar reasoning, inclusive of scarce resources, skilled manpower, reduced product life cycle, and increasing global competitions (Rahman, 2010).

This book has tried to incorporate applications and researches on open innovation concepts and strategies that are being applied to the enterprise sector belonging to SMEs by accommodating theoretical perspectives, researches as practices, and case studies. The book has 16 chapters divided into four sections, namely concepts, researchers, and practices; innovation marketing, communication, and growth management; entrepreneurships, strategies, and crisis management; and creativity, collaboration, and co-creation.

Where the Book Stands

Accommodating academic and research concepts that are being applied in real-life cases illustrating open innovation strategies for SMEs development, this book has the intent of being a common platform for raising knowledge on entrepreneurship and innovation. Covering diverse aspects surrounding the open innovation concepts in terms of policy, politics, economy, and culture, including opportunities, challenges, and risks, this book sets forth a benchmark in this field of interest. Furthermore, by incorporating scientific, theoretical, academic, and research concepts, as well as empirical and success cases focusing those concepts, this book will act as a guideline or flagship for those who are active in the aspect of open innovation paradigm for SMEs development. By far the book has the opportunity to be taken as a catalyst of knowledge repository in the field of business entrepreneurship for economic empowerment in a country, whether developed, developing, transitional, or under-developed.

Target Audience

The very basic nomenclature of open and collaborative innovation in terms of entrepreneurship leads to the enormous extent of periphery of the target groups that one could think of. If one thinks of an entrepreneurship in a country, it may accompany issues and contexts of the entire community, ranging from the government or policy initiators; entrepreneurs and enterprises; researchers, practitioners, and academics; users and intermediaries; and the population at large. Hence, the potential audience of this book may comprise of all these stakeholders within an economy. Furthermore, on regional and global scenarios, this book will target development partners, international agencies, and their mediators.

Organization of the Book

The book has been divided into four sections; concepts, researches, and practices; innovation marketing, communication and growth management; entrepreneurships, strategies and crisis management; and creativity, collaboration and co-creation. As a new field of research and practices on entrepreneurship, hardly one section or another could be separated very distinguishingly, but for the sake of readership, the intent is so. However, the book comprises 16 fine pieces of manuscripts focusing business development aspects of the small and medium sized enterprises sector covering almost every corner of research and practices of open innovation paradigm.

Innovation has become a major driver of economic growth in a country through sustained entrepreneurships. Moreover, recently coined "open innovation" is increasingly taking a lead in enterprise management in terms of sustained profitability. Foci of researchers and practitioners are revolving around innovation methods, processes, and strategies. Chapter 1, as the introductory chapter to the book, finds out open innovation researches and practices that are being carried out circumscribing development of SMEs through a longitudinal study. The chapter investigated researches that are being carried out by leading researchers and research houses across the globe, and at the same time, it also searched for open innovation practices that are being carried out for the development of SMEs. One may argue about the characteristics of researches or practices on open innovation in relevance to the SMEs development, this chapter sets forth the initiation of a new dialogue or a new dimension of research.

Open innovation has gained popularity in recent years. But a question arises: is this concept new or does it express old realities? Chapter 2, while trying to find out an answer in this aspect, discovered that the term "open innovation" is recent, and that its development has been facilitated largely by technological innovations. As a case study it collected data over a period of eight months from an agrifood SME in Quebec, Canada, and found that components associated with the basic open innovation model such as intellectual property, joint R&D, and co-product development are present there.

As a case of a business development company in the field of ICTs, Chapter 3 explores, describes, and examines the story and circumstances of building and managing innovative business ideas in an Egyptian small business enterprise. The chapter addresses specific circumstances, opportunities, and challenges with which a small enterprise pursuing innovation strategies is faced, and analyzes how effective deployment of innovative business ideas and bringing them to the market has contributed to strategic competitiveness. The study used qualitative research strategy in connection with theory to gain full understanding of a natural setting, and as a comprehensive method of substantiating or un-substantiating the context of achieving strategic competitiveness through new innovations and technologies, and provided insights into the nature of management of innovations, from which lessons can be drawn for other organizations both in small business sector organizations and generally.

With recent developments of open innovation concept as a formal management discipline, many organizations are thriving to raise internal competencies that can be leveraged to generate measured success. Chapter 4 presented a model with simple, but elegant structure necessary for the design and implementation of a successful open innovation program. The chapter explores the leading causes of failure in a new open innovation program, and offers guidelines and criteria that open innovation leaders and practitioners can use to avoid these pitfalls, and to establish a program that generates tangible returns.

With the advent of innovative technologies and multi-dimensional researches along the setting of entrepreneurship development philosophy, context and business management have been transformed from being traditional to innovative. Depending on diversity and nature of the transformation, innovation has been shifted from closed peripheries to open dimension. Looking at the immense benefits, ranging from small entrepreneurs to corporate to multinational business houses, all are adopting various innovation techniques. However, Chapter 5 argues that there is a research gap within the context of SMEs development through open innovation strategies. The chapter initiates a research model for carrying out researches on the development of SMEs through utilization of open innovation strategies. To set the research model, the chapter has developed an innovation opportunity framework, both at the policy level and at the entrepreneur level.

As mentioned earlier, innovation is vital to sustain and advance current activities, as well as it can be vital to growing new businesses. In this aspect, the challenge for organizations operating in a global environment is to meet the evolution of the marketplace, social needs and the needs of society. Open innovation allows organizations to draw from the global pool of knowledge to design products and business modes that provide value while meeting social needs. Chapter 6 looks at several SMEs in South America that are working in the entrepreneurship acting as social innovators by drawing on a range of technologies to create products and services or to commercialize existing products in a new way in order to meet pressing social needs around the world.

Chapter 7 presents a model of open innovation as a collaborative effort of firms in Israel, mostly SMEs that are managed under a government support programme with the assistive involvement of academia. While normally industry-academia cooperation is R&D focused, the chapter presents a model of open innovation not involving dedicated R&D. It deals with the process of assimilating existing technologies and methodologies. The model focuses on search, identification, and implementation phases to improve competitiveness through open innovation strategies. The model claims that specifically for SMEs, innovation not related to R&D, and especially open innovation, is hard to deal with alone.

Chapter 8 investigates a mechanism of organizational innovation serving to make sense of a maturing university community involving educational executives, academic staff, and students in the occasion of a new campus development in Macau SAR. The case study is aimed to investigate from the perspective of a learning enterprise, a reflective way of forward thinking to record the author's observation and interpretation of what is entailed in this process of upbringing a relatively young university in this age-old city, famous for its rich heritage of East (Chinese) meeting West (Europeans – Portuguese). The chapter examines the accountability framework for undergraduate curriculum reform, and by treating the electronic transformation (e-transformation) as one of the open innovation strategies, it explores the e-transformation of the university environment, based on holistic concerns of the campus community. The study argues that in practice, the lessons learned behind the e-transformation of the learning enterprise should benefit all walks of the local community, including the community of the SMEs.

Chapter 9 investigates and analyzes the sources of innovation and success in food service industry of Finland. It also discusses of what kind of network management capabilities were needed in these SMEs cases. The chapter demonstrates that the SMEs have a good chance to succeed if they can harness the external knowledge in their start up process. The chapter thus contributes to the prior strategic management and business networks research by analyzing the selected SME cases in the food service industry.

Recent scholarly discussions on open innovation put forward the notion that an organization's ability to internalize external knowledge and learn from various other sources in undertaking new product development is crucial to its competitive performance. However, little attention has been given to growth-oriented small firms in identifying and exploiting entrepreneurial opportunities related to such development, in an open innovation context, from a social learning perspective. Chapter 10, based on an instrumental case-firm in UK, demonstrates analytically how learning as entrepreneurial action takes place, drawing on situated learning theory. The chapter argues that such learning is dynamic in nature and is founded on specific organizing principles that foster both inter- and intra-communal learning.

Open innovation in entrepreneurships finds its acceptance at all levels of the business industry for adding value to the business. The value could be in the form of economic gain or enhancement of knowledge leading to a sustained financial base. Open innovation adopts various strategies to accomplish the task for enhancing the value gain. Depending on size, nature, pattern, or characteristics of the firm, various strategies are being adopted by enterprises. In recent years, open innovation is also becoming increasingly adopted in SMEs and the trend is rapidly increasing. However, despite the potency of open innovation strategies, most of the enterprises are yet to find a sustained business model, especially for the SMEs working at the periphery of that value chain, which is the focus of Chapter 11. The chapter is trying to formulate a business model incorporating partnership approach from academia, research houses, intermediaries and other stakeholders to be applicable for SMEs.

However, the nature of the relationships between business and risk factors in one country or another does not fit exactly into a "model" nor does it have a pure placebo effect. But, models' simplicity may appeal to managers and regulators in understanding important business risks and crisis related phenomena. Backed by this idea, Chapter 12 underpins a comparative study on SMEs handling risk and crisis management according to a new tailored model of a Balanced Scorecard. This new model of a risk and crisis management aims at improving both SMEs management adaptation and performance across all crises' stages, something not attempted so far in the literature. The application of such a Balanced-Scorecard comes from the author's experience as a banker financing various SME industries, as a bank consultant on risk management, as well as from the results of a survey performed on a sample of 48 Romanian and Cypriot SMEs, selected from the area of trading, manufacturing, and services.

Open source communities such as the ones responsible for Linux and Apache became well known for producing, with volunteer labor innovating over an open communication channel, high-quality software that has been widely adopted by organizations. In the web server market, Apache has dominated in terms of share for over 15 years, outperforming other corporations and research institutions. The resource-based view (RBV) of firms posits that an organization outperforms its competitors because it has valuable, rare, imperfectly imitable, and non-substitutable resources. Accordingly, one concludes that Apache possesses such resources to sustain its competitive advantage. However, one does not know what those resources are. This chapter is an effort to locate them. While trying to find an answer for the question, "What resources enable Apache to outperform its for-profit competitors consistently?" this research draws on the RBV to develop a series of propositions. Additionally, for each proposition developed, methods for their empirical validation are proposed.

Chapter 14 describes a new sustainable business model, Innovation 2.0, led and deployed by the author with the goal of increasing the innovation level at the telco service portfolio for the end-user, with thanks to the convergence of Internet and Telco worlds and the fresh and effervescent proposals created by start-ups. The model, created by a Spanish company leader in marketing and strategy issues, under the premises of fairness for the different players, requires an open-minded view of the businesses and, in fact, is a model for open innovation in telcos, which traditionally tend to address innovation from a prism more reactive than proactive. The model proposed has been deployed in Telefonica with very good results, deploying 12 commercial services, which have generated more than 10 million € in income and 8 million of savings in investments. Innovation 2.0 is a sustainable model, finding a balance for all parties in a win-win basis and constitutes an appropriate model for the twenty-first century telcos.

The nature of innovation is rapidly changing, and the concepts of open innovation and crowdsourcing need to be embraced. Innovation has quickly become not only the edge on which competition can occur but rather the characteristic of existence, the means to basic survival. Several forces are driving the rapid change in innovation as it is understood today. These forces tend to have a technological impetus, the very impetus that open innovation and crowdsourcing are built on. SMEs are facing a triple blow in the light of the evolving face of innovation. Survival is the eminent challenge that must be overcome. Chapter 15 elucidates on the major challenges faced by SMEs in crowdsourcing and the hindrance they face exposing to open innovation situations.

France has long been associated with a state-directed 'dirigiste' model of linear R&D focused on large programmes, such as development of the TGV high-speed train. Contemporary France has, however, largely left the state-centered model behind, introducing a range of devices since the turn of the 21st century aimed at opening French innovation to international and cross-sectoral collaboration for increased productivity and national competitiveness. Chapter 16, as a case study, traces the opening of the French innovation system and the way one new academic, industrial, and government collaboration aims to make use of new features of the system to accelerate development of an eco-innovation cluster focused on cities of the future.

CONCLUSION

Entrepreneurship, a crucial element of economic expansion for a country, essentially plays the same role in all spheres of social and democratic processes among developed, developing, and transitional economies. Similarly, innovation is the term newly evolved and at the same time equally adopted by governments, scientists, academics, and entrepreneurs to progress further for exploration of new knowledge and ventures for economic and value gain. In this aspect, open innovation perspective of business development, especially for the SMEs, deserves scholarly attention from every corner of research dimensions. Contribution of similar researches and practices has already been acclaimed by leading countries in the arena of knowledge economy, and a book of this nature will certainly find its position among the communities of innovation dynamics.

Hakikur Rahman
University of Minho, Portugal

Isabel Ramos
University of Minho, Portugal

REFERENCES

Chesbrough, H. W. (2003). *Open innovation: The new imperative for creating and profiting from technology*. Boston, MA: Harvard Business School Press.

Chesbrough, H. W. (2006). *Open innovation: The new imperative for creating and profiting from technology*. Boston, MA: Harvard Business School Press.

EIRMA. (2009). *Joining forces in a world of open innovation: Guidelines for collaborative research and knowledge transfer between science and industry*. EIRMA, EUA, EARTO, ProTon Europe, version 1.1 October 2009.

Nørager, M. (2009). *How to manage SMEs through the transformation from non innovative to innovative?* PhD Thesis, Doctoral School of Organisation and Management Studies, Copenhagen Business School, Denmark.

Rahman, H. (2010). Open innovation: Opportunities and challenges for SMEs. In Cruz-Cunha, M. M., & Varajão, J. E. (Eds.), *E-business issues challenges and opportunities for SMEs: Driving competitiveness* (pp. 87–100). Hershey, PA: IGI Global. doi:10.4018/978-1-61692-880-3.ch006

Acknowledgment

We are indebted to the editorial staff of IGI Global for offering us the opportunity to edit this book focusing on theoretical and scientific aspects of open innovation regarding small and medium enterprises. We are also grateful to Editorial Advisory Board Members and all contributing Authors of this book for their support, cooperation, contribution and collaboration. Our indebtedness extends to all Reviewers, especially Matthew C. Heim, Amiram Porath, and Simona Mihai Yiannaki for their extended support during the tiring review process.

We would not have completed this work unless continuous support came from our family members.

Finally, our regards to the colleagues at the Department of Information Systems of the University, especially João Alvaro Carvalho without whom we could not able to complete this research publication.

Hakikur Rahman
University of Minho, Portugal

Isabel Ramos
University of Minho, Portugal

Section 1
Concepts, Researches and Practices

Chapter 1
Research and Practices on Open Innovation:
Perspectives on SMEs

Hakikur Rahman
University of Minho, Portugal

Isabel Ramos
University of Minho, Portugal

ABSTRACT

Innovation has become a recognized driver of economic prosperity of a country through sustained growth of its entrepreneurships. Moreover, recently coined term open innovation is increasingly taking a lead in enterprise management in terms of sustained profitability. Foci of researchers and practitioners are revolving around innovation methods, processes, and strategies. This chapter seeks to find out open innovation researches and practices that are being carried out circumscribing development of small and medium enterprises (SMEs) through a longitudinal study. Along this context the study is investigating into researches that are being carried out by leading researchers and research houses across the globe, and at the same time, it also investigates open innovation practices that are being carried out for the development of SMEs. Before its conclusion, the chapter attempts to develop a framework for future research practices.

INTRODUCTION

Innovation is no more just a research topics but became a significant driver for prosperity, growth and sustained profitability to global entrepreneurships). Furthermore, due to opening up the innovation processes and combining internally and externally developed technologies and strategies to create economic value the innovation has crossed the boundary of closed innovation to open innovation (Chesbrough 2003a; 2003b; Rahman & Ramos, 2010), which gained interest of researchers and practitioners favoring special

DOI: 10.4018/978-1-61350-519-9.ch001

issue of publication, dedicated conferences and growing literature (Fredberg, Elmquist & Ollila, 2008).

As mentioned above and also found in contemporary literatures that innovation is shifting from the closed and controlled environment of the corporate entrepreneurs towards more open and flexible model, based on cooperation and coordination among various parties. Knowledge and new technologies are no longer remaining sole properties of major corporations. In this aspect, SMEs are playing important role in networking and making innovation clusters in association with universities and research houses, being recognized as major driving forces in the open innovation paradigm (European Union, 2005).

SMEs also play a crucial role in raising investments in R&D and making countries more competitive, which is true for not only the European Union but also in other countries (European Union, 2005). Moreover, the majority of the transitional economies have acknowledged that SMEs are potential engine of economic growth and source of sustainable development, which are essential for industrial restructuring, new job creation, and income generation of the population at large (Koyuncugil & Ozgulbas, 2009). However, as this research has found and being supported by many researchers, utilization of open innovation strategies for the development of small and medium enterprises (SMEs[1]) remains low in terms of researches and practices (Chesbrough 2003a; 2003b; ; Lichtenthaler & Ernst, 2009; Lindermann, Valcareel, Schaarschmidt & Von Kortzfleisch, 2009; Van de Vrande, de Jong, Vanhaverbeke & de Rochemont, 2008; 2009; West, Vanhaverbeke & Chesbrough, 2006). This study has tried to synthesize researches and practices in this aspect by carrying out a longitudinal study. In doing so, a thorough literature review has been conducted emphasizing researches conducted by leading researchers and practitioners, though hardly these could be recognized as cent percent contribution towards SMEs growth. While investigating into the open innovation aspects of SMEs, the study covered characteristics of individual firms, group of firms or taken at regional or national contexts. The intention is to serve as a basis to commence researches and or practices along the route to enrich knowledge of small and medium scale entrepreneurs cluster among the various business clusters segmented by recognized agencies or institutes engaged in this field of study.

BACKGROUND

Innovative entrepreneurship is a function, accountability, or task which can be fulfilled by an individual alone or by teaming up with one or more partners, or with the support of a small firm or similar venture. Or in a collaborative platform, even a large firm can function as an entrepreneur. The main point here is that the entrepreneur is the one who brings together the necessary resources (financial, logistic, managerial and personal) that the innovation calls for. The entrepreneur is the one who finds the place of application and directs the execution of the change. Sometimes a long time passes before a promising invention is taken up by a true entrepreneur. Probably it may happen that an invention or discovery and an entrepreneur do not find each other. Fortunately in the realm of technology advancement, it is quite frequent that the match is made easily. However, in most cases the Schumpeterian entrepreneur drives the innovation process during the first realization of the revolutionary innovation. Furthermore, the process following the pioneering innovation (also known as diffusion), is also mostly driven by entrepreneurs and majority of the initiative appears at the beginning of the entrepreneurship sequence (Kornai, 2010). The entire process can be familiarized as a process of innovation.

Innovation is essential for enterprises and entrepreneurships, especially surviving the current economic conditions or planning sustainable growth relative to their competitors, locally and globally. Though not plenty, but tools exist to assist the entrepreneurs to measure their propensity to innovate and increase their capability for innovation or their innovation performance. However, this research finds that the situation is scanty for SMEs, as there is a lack of awareness and capacity which causes apprehension about innovation, open innovation, intellectual property and other strategies (Gassmann, 2006; Van de Vrande, de Jong, Vanhaverbeke & de Rochemont, 2009). Furthermore, amongst SMEs who have been subjected to related research there is persuasive evidence that innovation tends to be a domestic affair with more developments coming from existing resources than from outside sources (Bevis and Cole, 2010).

In this context, most of the researches on open innovation remain restricted towards targeting common stakeholders through major global entrepreneurs or their alliances. In addition, it is a fact that a few of those global business houses are controlling the entire market or system of open innovation development through process modification and or diversification of resources. Moreover, despite immense potentiality to reach out the stakeholders at the grass roots through open ended demand, diversity of product variation, and scale of economic capacity major contemporary researches are confined towards generic pattern-oriented clients (Rahman & Ramos, 2010).

But, the scenario is rapidly changing in the recent decade. Innovation is no longer remains within a vertically integrated company with everything in-house. With the advent of open innovation concept, open and flexible cooperation among business houses, research centers and universities is regarded as the most beneficial approach for business development. In this new business model different actors are applying their principles in addition to other partners through interactive participations to bring out an acceptable outcome for value gain (Chesbrough, 2003a; Maijers, Vokurka, van Uffelen, & Ravensbergen, 2005; Wijffels, 2009).

As mentioned earlier, SMEs are the global drivers of technological innovation and economic development and represent the deep, broad and fertile platform that nourishes, sustains and regenerates the global economic ecosystem (Kowalski, 2009). At the same time, to engage the open innovation strategies, R&D (a key innovation indicator) is increasingly being outsourced to lower the cost of production (Dehoff & Sehgal, 2006; 2008). Figure 1 shows that there has been a shift over time towards the ever increasing cooperation in the area of innovations among businesses themselves, and between businesses and knowledge institutions.

Apart from the cooperation, coordination and collaboration, open innovation strategies involve vigorous networking with partner companies; interaction with start-up ventures, public research institutes, universities and external suppliers; sharing and accessing of outside information and technology; IP management; knowledge management; creative entrepreneurship thinking; and above all to be global visionary (Kowalski, 2009). The current section serves as a broad background on the concept of open innovation that has been seen by this research along with other researchers. However, the research hypothesis is to find out contemporary researches and practices following a respectable search from a dependable repository. The study has taken various approaches in this aspect, but mainly depended on contents from the ScienceDirect, a concern of Elsevier B.V. Furthermore, due to the subscription status of the researchers own institute, it was easy to obtain cross reference materials easily from other subscribed sources. The main literature review has been conducted across issues of researches and practices, as mentioned earlier and they have been presented in the next section. The next section has been divided into two threads; one on researches that have been carried out by contemporary researchers in the field of open innovation for the development of SMEs, and the other one on practices that have been adopted by global leaders in the field of open innovation for the development of SMEs.

Figure 1. Changes of patterns in R&D outsourcing (adopted from EIRMA, 2005; 2009)

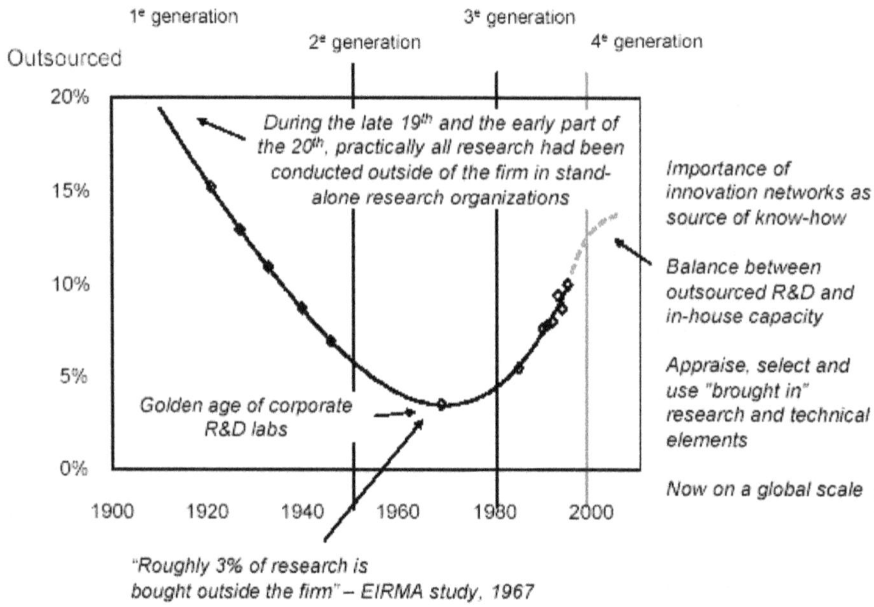

RESEARCHES AND PRACTICES: THE REVIEW

In the eyes of an open innovator on entrepreneurship development, economic prosperity is anticipated to result from exploiting innovation capacity, improving competitiveness, and enhancing productivity (BVCA, 2005). By far the open innovation and collaborative innovation are becoming the central topics in recent years surrounding business strategy and innovation (Huizingh, 2009) and open innovation is being claimed to be the new breed of innovation requiring enterprises to look beyond the boundaries of their organizations, thus using external and internal actors and knowledge for successful value creation (Thoben, 2008). However, as evident from this longitudinal study, open innovation has so far been adopted mainly in high-tech and multinational enterprises. Though open innovation has received increasing attention in the scientific research arena, but so far it has mainly been analyzed in corporate and high-tech multinational enterprises (MNEs) based on in-depth interviews and case studies (Chesbrough, 2003b; Kirschbaum, 2005; Lichtenthaler, 2010). Moreover, when looking for cases or examples, most of them focus on very specific industries, for example open source software (Henkel, 2006) or tabletop role-playing games (Lecocq & Demil, 2006) or crafts industries (Santisteban, 2006) or tourism (Novelli, Schmitz & Spencer, 2006; Hjalager, 2010). Even if a large sample of enterprises is being explored, the focus remains on specific issues rather than the full open innovation model (Van de Vrande, de Jong, Vanhaverbeke & de Rochemont, 2009). Perhaps, open innovation strategies depend on the very specific cases, applications and environments, and cannot be generalized, as such.

However, a few studies have demonstrated that open innovation also exists in smaller organizations (Van de Vrande, de Jong, Vanhaverbeke & de Rochemont, 2009) and this trend is increasing (Gassmann, Enkel & Chesbrough, 2010; Saarikoski, 2006; Dahlander & Gann, 2010; Fredberg, Elmquist & Ollila, 2008). To find out further about the researches and practices on open innovation for SMEs development, as mentioned before several searches were conducted among the contemporary researchers, their researches, research and research practices carried out by reputed organizations like OECD, European Union, European Commission and others in the field of open innovation, especially targeted for SMEs development. Following the search methodology of Saarikoski (2006:24) and supported by similar methodology on structured literature review of Fredberg, Elmquist and Ollila (2008:10), a search into the Internet with the search string 'open innovation AND SMEs development' (empirical setting of the research) was conducted and it yield 192,000 hits (though the string ["open innovation" AND "SMEs development"] resulted only 260 hits). These hits included contents on this aspect incorporating all those mentioned entities (researchers, researches, research organizations, and academia, national and international organizations). At the same time, Google scholar search for the same key words also yields 40,800 hits accommodating the mentioned entities. However, to keep the search less generalized and focused to specific search settings and foremost to have an overview on the contemporary research works including practices on open innovation, the search for this study was conducted on a content provider namely, ScienceDirect. In future as this research continues, similar search will be conducted on other content providers like, SCOPUS (subjected to be subscribed) and 'the ISI Web of Knowledge', or other subscribed entities utilizing additional search methodologies. This search has been carried out using the search formulae set_1 for researches on open innovation and search formula set-2 for practices on open innovation.

The Search Formulae set_1:

1. (open AND innovation for all fields and research AND SMEs for all fields) [All Sources(All Sciences)]
2. (open AND innovation for all fields and research AND SMEs for all fields) [All Sources(Business, Management and Accounting)]
3. (open AND innovation AND research for all fields and SMEs for all fields) [All Sources(Business, Management and Accounting)]
4. (open AND innovation for titles and research AND SMEs for all fields) [All Sources(Business, Management and Accounting)]

The search string (1) brings 1,694 counts; search string (2) brings 1,244 counts and search string (3) brings 1,079 counts; while search string (4) brings only 10 counts.

Search string (3) (with 1,079 counts) has been taken as the entry point of this longitudinal study and among them 940 were journal articles and 139 books. Table 1 shows their publication pattern considering major number of publication (here the minimum is 40) and Table 2 shows their years of publication (here the data has been given from 2000 till the date of the search, which is October 21, 2010). Noteworthy to mention that the search was conducted applying to all fields. But, when the search was modified with 'open innovation' in the title and 'research+smes' within the fields, the result returned only 10 journal articles and books.

The Search Formulae set_2:

5. (open AND innovation for all fields and practice AND SMEs for all fields) [All Sources(All Sciences)]
6. (open AND innovation AND practice for all fields and SMEs for all fields) [All Sources(All Sciences)]
7. (open AND innovation for all fields and practice AND SMEs for all fields) [All Sources(Business, Management and Accounting)]
8. (open AND innovation for titles and practice AND SMEs for all fields) [All

Table 1. Number of entries in different journals (with minimum count of 40)

Search string # 3 with minimum count of 40	Name of Journal	Number of entries
	Technovation	154
	Research Policy	144
	Technological Forecasting and Social Changes	50
	European Management Journal	45
	Industrial Marketing Management	41

Table 2. Number of publications in various years on OI researches for SMEs

Search string # 3	Year of publication	Number of publications
	2011	3
	2010	160
	2009	124
	2008	124
	2007	103
	2006	96
	2005	91
	2004	57
	2003	53
	2002 and earlier since 1991	268

(counts were taken since 2003 till the date of the search, which is October 21, 2010)

Sources(Business, Management and Accounting)]

Similarly, the search formulae set_2 gives the hit counts for the key words incorporating 'practices'. The search string (5) brings 1,447 counts; search string (6) brings 1,447 counts and search string (7) brings 937 counts; while the search result for string (8) returned only 8 journal articles and books for practices ('open innovation' in the title and 'practice+smes' within all fields). This evidently is confirming the low number of researches and practices in the area of open innovation, specifically for the SMEs. Furthermore, Table 3 shows their publication pattern considering major number of publication (here the minimum is 30) and table-4 shows their years of publication (here the data has been given from 2000 till the date of the search, which is October 21, 2010).

As evident from Tables 1 through 4 is that the trend of researches and practices using open innovation strategies for SMEs development is growing after the term 'open innovation' being coined by Prof. Henry Chesbrough. These tables also reveal that the trend of using OI strategies was there as number of articles before 2002 with the available data from the ScienceDirect is significant (see Table 2 and 4). However, the main purpose of this study is to find out thematic patterns or themes of researches and practices obtained from most relevant contents of these searches. To be more specific, the most relevant contents were separated from these search using search string (3) and (4). Also, the main notion

Table 3. Number of entries in different journals (with minimum count of 30)

Search string # 6 with minimum count of 30	Name of Journal	Number of entries
	Technovation	129
	Research Policy	117
	Technological Forecasting and Social Changes	46
	Industrial Marketing Management	40
	Journal of Business Venturing	39
	European Management Journal	38

of using ScienceDirect is to provide the first impression of freely available content without being subscribed, notwithstanding other arguments.

Another set of search was carried out among the publication of the leading researchers and practitioners in this field. Among them, the book, "Open Innovation: The New Imperative for Creating and Profiting from Technology" (Chesbrough, 2003a, being the most cited author on 'open innovation' (Fredberg, Elmquist & Ollila, 2008)); "Open Innovation: Practice, Trends, Motives and bottlenecks in the SMEs" (De Jong, 2006); "Open Innovation: Researching a new Paradigm" (Chesbrough, Vanhaverbeke & West, 2006); Journal articles written by Van de Vrande, de Jong, Vanhaverbeke & de Rochemont, 2008; 2009; De Jong, Vanhaverbeke, Kalvet & Chesbrough, 2008; Gassman, 2006; and Gasmann, Enkel & Chesbrough, 2010 were included. The first book was selected as the most cited book in this sector, the second article was selected as the most relevant search return each time made on search engines, and the rest were selected by the authors.

As a third check, contribution of forerunners in the concept of open innovation were also included in the categorizarion, such as Schumpeter (1934; 1942; 1950), Von Hippel (1986; 1988) and Davenport (1993a; 1993b). Noteworthy to mention that there were several others books (available in the references), but they are not being included here, as separate entities. And, unless they provide fundamental concepts on open innovation, literatures earlier than 2003 were less emphasized.

Table 4. Number of publications in various years on OI practices for SMEs

Search string # 6	Year of publication	Number of publications
	2011	2
	2010	142
	2009	111
	2008	113
	2007	91
	2006	85
	2005	79
	2004	48
	2003	47
	2002 and earlier since 1991	219

(counts were taken since 2003 till the date of the search, which is October 21, 2010)

Table 5. Synthesized research themes on OI carried out for/by the SMEs

Research themes	Literatures from the search
Conceptualization of open innovation	Chesbrough, 2003a; Lee, Park, Yoon & Park, 2010; Savioz & Blum, 2002; Amara, Landry, Becheikh & Ouimet, 2008
Establishment of research model	Chesbrough, 2003b; 2006; Raymond & St-Pierre, 2010; Edwards, Delbridge & Munday, 2005; Thorgren, Wincent & Örtqvist, 2009; Lawson, Longhurst & Ivey, 2006; Major & Cordey-Hayes, 2000; Rhee, Park & Lee, 2010
Development of business model	Chesbrough, 2006; De Jong, 2006; Partanen, Möller, Westerlund, Rajala & Rajala, 2008; Freel & De Jong, 2009; Cooke, 2005; Belussi, Sammarra & Sedita, 2010
Opportunities and challenges of open innovation	Groen & Linton, 2010; Knudsen & Mortensen, 2010; Hoffman, Parejo, Bessant & Perren, 1998; Tödtling & Trippl, 2005; Levy, Powell & Galliers, 1999; Del Brío & Junquera, 2003; Rahman & Ramos, 2010
Adoption of OI strategies and technologies	Van de Vrande, de Jong, Vanhaverbeke, & Rochemont, 2009; Zeng, Xie & Tam, 2010; Ferneley & Bell, 2006; Bougrain & Haudeville, 2002; Izushi, 2003; De Jong & Marsili, 2006; Lichtenthaler, 2010; Lee & Lan, 2011; Leiponen & Byma, 2009; Mention, 2010; Dickson, Weaver & Hoy, 2006; Hjalager, 2010; O'Regan, Ghobadian & Sims, 2006; Laforet, 2008; De Jong & Hippel, 2009
Measuring the impact of OI strategies	Massa & Testa, 2008; Woodhams & Lupton, 2009; Grupp & Schubert, 2010; Huang, Soutar & Brown 2004
Development of tools or instruments based on OI strategies	Kohn & Hüsig, 2006; Kaufmann & Tödtling, 2002

Finally, to avail the information about open innovation practices in SMEs, this study looked at various publications from international organizations like, OECD, European Union, and European Commission; individual organizations like, Vinnova, Vision Era-Net; portals like, OpenInnovation dot net; and articles from special issues from journals like, MIT Sloan Management Review, Technovation, Research Policy, and Harvard Business Review.

After several round of iterations, the following research themes and practice areas were taken into considereation for further exploration, pending further research impact on these themes and extended debate on their substances in relation to the development of smaller firms serving at the grass roots. Table 5 are showing the selected research themes and Table-6 are showing the practice areas, which are being discussed in the next sub-section in terms of their relevant in the mentioned references in these tables.

Researches and Researchers: Research Themes

Ranging from the conceptualization, establishment of research model to development of business model, adaptation of strategies, measurement of the impact and development of tools for use and dissemination of the strategies, this study emphasizes on seven distinct research themes. One may argue about this setting of the research theme, but this study observed that without the conceptualization of a process, it cannot be established, and similarly next stages are dependent on the previous stage, such as without learning about the opportunities and challenges behind the strategies, the methodologies cannot be adopted, and so forth. Another school of thoughts could be how much these themes are relevant to the SMEs. These researchers argue that adoption of open innovation strategies to small scale enterprises are yet to reach maturity in even developed nations or nations who are leaders in doing so, hence these themes are also need to be experimented at the

Table 6. Synthesized practice areas on OI carried out for/by the SMEs

Areas of practices	Literatures from the search
Venturing	Van de Vrande, de Jong, Vanhaverbeke & Rochemont, 2009; Kaivanto & Stoneman, 2007
Technology licensing	Bidault, 2004; Kollmer & Dowling, 2004: Grindley & Teece, 1997; Ziedonis, 2007; Lichtenthaler, 2007; Nagaoka & Kwon, 2006
Licensing of intellectual property (IP),	Van de Vrande, de Jong, Vanhaverbeke & Rochemont, 2009; Bianchi, Cavaliere, Chiaroni, Frattini & Chiesa, 2010; Lichtenthaler, 2010
Customer and supplier integration	Van de Vrande, de Jong, Vanhaverbeke & Rochemont, 2009; Van Hemel & Cramer, 2002; Bayraktar, Demirbag, Koh, Tatoglu & Zaim, 2009; Alshawi, Missi & Irani, 2010; Fernández-Viñé, Gómez-Navarro & Capuz-Rizo, 2010; De Jong & von Hippel, 2009
External networking and external participation	Van de Vrande, de Jong, Vanhaverbeke & Rochemont, 2009; Zeng, Xie & Tam, 2010; Mention, 2010; Tödtling & Trippl, 2005; Partanen, Möller, Westerlund, Rajala & Rajala, 2008; Van Hemel & Cramer, 2002; Spithoven, Clarysse & Knockaert, 2010; Jones & Macpherson, 2006

grass roots situation and call for extensive research, especially in marginal socio-economic situations.

Conceptualization of Open Innovation

Though the concept of open innovation was in the market for many years before the term was being popularized by Prof. Henry Chesbrough, this study finds many researchers are researching on modernizing the concepts of open innovation strategies or advancing the degree of novelty of innovation. With the adoption of OI strategies in the entrepreneurship researchers initiated extended studies to reach out to the market by breaking the boundary of the firm. Terms like, innovation merchants, innovation architects, innovation missionaries, innovation intermediaries, or outsourcing R&Ds, use of venture capital, technology intelligence, licensing management, intellectual property management started gaining their acceptance and popularity to fill an ever-existing gap between the producer and the user (Chesbrough, 2003a; Lee, Park, Yoon & Park, 2010; Savioz & Blum, 2002; Amara, Landry, Becheikh & Ouimet, 2008). Apart from the mentioned terms, 'Users-as-innovators', 'customers-as-innovators', 'suppliers-as-innovators' (von Hippel 1986; 1988), 'networked coordination' (Powell, 1990), 'co-opetition' (Brandeburger & Nalebuff 1996), 'communities of practice' (Wenger 1998), and the 'private-collective innovation model' (Von Hippel & Von Krogh 2003) are other concepts which have been developed to describe the changing modes of entrepreneurship coordination. At the same time the research dimension has also shifted from 'closed boundaries to networked paradigm' (Livieratos & Papoulias, 2009; Rahman & Ramos, 2010).

Establishment of Research Model

With the term 'open innovation' being popularized, majority of the researchers are trying to establish various research models, especially incorporating the role of external partners (SMEs, academia, research house, universities and intermediaries) to achieve improved performance and efficiency on product, process, service and organizational innovation. Along the path to the paradigm shift, as indicated by majority of the researchers, research models have tried to validate the role of internal R&D, effect of the firm's size, the link between R&D activities and innovation exposition, and various channels of open innovation thus mainstreaming open innovation research in the entrepreneurships through illustrating the conceptual arguments and conceptual frameworks incorporating various drivers of innovativeness (Chesbrough, 2003b; 2006; Raymond & St-Pierre, 2010; Edwards, Delbridge & Munday, 2005; Thorgren, Wincent & Örtqvist, 2009; Lawson, Longhurst & Ivey, 2006; Major & Cordey-Hayes, 2000; Rhee, Park & Lee, 2010).

Development of Business Model

Since the mid-1980s a new systemic model of innovation has emerged by incorporating a number of factors, such as externalities, transferability, modularity, network structure, and others, which are not included in the previously dominant linear model (Livieratos & Papoulias, 2009). In this respect, innovation is viewed as a systemic, path dependent and knowledge-centric social process influenced by the institutional environment (Chesbrough, 2003a). Livieratos and Papoulias (2009) argue that, within the open innovation model the innovation process becomes more complex and fragmented, actors are increasingly heterogeneous and more interdependent, and the period from conceptualization to commercialization is shorter. Livieratos and Papoulias (2009) further argue that, this model has created porous boundaries between the innovative company and its surrounding environment, changing the inter- and intra-organizational modes of coordination and triggering new answers to Coase's (1937) question as to, 'what determines the boundaries of the organization'.

While investigating the existence and the performance of an Open Regional Innovation System (ORIS model) characterized by the firms' adoption of an open innovation strategy Belussi, Sammarra and Sedita (2010) argue that in terms of adopting open innovation strategies, it overcomes not only the boundaries of the firms but also the boundaries of the region. Furthermore, Damaskopoulos and Evgeniou (2003) find that open innovation strategy based business models adopt frameworks comprising three interrelated levels of analysis, such as the level of the firm, the level of the market and industrial structures and the regulatory environment.

Literatures depicted adoption of framework or business model incorporating OI strategies, such as Triple Helix or ORIS, Open Business Model articulating value creation by emphasizing the role of social capital (Wang, Jaring & Wallin, 2009; West, 2006; Chesbrough, 2006; De Jong, 2006; Partanen, Möller, Westerlund, Rajala & Rajala, 2008; Freel & De Jong, 2009; Cooke, 2005; Belussi, Sammarra & Sedita, 2010).

Opportunities and Challenges on Adopting Open Innovation Strategies

The successful establishment and management of an effective innovation network often remain as a critical challenge for SMEs (Lazzarotti, Manzini & Pizzurno, 2008). SMEs are more open to open innovation because of their limited size and resources. At the same time, intense competition and more demanding customers are found to be the major motivation for open innovation. The most important bottleneck for open innovation is differences in organization and culture between the individual partners (De Jong, 2006).

While researchers are carrying out researches to develop innovation opportunity framework (Levy, Powell & Galliers, 1999; Wang, Jaring & Wallin, 2009; Rahman & Ramos, 2010), the review finds some other researchers find open innovation as throwing challenges for the SMEs development. Groen and Linto (2010) raised a challenge, as whether the term open innovation hindering growth in research and understanding and if so should the term be used as it is currently? Knudsen and Mortensen (2010) chart an unnoticed theme in the current debate on open innovation, as a foundational question whether increasing openness is beneficial? They further investigate that, with increasing degrees of openness the product development projects are slower than the average in the industry, slower than what is usual for the firm's projects and had higher cost than the average in the industry and the firm's usual projects.

Hoffman, Parejo, Bessant and Perren (1998) mention that, despite the strong commitment to support innovation within SMEs at both regional and local level, the actual processes whereby small firms undertake innovative activity remain unclear. In this context, Tödtling & Trippl (2005) argue that, there is no "ideal model" for innovation policy as innovation activities differ strongly between central, peripheral and old industrial areas.

Strategies and Technologies

Majority of the research documents discuss on adoption of open innovation strategies in the form of practices or applications incorporating innovation technologies. They focus on inter-firm cooperation, cooperation with intermediary institutions, cooperation with research organization (Zeng, Xie & Tam, 2010; Izushi, 2003; Leiponen & Byman, 2009; Mention, 2010); management attitude, planning and external orientation (De Jong & Marsili, 2006); R&D alliances (Dickson, Weaver & Hoy, 2006) and find these are providing significant impact on the innovation performances for SMEs. There are researches on the introduction of tools, for example, bricolage (Ferneley & Bell, 2006); or taxonomies, for example, fruit flies approach (De Jong & Marsili, 2006); or terminologies, for example, technology exploitation (van de Vrande, de Jong, Vanhaverbeke, & Rochemont, 2009; Lichtenthaler, 2010) or technology exploration (van de Vrande, de Jong, Vanhaverbeke, & Rochemont, 2009). In these contexts, O'Regan, Ghobadian and Sims (2006) state that close association between strategy, organizational culture, leadership and innovation plays important role in achieving successful innovation.

Lee and Lan (2011) argue that adoption of knowledge management is becoming an emerging agenda in developing business strategies. They further argue that, implementation depends on a harmonious amalgamation of infrastructure and process capabilities, including technology, culture and organizational structure. However, based on a random sample of 500 South Yorkshire non-hitech manufacturing SMEs, Laforet (2008) finds that the size, strategy and market orientation are associated with innovation.

Measuring the Impact of OI Strategies

As the adoption of OI strategies are being increased at all levels of the entrepreneurships, especially for the SMEs, researchers engaged themselves in finding the ultimate benefits of utilizing OI strategies by measuring their impact. Messa and Testa (2008) points on various indicators of the innovation measurements. Rejeb, Morel-Guimarães, Boly and Assiélou (2009) argue that innovation, as a competitive economic factor, is a process that compels a continuous, evolving and mastered management. Therefore, innovative companies need to measure their innovation capacity. In this aspect, innovation indicators have been used largely by "innovation scholars", a community that consists of researchers from a variety of disciplines (ranging from engineering to sociology and political science), who have a common research focus on technological innovation (Grupp & Schubert, 2010). However, Huang, Soutar and Brown (2004) indicated that four factors underline the commonly used success measurement: financial performance, objective market acceptance, subjective market acceptance and product-level measures. They also mention that these four factors are related to each other and can be used to well predict the overall measurement.

Tools and Instruments

Regarding employment of software in the innovation process in SMEs, Kohn and Hüsig (2006) during their investigation have found that a large variety of software products are available in the market. Their research while trying to address the question of how far these products are specifically used in practice, they find that these software products are rarely used to support the innovation process in German SMEs. Kaufmann and Tödtling (2002) mention that, the problem that most SMEs hardly interact with knowledge providers from outside the business sector (for example, universities or intermediaries) and the interaction is not reduced by the support instruments. SMEs perform insufficiently the function of interfaces to innovation-related resources and information from outside the environment. Kaufmann and Tödtling (2002) also argue that, there is a lack of proactive consultancy concerning strategic, organizational and technological weaknesses which is necessary because most of the time, the firms are not aware of such deficiencies within themselves.

Practices and Practitioners: Areas of Practices

Despite the emergence of outsourcing R&D and it's emphasize by early introducers of open innovation in the entrepreneurship enhancement, this study does not find significant contributions from the researchers or practitioners illustrating the application of R&D outsourcing for SMEs development. There is a visible gap in applying OI strategies in practice, especially for SMEs. Van de Vrande, de Jong, Vanhaverbeke and Rochemont (2009) conducted a study, which they claim as a first, explorative one to address this gap by focusing on SMEs. Their study tried to measure the extent of application of OI practices by SMEs and find out whether there is a positive trend on adoption of OI model over the time. However, by looking up at the most frequently applied areas of researches from the most relevant literatures encountered during the review, the following areas are being put forward by this study:

Venture

Venturing implies to starting up of new organization through spin-off or spin-out processes, including getting support from the parent organization in the form of finance, human capital, legal advice or administrative service. But, most OI studies have primarily targeted to venture activities in large enterprises (Van de Vrande, de Jong, Vanhaberbeke & Rochemont, 2009). Van de Vrande, et al. (2009) in their study finds a stable perceived trend of venturing among their surveyed SMEs. However, a distinctive and encouraging fact that governments (in Europe and elsewhere) get involved to stimulate innovation in the SME sector and as SMEs face financial constraints predominantly, governments encourage the provision of debt and equity (venture capital) finance to such firms (Kaivanto & Stoneman, 2007).

Technology Licensing

Technology licensing usually provides monetary and strategic benefits (Bidault, 2004; Kollmer & Dowling, 2004). The monetary benefits refer to generating licensing revenues, and equally important are the strategic benefits from technology licensing that, the firm does not attempt to directly generate licensing revenues, but tries to improve its competitive position, which indirectly affects its financial performance (Grindley & Teece, 1997; Ziedonis, 2007). The strategic benefits fall into two categories: licensing to reinforce a firm's product market position and licensing to augment a firm's technological position (Lichtenthaler, 2007; Nagaoka & Kwon, 2006). Lichtenthaler (2010) shows that technology licensing offers important strategic benefits beyond generating licensing revenues, which underscore the need for an integrated management of technology licensing activities. However, as this study observes, there is a research gap to find out the impact of technology licensing in SMEs.

Licensing of Intellectual Property

Emerged from the very basic notion of open innovation, thus being coined as the result of purposive inflows and outflows of knowledge, and can be termed as the flow of intellectual property (IP). Van de Vrande et al. (2009) defined this as selling or offering licenses or royalty agreements to other organizations to gather profit from the intellectual property, such as patents, copyrights or trademarks. Similarly, they mentioned about buying or using IP of other organizations to benefit from external knowledge. Van de Vrande et al. (2009) find stable perceived trend among their surveyed SMEs, with increased number of inward IP licensing than outward IP licensing. To cite another example, though hardly be treated as SMEs, but Bianchi et al. (2010) find in their study that bio-pharmaceutical firms are successfully using licensing agreements aiming to acquire (inbound OI) or commercially exploit (outbound OI) technologies and knowledge.

Customer and Supplier Engagement

Customer integration has been found as the most influential external stimulus of open innovation (Van Hemel & Cramer, 2002). Empirical research by innovation scholars has clearly documented that many of the innovative products are in fact developed, prototyped, tested and improved by "lead users" (De Jong & von Hippel, 2009).

Van Hemel & Cramer (2002) find that among others, customer demand is the most influential external stimuli for ecodesign of product development. In this aspect, close relationship with customers based on mutual trust is an essential element for developing eco-efficiency of SMEs (Fernández-Viñé, Gómez-Navarro & Capuz-Rizo, 2010). Firms may benefit from their customers' ideas and innovations by proactive market research, providing tools to experiment with and/or develop products similar to the ones that are currently offered in the market, or by producing products based on the designs of customers and evaluating what may be learned from general product development (Van de Vrande, de Jong, Vanhaberbeke & Rochemont, 2009).

Alshawi, Missi and Irani (2010) referred that, there is a shift from product-oriented business strategy to customer focused relationship strategy has been identified as a major change agent in companies. Alshawi, Missi and Irani (2010) further argue that like large organizations, many SMEs have implemented Customer Relationship Management (CRM), so that they can compete effectively in today's highly changeable economic and market climate. Bayraktar, Demirbag, Koh, Tatoglu and Zaim (2009) support the previous argument by indicating that the forward-looking enterprises today are dynamic; they work in partnership with suppliers, customers and even with competitors; share information and knowledge aiming to create a shared supply chain that is capable of competing if not leading a particular industry.

External Networking and External Participation

Innovation is perceived as a broad and multi-dimensional concept, and can be differentiated as the capacity to innovate now, also in the future, along the entire innovation process of ongoing learning, searching and exploring, resulting in making of new products and processes, establishment of new forms of organization and creation of new markets (Lundvall, 1995). More specifically, as innovation intrinsically involves doing something different, successful innovation is increasingly likely to require sources of complementary competence that lie outside the innovating firm (Freel, M. & de Jong, J.P.J. (2009). Thereby, the place of innovation is not for any single company alone, but for more and more companies to be embedded in a larger network (Kühne, Vanhonacker, Gellynck & Verbeke, 2010), and the complexity of innovation processes led to a tremendous growth in the use of external networks by SMEs (Zeng, Xie & Tam, 2010).

The ability to absorb external knowledge has become a major driver for competition (Spithoven,

Clarysse & Knockaert, 2010), and external organizations can play an active role by 'intertwining' knowledge to support the development of processes, systems and routines that distribute and institutionalize learning throughout the organization (Jones & Macpherson, 2006).

External networking may need to focus on accumulating external knowledge or human capital, leading to external participation through equity investment in order to gain access to the external knowledge or to obtain other synergies (Van de Vrande, de Jong, Vanhaberbeke & Rochemont, 2009).

An analysis made by Mention (2010) shows that firms provided with information from market sources and from internal sources as well as firms involved in science-based collaboration for their product innovations are more likely to introduce new market innovations. Moreover, with respect to the innovation networks of firms, there is a widespread consensus nowadays that local connections do not suffice to sustain innovativeness (Tödtling & Trippl, 2005). Evidence from SMEs and entrepreneurship studies suggests that a key condition for small firms to be innovative and grow is that they should have network mobilization capability; i.e. the ability to establish interactive networks of partners as external stimuli and proactively seek for external assistance (Van Hemel & Cramer, 2002; Jones & Macpherson, 2006; Partanen, Möller, Westerlund, Rajala & Rajala, 2008; Spithoven, Clarysse & Knockaert, 2010).

FUTURE RESEARCH DIRECTIONS

Along the practice areas of open innovation, the timely identification of opportunities for out-licensing of a firm's technologies outside the core business process is essential for its business success. This is particularly challenging for SMEs due to their lack of specialized knowledge base and also limited financial resources that can be devoted

Table 7. Summarization of future research framework

Future research framework
Future researches may need to improve the current understanding of open innovation in SMEs
Focus of future researches may be in the nature of innovation and the extent to which open innovation is embedded in SMEs
Future researches may investigate how entrepreneurs engage in open innovation during their growth phases, and particularly what managerial implication can be derived
Future researches may find out the characteristics of SMEs that are more likely to get benefit from collaboration, particularly via an intermediary
Future researches could be carried out by minimizing the screening questions which may opt out, especially start-ups and micro-enterprises, as these enterprises have been identified as the sources of breakthrough innovations and challengers of contemporary innovation actors
Future researchers could attempt to survey open innovation in broader samples of enterprises in more detailed and exploratory way
Focus of future research could be on the requirement of OI on differences in culture, structure and decision making among partners of different sizes and sectors
Future studies may incorporates findings of OI strategies in improving innovation cooperation for SMEs in emerging economies and developing countries and extend the generalizations of the findings
Future research could be focused on identifying different segments within the population of every stakeholder (the entrepreneurs could be segmented by industry or geographic location and the academics by discipline)
Future researches could pay more attention to the outflows of knowledge, which is intellectual property management
Ref: Van dc Vrande, de Jong, Vanhaberbeke & Rochemont, 2009; Lee, Park, Yoon & Park, 2010; Zeng, Xie & Tam, 2010; Massa & Testa, 2008

to innovation activities (Bianchi, Campodall'Orto, Frattini & Vercesi, 2010).

Despite the vast growth in research on open innovation, there are several openings of further research that this study have observed; such as linking open innovation research with other management areas, like marketing, human resources, change process, etc (Van de Vrande, Vanhaverbeke & Gassman, 2010). Moreover, to match the global demand and supply of innovation, businesses need to internationalize their innovation activities through collaboration with external partners (for example, customers, suppliers, universities, and intermediaries) (De Backer & Cervantes, 2008). Foremost, the understanding of open innovation at the outer periphery of the business chain, driving factors of global innovation networking across different SMEs sub-sectors, accessibility and relationship of open innovation strategies with the implementing firms deserves further attention and research.

As mentioned in the problem statement, and also within the texts in the literature review, it is worthy to mention for sake of future research that the concentration on open innovation researches are primarily focused to corporate businesses and it is unclear whether these findings can be generalized to SMEs (Pedersen, Sondergaard & Esbjerg, 2009). A small size pilot project may be initiated in a few countries of similar social and economic patterns to learn about the incubation of OI strategies before making a further leap to standardize a common platform for a larger number of countries in a region.

Based on the study findings, this research would like to establish a future research framework involving products, processes, services and organizational transformations. Table 7 summarizes the framework.

CONCLUSION

The small firms that do innovate successfully increase their chances of survival (Cofis & Marsili, 2003) and growth (De Jong, Vermeulen & O'Shaughnessy, 2004). The behavior of small firms can vary substantially. Some small firms survive by competing in a market niche, while others pursue more radical innovations and eventually, themselves become market leaders (De Jong & Marsili, 2006).

This study finds the evidence of clustered researches in various segments of SMEs sectors, but a continued research covering major categories of SMEs sectors is demanding. Along the study, the significant research themes and their practices have been explored, including their nature, number, exposition and potency. The study concludes that a research gap exists at the periphery where most of the entrepreneurship nourishes among the developed economies, which are the SMEs. The study also concludes that there is a broadening gap in terms of applying researches in the form of practices aiming at SMEs in the arena of open innovation.

REFERENCES

Alshawi, S., Missi, F., & Irani, Z. (2010). (in press). Organisational, technical and data quality factors in CRM adoption — SMEs perspective. *Industrial Marketing Management.*

Amara, N., Landry, R., Becheikh, N., & Ouimet, M. (2008). Learning and novelty of innovation in established manufacturing SMEs. *Technovation, 28*(7), 450–463. doi:10.1016/j.technovation.2008.02.001

Bayraktar, E., Demirbag, M., Koh, S. C. L., Tatoglu, E., & Zaim, H. (2009). A causal analysis of the impact of information systems and supply chain management practices on operational performance: Evidence from manufacturing SMEs in Turkey. *International Journal of Production Economics, 122*(1), 133–149. doi:10.1016/j.ijpe.2009.05.011

Belussi, F., Sammarra, A., & Sedita, S. R. (2010). Learning at the boundaries in an "Open Regional Innovation System": A focus on firms' innovation strategies in the Emilia Romagna life science industry. *Research Policy*, *39*(6), 710–721. doi:10.1016/j.respol.2010.01.014

Bevis, K. I., & Cole, A. (2010). *Open innovation readiness: A tool.* ISPIM XXIth International Conference, Bilbao, June 2010, International Society for Professional Innovation Management.

Bianchi, M., Campodall'Orto, S., Frattini, F., & Vercesi, P. (2010). Enabling open innovation in small- and medium-sized enterprises: How to find alternative applications for your technologies. *R & D Management*, *40*(4), 414–431. doi:10.1111/j.1467-9310.2010.00613.x

Bianchi, M., Cavaliere, A., Chiaroni, D., Frattini, F., & Chiesa, V. (2010). (in press). Organisational modes for open innovation in the bio-pharmaceutical industry: An exploratory analysis. *Technovation*.

Bidault, F. (2004). Global licensing strategies and technology pricing. *International Journal of Technology Management*, *27*(2/3), 295–305. doi:10.1504/IJTM.2004.003959

Bougrain, F., & Haudeville, B. (2002). Innovation, collaboration and SMEs internal research capacities. *Research Policy*, *31*(5), 735–747. doi:10.1016/S0048-7333(01)00144-5

Brandenburger, A., & Nalebuff, B. (1996). *Co-opetition*. New York, NY: Doubleday.

BVCA. (2005). *Creating success from university spin-outs. A Review: BVCA (British Venture Capital Association)*. Library House.

Chesbrough, H. W. (2003a). The era of open innovation. *MIT Sloan Management Review*, *44*(3), 35–41.

Chesbrough, H. W. (2003b). *Open business models: How to thrive in the new innovation landscape*. Boston, MA: Harvard Business School Press Books.

Chesbrough, H. W. (2006). *Open innovation: The new imperative for creating and profiting from technology*. Boston, MA: Harvard Business School Press.

Chesbrough, H. W., Vanhaverbeke, W., & West, J. (Eds.). (2006). *Open innovation: Researching a new paradigm*. London, UK: Oxford University Press.

Coase, R. (1937). The nature of the firm. *Economica*, *4*(16), 386–405. doi:10.1111/j.1468-0335.1937.tb00002.x

Cofis, E., & Marsili, O. (2003). *Survivor: The role of innovation in firm's survival*. No. 03-18. WPT. Koopmans Institute, USE, Utrecht University.

Cooke, P. (2005). Regionally asymmetric knowledge capabilities and open innovation: Exploring Globalisation 2 - A new model of industry organisation. *Research Policy*, *34*(8), 1128–1149. doi:10.1016/j.respol.2004.12.005

Dahlander, L., & Gann, D. M. (2010). How open is innovation? *Research Policy*.

Damaskopoulos, P., & Evgeniou, T. (2003). Adoption of new economy practices by SMEs in Eastern Europe. *European Management Journal*, *21*(2), 133–145. doi:10.1016/S0263-2373(03)00009-4

Davenport, T. H. (1993a). *Process innovation*. Boston, MA: Harvard Business School Press.

Davenport, T. H. (1993b). *Process innovation: Reengineering work through information technology. Ernst & Young*. Harvard Business School Press.

De Backer, K., & Cervantes, M. (2008). *Open innovation in global networks*. OECD.

De Jong, J. P. J. (2006). *Open innovation: Practice, trends, motives and bottlenecks in the SMEs (Meer Open Innovatie: Praktijk, Ontwikkelingen, Motieven en Knelpunten in het MKB)*. Zoetermeer: EIM.

De Jong, J. P. J., & Marsili, O. (2006). The fruit flies of innovations: A taxonomy of innovative small firms. *Research Policy, 35*, 213–229. doi:10.1016/j.respol.2005.09.007

De Jong, J. P. J., Vanhaverbeke, W., Kalvet, T., & Chesbrough, H. (2008). *Policies for open innovation: Theory, framework and cases*. Research project funded by VISION Era-Net, Helsinki, Finland.

De Jong, J. P. J., Vermeulen, P. A. M.. & O'Shaughnessy, K. C. (2004). Effects of innovation in small firms. *M & O, 58*(1), 21-38.

De Jong, J. P. J., & von Hippel, E. (2009). Transfers of user process innovations to process equipment producers: A study of Dutch high-tech firms. *Research Policy, 38*(7), 1181–1191. doi:10.1016/j.respol.2009.04.005

Dehoff, K., & Sehgal, V. (2006). Innovators without borders. *strategy+business*, Autumn 2006.

Dehoff, K. & Sehgal, V. (2008). Beyond borders: The global innovation 1000. *strategy+business*, Winter 2008.

Del Brío, J. Á., & Junquera, B. (2003). A review of the literature on environmental innovation management in SMEs: Implications for public policies. *Technovation, 23*(12), 939–948. doi:10.1016/S0166-4972(02)00036-6

Dickson, P. H., Weaver, K. M., & Hoy, F. (2006). Opportunism in the R&D alliances of SMES: The roles of the institutional environment and SME size. *Journal of Business Venturing, 21*(4), 487–513. doi:10.1016/j.jbusvent.2005.02.003

Edwards, T., Delbridge, R., & Munday, M. (2005). Understanding innovation in small and medium-sized enterprises: A process manifest. *Technovation, 25*(10), 1119–1127. doi:10.1016/j.technovation.2004.04.005

EIRMA. (2005). *Responsible partnering: A guide to better practices for collaborative research and knowledge transfer between science and industry*. EIRMA, EUA, EARTO, ProTon Europe, version 1.0 January 2005.

EIRMA. (2009). *Joining forces in a world of open innovation: Guidelines for collaborative research and knowledge transfer between science and industry*. EIRMA, EUA, EARTO, ProTon Europe, version 1.1 October 2009.

European Union. (2005). *EU-Summary*. The joint Japan-EU Seminar on R&D and Innovation in Small and Medium size Enterprises (SMEs), Tokyo, November 18, 2004.

Fernández-Viñé, M. B., Gómez-Navarro, T., & Capuz-Rizo, S. F. (2010). Eco-efficiency in the SMEs of Venezuela. Current status and future perspectives. *Journal of Cleaner Production, 18*(8), 736–746. doi:10.1016/j.jclepro.2009.12.005

Ferneley, E., & Bell, F. (2006). Using bricolage to integrate business and information technology innovation in SMEs. *Technovation, 26*(2), 232–241. doi:10.1016/j.technovation.2005.03.005

Fredberg, T., Elmquist, M., & Ollila, S. (2008). *Managing open innovation: Present findings and future directions. Report VR 2008:02*. VINNOVA - Verket för Innovationssystem/Swedish Governmental Agency for Innovation Systems.

Freel, M., & De Jong, J. P. J. (2009). Market novelty, competence-seeking and innovation networking. *Technovation, 29*(12), 873–884. doi:10.1016/j.technovation.2009.07.005

Gassmann, O. (2006). Opening up the innovation process: Towards an agenda. [Blackwell Publishing Ltd.]. *R & D Management, 36*(3), 223–228. doi:10.1111/j.1467-9310.2006.00437.x

Gassmann, O., Enkel, E., & Chesbrough, H. (2010). The future of open innovation. *R & D Management, 40*(3), 213–221. doi:10.1111/j.1467-9310.2010.00605.x

Grindley, P. C., & Teece, D. J. (1997). Managing intellectual capital: licensing and cross-licensing in semiconductors and electronics. *California Management Review, 39*, 1–34.

Groen, A. J., & Linton, J. D. (2010). Is open innovation a field of study or a communication barrier to theory development? *Technovation, 30*(11-12), 554. doi:10.1016/j.technovation.2010.09.002

Grupp, H., & Schubert, T. (2010). Review and new evidence on composite innovation indicators for evaluating national performance. *Research Policy, 39*(1), 67–78. doi:10.1016/j.respol.2009.10.002

Henkel, J. (2006). Selective revealing in open innovation processes: The case of embedded Linux. *Research Policy, 35*, 953–969. doi:10.1016/j.respol.2006.04.010

Hjalager, A.-M. (2010). A review of innovation research in tourism. *Tourism Management, 31*(1), 1–12. doi:10.1016/j.tourman.2009.08.012

Hoffman, K., Parejo, M., Bessant, J., & Perren, L. (1998). Small firms, R&D, technology and innovation in the UK: A literature review. *Technovation, 18*(1), 39–55. doi:10.1016/S0166-4972(97)00102-8

Huang, X., Soutar, G. N., & Brown, A. (2004). Measuring new product success: An empirical investigation of Australian SMEs. *Industrial Marketing Management, 33*, 117–123. doi:10.1016/S0019-8501(03)00034-8

Huang, X., Soutar, G. N., & Brown, A. (2004). Review and new evidence on composite innovation indicators for evaluating national performance. *Research Policy, 39*(1), 67–78.

Huizingh, K. R. E. (2009). The future of innovation. *Proceedings of The XX International Society for Professional Innovation Management* - ISPIM Conference 2009, (p. 21).

Izushi, H. (2003). Impact of the length of relationships upon the use of research institutes by SMEs. *Research Policy, 32*(5), 771–788. doi:10.1016/S0048-7333(02)00085-9

Jones, O., & Macpherson, A. (2006). Inter-organizational learning and strategic renewal in SMEs: Extending the 4I framework. *Long Range Planning, 39*(33), 155–175. doi:10.1016/j.lrp.2005.02.012

Kaivanto, K., & Stoneman, P. (2007). Public provision of sales contingent claims backed finance to SMEs: A policy alternative. *Research Policy, 36*(5), 637–651. doi:10.1016/j.respol.2007.01.001

Kaufmann, A., & Tödtling, F. (2002). How effective is innovation support for SMEs? An analysis of the region of Upper Austria. *Technovation, 22*(3), 147–159. doi:10.1016/S0166-4972(00)00081-X

Kirschbaum, R. (2005). Open innovation in practice. *Research on Technology Management, 48*, 24–28.

Knudsen, M. P., & Mortensen, T. B. (2010). (in press). Some immediate – but negative – effects of openness on product development performance. *Technovation*.

Kohn, S., & Hüsig, S. (2006). Potential benefits, current supply, utilization and barriers to adoption: An exploratory study on German SMEs and innovation software. *Technovation, 26*(8), 988–998. doi:10.1016/j.technovation.2005.08.003

Kollmer, H., & Dowling, M. (2004). Licensing as a commercialization strategy for new technology-based firms. *Research Policy, 33*, 1141–1151. doi:10.1016/j.respol.2004.04.005

Kornai, J. (2010). *Innovation and dynamism: Interaction between systems and technical progress*. Working Paper No. 2010/33, UNU-WIDER.

Kowalski, S. P. (2009). *SMES, open innovation and IP management: Advancing global development.* A presentation at the WIPO-Italy International Convention on Intellectual Property and Competitiveness of Micro, Small and Medium-Sized Enterprises (MSMEs), Rome, Italy, December 10-11, 2009.

Koyuncugil, A. S., & Ozgulbas, N. (2009). Risk modeling by CHAID decision tree algorithm. *ICCES, 11*(2), 39–46.

Kühne, B., Vanhonacker, F., Gellynck, X., & Verbeke, W. (2010). Innovation in traditional food products in Europe: Do sector innovation activities match consumers' acceptance? *Food Quality and Preference, 21*(6), 629–638. doi:10.1016/j.foodqual.2010.03.013

Laforet, S. (2008). Size, strategic, and market orientation affects on innovation. *Technovation, 25*(10), 1119–1127.

Laursen, K., & Salter, A. (2006). Open for innovation: The role of openness in explaining innovation performance among UK manufacturing firms. *Strategic Management Journal, 27*, 131–150. doi:10.1002/smj.507

Lawson, C. P., Longhurst, P. J., & Ivey, P. C. (2006). The application of a new research and development project selection model in SMEs. *Technovation, 26*(2), 242–250. doi:10.1016/j.technovation.2004.07.017

Lazzarotti, V., Manzini, R., & Pizzurno, E. (2008). Managing innovation networks of SMEs: A case study. *Proceeding of the International Engineering Management Conference: Managing Engineering, Technology and Innovation for Growth (IEMC Europe 2008)*, Estoril, Portugal, (pp. 521-525).

Lecocq, X., & Demil, B. (2006). Strategizing industry structure: The case of open systems in low-tech industry. *Strategic Management Journal, 27*, 891–898. doi:10.1002/smj.544

Lee, M. R., & Lan, Y.-C. (2011). Toward a unified knowledge management model for SMEs. *Expert Systems with Applications, 38*(1), 729–735. doi:10.1016/j.eswa.2010.07.025

Lee, S., Park, G., Yoon, B., & Park, J. (2010). Open innovation in SMEs—An intermediated network model. *Research Policy, 39*(2), 290–300. doi:10.1016/j.respol.2009.12.009

Leiponen, A., & Byma, J. (2009). If you cannot block, you better run: Small firms, cooperative innovation, and appropriation strategies. *Research Policy, 38*(9), 1478–1488. doi:10.1016/j.respol.2009.06.003

Levy, M., Powell, P., & Galliers, R. (1998). Assessing information systems strategy development frameworks in SMEs. *Information & Management, 36*(5), 247–261. doi:10.1016/S0378-7206(99)00020-8

Lichtenthaler, U. (2007). The drivers of technology licensing: an industry comparison. *California Management Review, 49*, 67–89.

Lichtenthaler, U. (2010). Technology exploitation in the context of open innovation: Finding the right job for your technology. *Technovation, 30*(7-8), 429–435. doi:10.1016/j.technovation.2010.04.001

Lichtenthaler, U., & Ernst, H. (2009). Opening up the innovation process: The role of technology aggressiveness. *R & D Management, 39*(1), 38–54. doi:10.1111/j.1467-9310.2008.00522.x

Lindermann, N., Valcareel, S., Schaarschmidt, M., & Von Kortzfleisch, H. (2009). SME 2.0: roadmap towards Web 2.0- based open innovation in SME-network- A case study based research framework. In Dhillon, G., Stahl, B. C., & Baskerville, R. (Eds.), *CreativeSME2009, IFIP International Federation for Information Processing, IFIP AICT 301* (pp. 28–41).

Livieratos, A. D., & Papoulias, D. B. (2009). *Towards an open innovation growth strategy for new, technology-based firms*. National Technical University of Athens. Retrieved October 30, 2010, from http://www.ltp.ntua.gr/uploads/GJ/eK/GJeKw8Jf5RqjCO9CWapm4w/Growth.pdf

Lundvall, B. (1995). *National systems of innovation: Towards a theory of innovation and interactive learning*. London, UK: Biddles Ltd.

Maijers, W., Vokurka, L., van Uffelen, R., & Ravensbergen, P. (2005). Open innovation: Symbiotic network, knowledge circulation and competencies for the benefit of innovation in the horticulture delta. *Presentation IAMA Chicago 2005.*

Major, E. J., & Cordey-Hayes, M. (2000). Engaging the business support network to give SMEs the benefit of foresight. *Technovation, 20*(11), 589–602. doi:10.1016/S0166-4972(00)00006-7

Massa, S., & Testa, S. (2008). Innovation and SMEs: Misaligned perspectives and goals among entrepreneurs, academics, and policy makers. *Technovation, 28*(7), 393–407. doi:10.1016/j.technovation.2008.01.002

Mention, A.-L. (2010). (in press). Co-operation and co-opetition as open innovation practices in the service sector: Which influence on innovation novelty? *Technovation*.

Nagaoka, S., & Kwon, H. U. (2006). The incidence of cross-licensing: A theory and new evidence on the firm and contract level determinants. *Research Policy, 35*(9), 1347–1361. doi:10.1016/j.respol.2006.05.007

Novelli, M., Schmitz, B., & Spencer, T. (2006). Networks, clusters and innovation in tourism: A UK experience. *Tourism Management, 27*(6), 1141–1152. doi:10.1016/j.tourman.2005.11.011

O'Regan, N., Ghobadian, A., & Sims, M. (2006). Fast tracking innovation in manufacturing SMEs. *Technovation, 26*(2), 251–261. doi:10.1016/j.technovation.2005.01.003

Partanen, J., Möller, K., Westerlund, M., Rajala, R., & Rajala, A. (2008). Social capital in the growth of science-and-technology-based SMEs. *Industrial Marketing Management, 37*(5), 513–522. doi:10.1016/j.indmarman.2007.09.012

Pedersen, M., Sondergaard, H. A., & Esbjerg, L. (2009). Network characteristics and open innovation in SMEs. In K. R. E. Huizingh, S. Conn, M. Torkkeli, & I. Bitran (Eds.), *Proceedings of The XX ISPIM Conference*.

Powell, W. (1990). Neither market nor hierarchy: Network forms of organization. In Stow, B., & Cummings, L. L. (Eds.), *Research in organizational behavior*. Greenwich, CT: JAI Press.

Rahman, H., & Ramos, I. (2010). Open innovation in SMEs: From closed boundaries to networked paradigm. *Issues in Informing Science and Information Technology, 7*, 471–487.

Raymond, L., & St-Pierre, J. (2010). R&D as a determinant of innovation in manufacturing SMEs: An attempt at empirical clarification. *Technovation, 30*(1), 48–56. doi:10.1016/j.technovation.2009.05.005

Rejeb, H. B., Morel-Guimarães, L., Boly, V., & Assiélou, N. G. (2009). Measuring innovation best practices: Improvement of an innovation index integrating threshold and synergy effects. *Technovation*, *28*, 838–854. doi:10.1016/j.technovation.2008.08.005

Rhee, J., Park, T., & Lee, D. H. (2010). Drivers of innovativeness and performance for innovative SMEs in South Korea: Mediation of learning orientation. *Technovation*, *30*(1), 65–75. doi:10.1016/j.technovation.2009.04.008

Saarikoski, V. (2006). *The odyssey of the mobile Internet- The emergence of a networking attribute in a multidisciplinary study*. Academic dissertation, TIEKE, Helsinki.

Santisteban, M. A. (2006). Business systems and cluster policies in the Basque country and Catalonia (1990-2004). *European Urban and Regional Studies*, *13*(1), 25–39. doi:10.1177/0969776406059227

Savioz, P., & Blum, M. (2002). Strategic forecast tool for SMEs: How the opportunity landscape interacts with business strategy to anticipate technological trends. *Technovation*, *22*(2), 91–100. doi:10.1016/S0166-4972(01)00082-7

Schumpeter, J. A. (1934). *The theory of economic development*. Cambridge, MA: Harvard University Press.

Schumpeter, J. A. (1942). *Capitalism, socialism, and democracy*. New York, NY: Harper & Row.

Schumpeter, J. A. (1950). *Capitalism, socialism, and democracy* (3rd ed.). New York, NY: Harper & Row.

Spithoven, A., Clarysse, B., & Knockaert, M. (2010). Building absorptive capacity to organize inbound open innovation in traditional industries. *Technovation*, *30*(2), 130–141. doi:10.1016/j.technovation.2009.08.004

Thoben, K. D. (2008). A new wave of innovation in collaborative networks. *Proceedings of the 14th International Conference on Concurrent Enterprising: ICE 2008*. Lisbon, Portugal, 23-25 June 2008, (pp. 1091-1100). Nottingham, UK: University of Nottingham, Centre for Concurrent Enterprise.

Thorgren, S., Wincent, J., & Örtqvist, D. (2009). Designing interorganizational networks for innovation: An empirical examination of network configuration, formation and governance. *Journal of Engineering and Technology Management*, *26*(3), 148–166. doi:10.1016/j.jengtecman.2009.06.006

Tödtling, F., & Trippl, M. (2005). One size fits all? Towards a differentiated regional innovation policy approach. *Research Policy*, *34*(8), 1203–1219. doi:10.1016/j.respol.2005.01.018

Van de Vrande, V., de Jong, J. P. J., Vanhaverbeke, W., & de Rochemont, M. (2008). *Open innovation in SMEs: Trends, motives and management challenges. SCALES-initiative (SCientific AnaLysis of Entrepreneurship and SMEs): SMEs and Entrepreneurship Programme*. Zoetermeer: Netherlands Ministry of Economic Affairs.

Van de Vrande, V., de Jong, J. P. J., Vanhaverbeke, W., & de Rochemont, M. (2009). Open innovation in SMEs: Trends, motives and management challenges. *Technovation*, *29*(6-7), 423–437. doi:10.1016/j.technovation.2008.10.001

Van de Vrande, V., Vanhaverbeke, W., & Gassman, O. (2010). Broadening the scope of open innovation: Past research, current state and future directions. *International Journal of Technology Management*, *52*(3-4), 221–235. doi:10.1504/IJTM.2010.035974

Van Hemel, C., & Cramer, J. (2002). Barriers and stimuli for ecodesign in SMEs. *Journal of Cleaner Production*, *10*(5), 439–453. doi:10.1016/S0959-6526(02)00013-6

Von Hippel, E. (1986). Lead users: A source of novel product concepts. *Management Science, 32*(7), 791–805. doi:10.1287/mnsc.32.7.791

Von Hippel, E. (1988). *The sources of innovation*. New York, NY: Oxford University Press.

Von Hippel, E., & Von Krogh, G. (2003). Open source software and the private–collective innovation model: Issues for organization science. *Organization Science, 14*(2), 209–223. doi:10.1287/orsc.14.2.209.14992

Wang, L., Jaring, P., & Wallin, A. (2009). Developing a conceptual framework for business model innovation in the context of open innovation. *Proceedings of the Third IEEE International Conference on Digital Ecosystems and Technologies (IEEE DEST 2009)*, (pp. 460-465).

Wenger, E. (1998). *Communities of practice: Learning, meaning and identity*. Cambridge University Press.

West, J. (2006). Does appropriability enable or retard open innovation? In H. Chesbrough, W. Vanhaverbeke & J. West (Eds.), *Open innovation: Researching a new paradigm*. London, UK: Oxford University Press.

West, J., Vanhaverbeke, W., & Chesbrough, H. (2006). Open innovation: A research agenda. In H. Chesbrough, W. Vanhaverbeke & J. West (Eds.), *Open innovation: Researching a new paradigm*. London, UK: Oxford University Press.

Wijffels, H. (2009). *Leadership, sustainability and levels of consciousness*. A speech at the Conference on the Leadership for a Sustainable World, June 5, 2009, Grote Kerk, Den Haag, the Netherlands.

Woodhams, C., & Lupton, B. (2009). Analysing gender-based diversity in SMEs. *Scandinavian Journal of Management, 25*(2), 203–213. doi:10.1016/j.scaman.2009.02.006

Zeng, S. X., Xie, X. M., & Tam, C. M. (2010). Relationship between cooperation networks and innovation performance of SMEs. *Technovation, 30*(3), 181–194. doi:10.1016/j.technovation.2009.08.003

Ziedonis, A. A. (2007). Real options in technology licensing. *Management Science, 53*, 1618–1633. doi:10.1287/mnsc.1070.0705

ADDITIONAL READING

Chesbrough, H.W. (2003) The Era of Open Innovation. *Sloan Management Review*, 44, 3 (Spring): 35-41.

Chesbrough, H.W. (2003) "Open Innovation: How Companies Actually Do It," *Harvard Business Review*, 81, 7 (July): 12-14.

Chesbrough, H.W. (2003) "Open Platform Innovation: Creating Value from Internal and External Innovation," *Intel Technology Journal*, 7, 3 (August): 5-9.

Chesbrough, H.W. (2004), "Managing Open Innovation: Chess and Poker," Research-Technology Management, 47, 1 (January): 23-26.

Chesbrough, H. W. (2006). Open Innovation: A New Paradigm for Understanding Industrial Innovation. In Chesbrough, H., Vanhaverbeke, W., & West, J. (Eds.), *Open Innovation: Researching a New Paradigm* (pp. 1–12). Oxford: Oxford University Press.

Chesbrough, H. W. (2006). New Puzzles and New Findings. In Chesbrough, H., Vanhaverbeke, W., & West, J. (Eds.), *Open Innovation: Researching a New Paradigm* (pp. 15–34). Oxford: Oxford University Press.

Chesbrough, H. W. (2006). *Open Business Models: How to Thrive in the New Innovation Landscape*. Boston: Harvard Business School Press.

Chesbrough, H.W. (2007) "The market for innovation: implications for corporate strategy." California Management Review, 49, 3 (Spring): 45–66.

Chesbrough, H. W. (2007). Business model innovation: it's not just about technology anymore. *Strategy and Leadership*, *35*(Iss: 6), 12–17. doi:10.1108/10878570710833714

Chesbrough, H.W. and A. K. Crowther (2006) "Beyond high tech: early adopters of open innovation in other industries," R&D Management, 36, 3 (June): 229-236

Davenport, T. H. (1994, July). Reengineering: Business Change of Mythic Proportions? *Management Information Systems Quarterly*, 121–127. doi:10.2307/249760

Davenport, T.H. (1998) Putting the enterprise into the enterprise system, *Harvard Business Review*, July–August, 76(4): 121–131

Davenport, T. H., & Beers, M. C. (1995). Managing Information About Processes. *Journal of Management Information Systems*, *12*(1), 57–80.

Davenport, T. H., & Harris, J. G. (2005). Automated decision making comes of age. *MIT Sloan Management Review*, *46*(4), 83–89.

Davenport, T. H., & Short, J. E. (1990, Summer). The New Industrial Engineering: Information Technology and Business Process Redesign. *Sloan Management Review*, 11–27.

Tilley, F., & Tonge, J. (2003). Introduction. In Jones, O. & Tilley, F. (Eds.). *Competitive Advantage in SME's: Organising for Innovation and Change*, John Wiley & Sons, pp. 1-12.

KEY TERMS AND DEFINITIONS

Open Innovation Concepts: Concept that uses the purposive inflows and outflows of knowledge to accelerate internal innovation and expand the markets for external use of innovation.

Open Innovation Practices: Practices carried out by the entrepreneurships incorporating open innovation researches.

Open Innovation Researches: Researches utilizing open innovation concepts and strategies.

Open Innovation Strategies: Strategies that incorporate fresh perspectives, knowledge and inspiration from the inside and outside of an entrepreneurship, thus allowing to go beyond day-to-day thinking and opens up the way to entirely new possibilities.

SMEs Development: Development of the small and medium enterprises in terms of economic, knowledge, human skills and other value additions.

ENDNOTE

1 Small and Medium scale Enterprises (SMEs) by definition are small firms with a small number of headcounts and the turnover fall below a certain limit. In terms of headcounts, they vary from country to country and regions to regions. In Europe, the headcounts are less than 250 and the turnover is less than 50 million Euros; http://ec.europa.eu/enterprise/policies/sme/facts-figures-analysis/sme-definition/index_en.htm. The study has adopted this definition.

Chapter 2
Open Innovation and Organizational Capacities:
Case Study of an SME

Denis Remon
Université du Québec à Trois-Rivières, Canada

ABSTRACT

Open innovation has gained popularity in recent years. But is the concept new or does it express old realities? The literature review of this case study found that the term "open innovation" is recent and that its development has been facilitated by technological innovations. The case study collected data over a period of eight months from an agrifood SME in Quebec, Canada. The aim was to go beyond the basic model of open innovation and integrate dynamic, absorptive and appropriative capacities into a new working open innovation model. Initial results show that components associated with the basic open innovation model such as intellectual property, joint R&D and co-product development are present as well as certain organizational capacities. The difficulty of applying the concept is due to its interactions rather than its components taken individually. The study concludes that further work is needed to extend the applicability and the foundations of open innovation.

INTRODUCTION

The concept of "open innovation" has laid the groundwork for new perspectives on business innovation, starting first from case studies of major U.S. companies (Chesbrough, 2003). Chesbrough (2003), who first developed the concept, has highlighted how some companies like IBM, Xerox and Intel have used the knowledge and expertise of other companies to (re)define their business model and reposition themselves in the market, thus becoming more competitive and profitable.

Until very recently, open innovation was essentially described in terms of technological innovation (Chesbrough, 2003, 2006; Chesbrough, Vanhaverbeke & West, 2006; Van der Meer, 2007)

DOI: 10.4018/978-1-61350-519-9.ch002

and was geared primarily towards professionals and practitioners (Fredberg, Elmquist & Ollila, 2008; Lee, Park, Yoon & Park, 2010). The underlying theories have been little developed (Elmquist, Fredberg & Ollila, 2009; Stoyanov, 2008) and only a limited number of empirical studies have been published (Huang & Rice, 2009; Van de Vrande, de Jong, Vanhaverbeke & Rochemont, 2008, 2009; Van de Vrande, Vanhaverbeke & Gassman, 2010). However, with time, there has been an expansion of the original, more limited frame of reference. Indeed, some authors (Lichtenthaler & Lichtenthaler, 2009; Teece, 2007; Van de Vrande et al., 2009) have shown how some organizational concepts, especially in terms of dynamic, absorptive and appropriative capacities, can be usefully integrated into the concept of open innovation, providing a more solid foundation for the concept and facilitating its application. This expansion of the concept seems warranted, especially as these capacities had been largely overlooked until then (Chesbrough et al., 2006).

In light of the foregoing, the author has chosen to make a case study of a small company with fewer than 10 workers. The aim of the study was to describe the process of open innovation and validate the presence of organizational capacities. The research goal is, therefore, to empirically validate a model of open innovation which incorporates organizational capacities.

The research has been conducted using the basic model of open innovation (Chesbrough, 2003) and has integrated dynamic (Teece, 1982; Teece, Pisano & Shuen, 1997), absorptive (Cohen and Levinthal, 1990) and appropriative capacities (Bonnin, 2006; Kolk & Püümann, 2008). The author has worked out on various aspects of this new model with the owners of the SME over a period of eight months. The chapter discusses about the frame of reference, the research method that has been used, and the results, including the discussion on the results and the direction for future research before the conclusion.

BACKGROUND

The current literature on open innovation provides little theoretical structuring (Van de Vrande et al., 2010). It first focuses on describing the contexts which show how some companies like IBM, Xerox and others[1] were able to make a comeback at a crucial point in their history (Chesbrough, 2003) in the United States. Then, the literature has moved forward to Europe (Lazzarotti, Manzini & Pellegrini, 2010; Lichtenthaler, 2008; Stoyanov, 2008) and other places around the world. Authors such as Chesbrough et al. (2006) have called open innovation a "new paradigm" as they have tried to define the economic realities associated with it. But open innovation itself remains as a question, what is it?

Open innovation (OI), largely defined as "a wider set of contributions into a funnel of corporate development" (Leadbeater, 2010), can be traced back to the 19th Century, sharing to some extent the label of *Collective Invention* (Allen, 1983). This term has been used to express how the Iron industry, as a collective unit, has worked out its development up the World War II, when the R&D laboratories have then emerged. In 20th Century, one has seen very close parallels to OI such as the R&D generations (Wang et Kleiner, 2005) and innovation networks (Julien, 2003).

From 2003 to 2008, OI has been essentially a description of a phenomenon related to innovation in a few large American organizations like IBM, Intel and Xerox (Chesbrough, 2003; Stoyanov, 2008) with a marketing label strong enough to open and strengthen imagination but without any strong scientific foundations (Watson, 2008). Although some organizational capacities such as dynamic, absorptive and appropriative capacities have been referred to in Chesbrough et al. (2006), they have not been yet integrated and validated in a working model. A logical next step is, therefore, to better elaborate the theoretical underpinnings of OI, so as to facilitate its application and further clarify its meaning. First off, the study looks at some societal and technological changes which allow for open innovation.

Societal Change

As mentioned above, the open innovation label has been coined by Chesbrough in 2003. But the concept of being "open" is not new. Among the first authors to write about openness is Bergson (1932). He drew a distinction between a "closed society" and an "open society", in terms of moral and religious norms. Closed societies are typified by a static caste structure (la "cité") while open societies pertain to the entire human species; they are dynamic. But that dynamism is only alive for a while. Once integrated to the existing reality, he said, "it becomes molded with the inherent culture" (p. 144). So the new and "open" become old and static over time. Of course, this idea is not new as the Protestant Reformation, back in the 16th Century, was already cultivating renewal from within as their motto was "*Ecclesia reformata semper reformanda*", meaning that the Reformed Church must continually renew itself.

Following Bergson, Popper (1962; 1966) has also written about the "open society" and its enemies. In his books he makes reference to Plato, Hegel and Marx, arguing that a "closed" society, such as tribal societies, are characterized by habits arising out of an appetite to survive and compete against each other. On the contrary, the "open" society is characterized by its capacity to reason and exert freedom in each and every life decision. Greeks in the B.C. period were expressing both closed and open attitude. Popper finds the origins of the latter in the quest for justice, wisdom, truth and beauty (p. 139). Greeks were exercising "Θεωρια" (theoria) or contemplation. This contemplation contains the idea of "openness".

While Bergson and Popper were writing about "openness" in a variety of fields, during the same period, the label "open" was also applied to education, under the name of "New School"[2], led by Ferrière, Montessori, Freinet and Decroly (1994) in Europe, for instance, and Dewey (1975) in the United States. In the 1970's, Paquette (1976) with his "pédagogie ouverte" (open pedagogy) and "pédagogie fermée" (closed pedagogy) has been among the most influential in Quebec. In fact, that "open" paradigm in Education was already taking form under Rousseau with his seminal book, *Émile, ou de l'Éducation* (1762). Thereafter, a new paradigm in Education was articulated and numerous books and articles have been written to explore dichotomies such as "nurture or culture" or "traditional or person centered education".

Given this human quest for openness, which can be traced back to the Ancient Greeks, one can ask if the label "open innovation paradigm" suggested by (Chesbrough et al., 2006) is fully justified. Once analyzed and understood, the expression of "open innovation" is, in reality, a dichotomous notion: that of "openness" versus "closedness", each an unattainable absolute. In other words, it appears impossible to be one hundred percent "open", and it is also impossible to be one hundred percent "closed" as the human spirit is, in its very essence "open" by virtue of hope and love. Clearly, when it comes to innovation, cultural environment is a key factor for "opening" the mind.

Technical Change

To better understand open innovation, one must examine how technology has changed since World War II (WW2) and how it has prepared firms for "openness". According to De Bandt (2002), the "trente glorieuses" (meaning 30 years post WW2) reflects an exceptional time. In that period, and since then, technological change has played such an important role that it became a paradigm in its own right (De Bandt, 2002) influencing even the school system (Bertrand and Valois, 1992).

Using the Kondratiev cycle, which is used to illustrate long economic cycles such as the First Industrial Revolution, the Railroad and Steam Engine era, the Age of Steel, and Post World War

II, for instance, De Bandt (2002) describes how different factors have contributed to an enormous transition, whose effects can sometimes only become fully visible over the course of about 50 years. Of the 50 social, political and economic transformations he lists, the following have apparently been the most influential in the opening process:

- Rapid growth of multinationals since 1970.
- Oil crisis (1973-1974; 1981).
- Progressive liberalization of exchange and rapid growth of sector-based exchanges.
- The development of scientific and technological politics and innovation since 1970.
- Rapid expansion of the new information and communication technologies since 1980.
- Appearance and development of cooperative relationships between rival firms since 1980.
- Deregulations in many different industries since 1980.
- Informational decentralization since the mid '80s.
- Outsourcing since the mid '80s.
- National systems of innovation since the end of the 1980's.
- Networking development since the end of the 1980's.
- Emergence and rapid development of globalization since 1990.
- Local innovation systems since the mid '90s.
- Explosion of the new economy since the mid '90s.

As one considers De Bandt (2002), the key word here seems to be "transition" as the change has been progressive, rather than abrupt. Because various factors are interrelated, it can be difficult to isolate each variable and understand its particular role and influence. This helps explain how Chesbrough (2003) came to describe the cases of IBM and Xerox in his book[3] and how these companies have evolved during that transition period. What is clear is that these companies did not initiate open innovation nor have they been its champion. Indeed, they have been followers of social, political and economic trends, not "early adopters". For instance, IBM only ventured into new innovation venues when its crisis was at its peak[4]. This contribution provides a more integrative comprehension of OI, providing insight into how it became so relevant to modern business culture.

Amable, Barré and Boyer (1997) have also studied carefully the economic and innovation changes since WW2, and recently, Elci (2009) has identified external drivers of OI such as globalization, the advent of an economic downturn and the transition to a knowledge economy. With these changes at the macroeconomic level, certain adaptations in terms of behavior, rules and policies became possible and new opportunities for development appeared. In other words, from the 1970s up to 2000, it has been observed that new conditions have emerged, forcing or creating new sets of relations between firms, new challenges in terms of adaptation, new possibilities for innovation: creating a "perfect storm" for increasing "openness". In light of these changes, one can deepen the understanding by looking at open innovation from a systemic point of view.

OPEN INNOVATION NEEDS A SYSTEMIC UNDERSTANDING

Open innovation requires a systemic understanding for it is more complex than one assumes at first glance. Such questions as how to implement open innovation into SMEs bring attention on how social, political and technological changes are intertwined. In this section, the study looks at OI from a systemic perspective, while the next one considers how organizational capacities are integrated into the basic OI model.

Innovation System

Maula, Keil and Salmenkaita (2006) have used a systemic innovation framework to better understand OI relationships. This framework highlights the differences between autonomous and systemic innovations. The former applies to "closed" innovation, while the latter integrates "suppliers, customers, partners, developer communities and competitors" (p. 249) and leads to "open innovation processes" (p. 254). This type of systemic approach has value in helping to organize the various actors interacting with one's firm, so as to optimize the allocation of resources, including time, in the process leading up to commercialization.

But another form of systemic understanding can be useful for deepening the understanding of OI. Even though systemic comprehension of innovation (Maula et al., 2006) and value networks (Vanhaverbeke & Cloodt, 2006) are important components in OI, this analysis points to innovation systems (IS) being derived from national, regional, and local innovation agendas.

Héraud and Nanopoulos (1994) have studied innovation in Alsace (France), and have found that firms' innovation often has its origins upstream rather than from within the firm. That means that social, political, and economic factors as well as rules and laws are key determinants of innovation. This has also been reported from SMEs in the Eastern Townships and Beauce (Quebec) through the work of Doloreux and Melançon (2007). Both have found that these influences are geographically dispersed but play a significant role in the whole innovation process.

Considering the variety of contexts that have been described before including globalization, innovation systems (IS) need to be understood when examining OI. This study makes a distinction here between what Maula et al. (2006) and Bertalanffy (1973) wrote about systemic approaches. While the former applies systemic innovation to a group of firms having a common value proposition and sharing value capture, the latter integrates other systems, be they social, political, economic or any other that have a significant influence. It is from this macro level perspective that one best determine whether OI is a new paradigm or not (Chesbrough et al., 2006). If one considers that OI is a set of relationships between organizations working to deliver some innovation to the market, then hardly it can be called as a new paradigm. Following this logic, the study sees OI as a model of innovation coming out of a paradigm, not a paradigm in itself.

Organizational Capacities

Dynamic (Teece, 1982), absorptive (Cohen and Levinthal, 1990), and appropriative capacities (Bonnin, 2006; Kolk and Püümann, 2008) have been discussed over the past 25 years and are widely recognized. Teece, Pisano and Shuen (1997) define the dynamic capabilities as "the firm's ability to integrate, build, and reconfigure internal and external competences to address rapidly changing environments" (p. 516). The term *dynamic* refers to the "capacity to renew competences so as to achieve congruence with the changing environment" [...] while the term *capabilities* "emphasizes the key role of strategic management in appropriately adapting, integrating, and reconfiguring internal and external organizational skills, resources, and functional competences to match the requirements of a changing environment" (p. 515).

For Fosfuri and Tribo (2006), but especially for Cohen and Levinthal (1990), absorptive capacities mean the "ability of a firm to recognize the value of new, external information, assimilate it, and apply it to commercial ends" (p. 128). In order to recognize that new knowledge, one has to better develop adequate prior knowledge. For example, companies which have a strong knowledge-base in R&D will be better placed to understand the R&D of competitors and eventually to adapt it for their own context. It follows that the definition of appropriative capacities, proposed by Laffey (2004, *in*Bonnin, 2006), describes observed practice: it is "the process of taking something that belongs to others and making it one's own" (p. 50). This applies to technological innovation as well as to innovation processes.

In this research, the author has used and integrated those three organizational capacities in an attempt to enhance the basic technological OI model, on the one hand, and to get a better workable OI model, appropriate for applications with small and medium size businesses, on the other. Figure 1 reflects that attempt. As one can see, the characteristics of technological OI are mentioned on the left side of the figure while the three organizational capacities are represented on the right. Using the technological innovation in the figure shows how a business model can look or how new types of relationship with other actors can be developed.

DEFINITION OF CONCEPTS

Innovation has multiple definitions, ranging from invention-innovation-diffusion (*cf.* Schumpeter, 1947) to a set of 22 types of innovations under the umbrella of the world/the industry/the firm (Garcia & Calantone, 2002). It also ranges from technological and organizational innovations to innovation of commercialization (OECD, 2005). In most cases, innovation means "offering new products or services to consumers". With that in mind, Chesbrough (2003) defines open innovation as

> ... *a paradigm that assumes that firms can and should use external ideas as well as internal ideas, and internal and external paths to market, as the firms look to advance their technology. Open Innovation combines internal and external ideas into architectures and systems whose requirements are defined by a business model. The business model utilizes both external and internal ideas to create value, while defining internal mechanisms to*

Figure 1. Theoretical OI model integrating organizational capacities

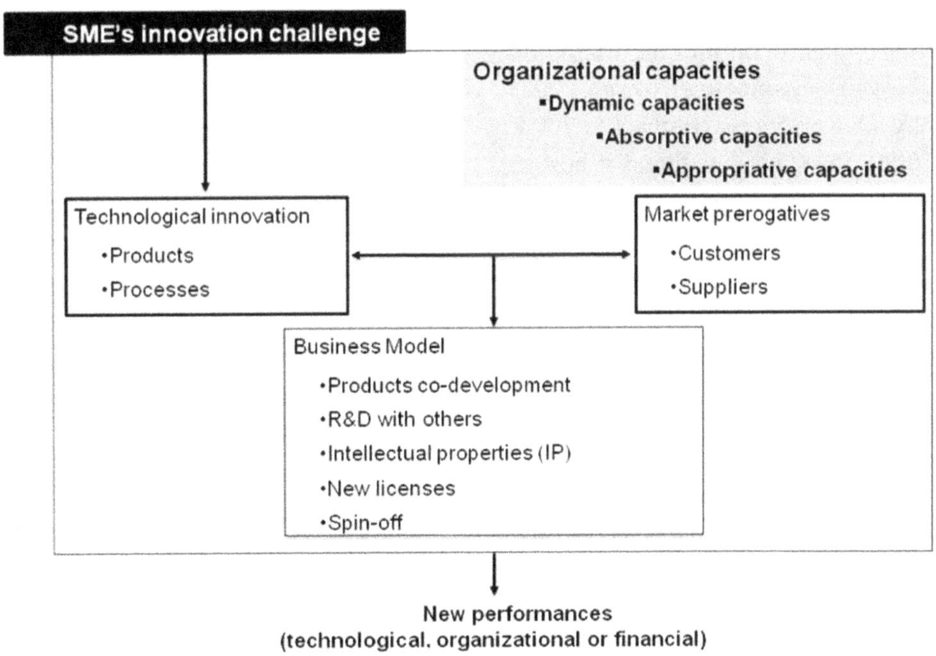

claim some portion of that value. Open Innovation assumes that internal ideas can also be taken to market through external channels, outside the current businesses of the firm, to generate additional value (p. xxiv).

Since Chesbrough (2003), some authors have offered other definitions. For example, Chesbrough *et al.* (2006) are saying that OI is "both a set of practices for profiting from innovation, and also a cognitive model for creating, interpreting, and researching those practices" (p. 286). For Leadbeater (2010), open innovation has two faces: an "IN" component according to which basic ideas originate from different sources ("crowdsourcing"). This type of open innovation is characterized by a funnel-model, that is to say that the ideas gathered converge toward a goal of business performance. The 'OUT' component, for its part, represents a type of common platform where it is possible both to contribute and to collect benefits. This definition includes the collective knowledge of individuals ("open source").

For the purposes of this research, the author has adopted the definition of Leadbeater (2010) with the use of the "IN" component, which represents a basic OI model. Moreover, instead of adopting the concept of paradigm or limiting the discussion to emerging best business practice (Rowell, 2008) or philosophy and practice (Hastbacka, 2004; Innovaro, 2006), the study has found it best to conceive of open innovation as a model.

In terms of organizational capacities, the study has adopted the following definitions from the literature review: (a) dynamic capacities: "resources which have been organized, coordinated and connected by economic, cultural and technical structures so as to bring these resources into use by new production entities" (Maillat, Quevit & Senn, 1993, p. 9), (b) absorptive capacity: "ability of a firm to recognize the value of new, external information, assimilate it and apply it to commercial ends" (Cohen & Levinthal, 1990, p. 128), and (c) capacity of appropriation: " that the person has power over the environment through the sensorial, perceptual and motor activity." (Bonnin, 2006, p. 53)

THE TECHNOLOGICAL AND ORGANIZATIONAL OPEN INNOVATION MODEL

The open innovation model of type IN or basic model of open innovation contains several elements such as co-product development, joint research and development (R&D), intellectual property (IP), new licenses and spin-off (Chesbrough, 2006). By integrating the three organizational capacities mentioned above (dynamic, absorptive and appropriative), one obtains a model of open innovation with explicit organizational capacities. Lichtenthaler and Lichtenthaler (2009), Teece (2007) and Van de Vrande et al. (2009) are among the first to incorporate organizational capacities in a theoretical model of open innovation.

Generally, these organizational capacities are distinguished from each other in that (a) the dynamic capacities are associated with the reorganization of human resources based on business strategies, (b) the absorptive capacities facilitate the acquisition of knowledge from outside of their application, and (c) the appropriative capacity represents the contribution through personal commitment of employees. These organizational capacities can sustain innovation in the world of business. By expanding the basic model of open innovation to include organizational capacities, business performance, whether technological, organizational or financial, is enhanced (Evan, 2009; Hastbacka, 2004).

According to Lichtenthaler and Lichtenthaler (2009), Teece (2007) and Van de Vrande et al. (2009), the more present and well-developed these capacities are in a given enterprise, the more likely it is that open innovation can be successfully implemented in that enterprise. In contrast, enterprises which lack organizational capacities will have limited possibilities for technological innovation. In the context of globalization, these capacities are of utmost importance. It is, therefore, necessary to accord them special attention. Indeed, could a company succeed in multi-faceted innovation without demonstrating strong organizational capacities? The challenge here is eminently practical, especially if ongoing innovation is desired. In light of this principle, one wonders how innovative SMEs are able to deal with these capacities. Are they present? How do they manage them? How do these capacities influence their business model?

RESEARCH METHOD

This research has used the basic open innovation model and enhanced it with organizational capacities built into the framework of an eight-month long case study (Eisenhardt, 1989) of a small agrifood SME. The study has used the methodological framework provided by Eisenhardt (1989) and Miles and Huberman (1984) because it facilitates the development of theoretical analyses. Of course, some approaches and operational realities of SMEs are different from those of larger companies. It was thus important to gather evidence that would be relevant to their dynamics of open innovation. In doing so, the study thus adopted a constructivist approach (Dyer and Wilkins, 1991), which facilitated the shift from our structural model (Figure 1) to a procedural model (Figure 2). The description of the company, its context and the tools used for data collection are being discussed next.

Description of Business

The company that has been analyzed is called Gaspor, Fermes St-Canut Farms. Gaspor is a spin-off of 347 Canada Inc.[5], which is a family-owned company, producing pork. Both companies are located in St-Canut, Mirabel, Quebec. 347 Canada Inc. produces over 16.000 pigs annually for the consumer market. In 2005, Gaspor won the Young Entrepreneurs of the Year Prize in its region and, in 2007, the Research and Development Award – for emerging innovation. Gaspor is

Figure 2. Gaspor's open innovation process model

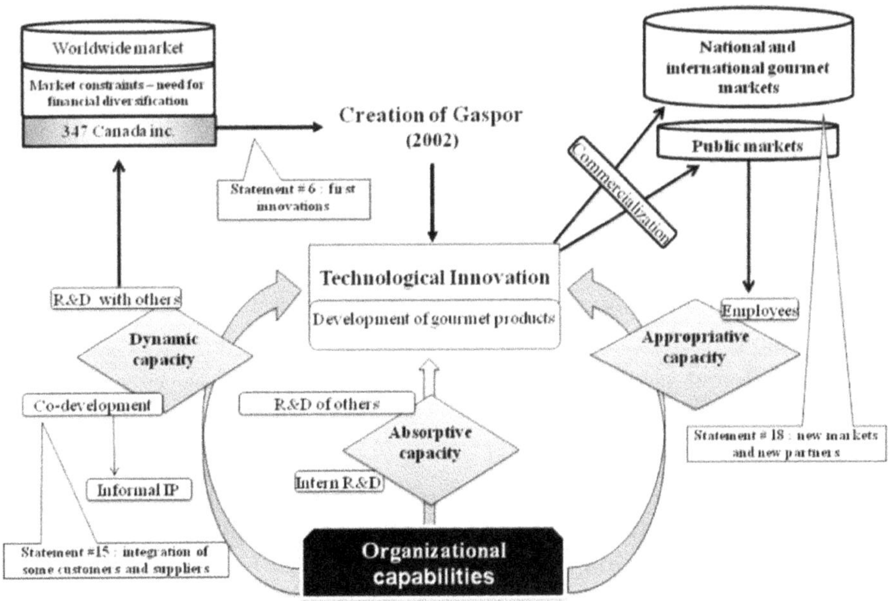

located on the same site as the mother company and takes advantage of the proximity of its facilities and amenities.

Gaspor is an offshoot of 347 Canada Inc. with a mandate to diversify, in a crisis situation where profitability was a key factor. The goal was to produce milk-fed piglets for gourmet markets. Technological innovations were required on the part of both companies in order to develop a new set of products of superior organoleptic qualities. These products were developed using R&D, which is confirmed by Canada Revenue Agency (CRA) having acknowledged the development under its program of scientific research and experimental development (SR & ED). Gaspor also distinguished itself by entering into co-product development with some of the greatest culinary chefs in Quebec.

BACKGROUND RESEARCH

This research with Gaspor took place from January to August 2009. During this period, there were about nine meetings of three hours each, including semi-structured interviews and data collection with the three owners. These were followed by a series of telephone conversations totaling about one hour each week. The topics covered were the following:

- Initial contact with the company and presentation of the research objectives.
- Description of business parameters and needs in terms of innovation.
- Presentation of open innovation framework and discussion.
- Presentation of different business models and discussion.
- Semi-structured interviews and collecting data on business models.
- Collecting data on the characteristics of Gaspor's model of open innovation.
- Proposal of a business model based on open innovation.

Data Collection

The Positioning of the Business Model

The objective of the data collection was twofold. First, the study has used the six categories of the business models of Chesbrough (2006). The categories consist of 27 statements (Appendix 1) divided among six models as follows: (1) undifferentiated, (2) differentiated, (3) segmented (4) externally aware (5) integrated with business model (6) platform player shapes markets. It was interesting to see what relevance these concepts would have for a small enterprise, in particular, for they were initially developed with reference to large U.S. companies. Was it possible to directly apply those statements to a SME?

Business Review Based on Open Innovation with Organizational Capacities

For the second step, the study has developed a semi-structured interview guide (Appendix 2) based on the basic model of open innovation and the three organizational capacities identified earlier. For methodological rigor, it used the criteria of Whittemore, Chase and Mandle (2001), which are primary (credibility, authenticity, integrity, critical thinking) and secondary (explicitness, clarity, creativity, completeness, consistency and sensitivity). Later on each owner of the company were interviewed.

RESULTS

The results obtained from all the data collected are being presented next.

Results from Analyzing the Business Models

As mentioned earlier, the study presented the 27 statements to the three owners of Gaspor to collect their responses on each one of the six models. Each respondent had the opportunity to check "A" or "B" or both (Table 1). An "A" indicated that the owner recognized the statement as being a part of their day-to-day operations while a "B" indicated the ideal business model one might dream of. When both options were checked, the "A" indicated the statement reflected the company's current business model, although to a limited degree, while the "B" indicated the statement described a practice that could be developed further in the future. This was the case with two of the models in particular: i.e. "externally aware" and "platform player shapes markets". The time dimension has not been taken into account in the list in order to reflect the fact that an ideal model might be desired, but might never be attained.

In Table 1 below, the six business models are represented, along with the number of statements by model. The possible total in theory is calculated by multiplying the number of utterances by the number of respondents. So, for undifferentiated and differentiated models we have respectively 12 (4×3) and nine (3×3) possible answers, and so on. As for the actual total, it is the cumulative of "A" and "B". Note the perfect agreement between theoretical and the actual total for the "segmented" business model. Thus, respondents clearly distinguish between statements representing their current business model and a business model which might be possible in the near future.

The distribution of responses to "A" and "B" to the statements could also indicate a positive discrimination. Indeed, while there was little recognition of the undifferentiated model as applicable to their small company (Current model [A], n = 1), they also saw little interest in that model for the future (Ideal model [B], n = 1). Similarly, we note that they do not see themselves as having a strong externally aware business model (Current model [A], n = 3), but they express an interest in this model as an ideal to strive towards (Ideal model [B], n = 13).

Results from Interviews

The three owners were then interviewed for close to an hour each to find out about their understanding of the open innovation model enhanced with organizational capacities. It has been found that, despite its small size, Gaspor applies several aspects of the basic model of open innovation and certain organizational capacities foster its ongoing development.

Table 1. Current and ideal Gaspor's business model

Model names	Type 1 Undifferentiated	Type 2 Differentiated	Type 3 Segmented	Type 4 Externally aware	Type 5 Integrated with business model	Type 6 Platform player shapes markets
Statements per model	4	3	5	5	5	5
Number of respondents	3	3	3	3	3	3
Theoretical total	12	9	15	15	15	15
Current model (A)	1	4	5	3	5	7
Ideal model (B)	1	5	10	13	12	11
Total responses	2	9	15	16	17	18

Joint R&D and Co-Product Development

As seen before, Gaspor was established when its parent company, 347 Canada Inc., wanted to produce a milk-fed piglet for the finest restaurants in Quebec, Canada, United States of America and Japan. The two companies have conducted research and development (R&D) and have developed their experimental design jointly. 347 Canada Inc. developed a dependable product, the piglets fed only with milk, while Gaspor has developed the meat cuts and the organoleptic qualities needed for competition in the world class market. With respect to co-development of products, Gaspor had suffered some failures in the past, but had succeeded with others. In the present case, several of the best chefs of Quebec contributed to the co-development of the product while joint R&D has been done with 347 Canada Inc.

Intellectual Property Rights

Intellectual property (IP), such as patents, are rather rare in the area of meat cutting and the development of organoleptic properties, which limits Gaspor's opportunities in this area. The results of their R&D were primarily aimed at characterizing certain cuts of the piglet. It was, therefore, difficult to target a specific technology for designation as a protected format, which would confer a competitive advantage. In this case, the term "informal intellectual property" might be more appropriate. At the time of collecting data, no IP had yet been purchased nor sold by the company.

Organizational Capacity

The interviews revealed that some aspects of dynamic, absorptive and appropriative capacities underpin Gaspor's technological innovation. The management roles of the owners had been defined from its inception in 2002; however, those affecting R&D, exportation, transportation and public communication have been redesigned since then in response to challenges as they arose. Thus the research sees that even on a small scale, roles are updated and the desire to have the best person in the right location is constant. This is reflected throughout the interviews on dynamic capabilities.

With respect to absorptive capacity, the company said that since it has been engaged in R&D, it is better able to decode R&D done elsewhere and incorporate it into their operations. One owner stated: "No doubt about it... Whatever the product, I tell people [...], there is a lot of potential. I knew nothing about research and development before, but now, I see companies benefiting from it. They are constantly evolving".

Finally, some appropriative capacities are present in that individual employees hired to sell in local public markets freely communicate their ideas about innovation. These ideas are discussed, approved and implemented, when appropriate, as part of a process where employees can take on new sets of responsibilities.

The study has summarized the different results in Figure 2. Initially, given the state of the global pork market, 347 Canada Inc. was obliged to find ways to increase income diversification. This established the rationale for creating a spinoff company that has been mandated to market specialty products for the gourmet market. The R&D for these new products was accomplished jointly with the mother company, 347 Canada Inc., in co-development with some well-known chefs. As these products were sold to specialty restaurants in North America and Japan, some of these restaurants were then integrated into the continuous creative process. By conducting R&D, Gaspor acquired some knowledge and developed certain skills that allowed them to better understand the R&D done elsewhere, all of which had a positive impact in the company. At this stage of its development, the only new hiring is for employees who work in public markets. Nevertheless, they are given a role in the company's ongoing innovation which allows them to contribute to improving its technological innovation. Finally, some links in the Figure 2 are supported by some statements of the list mentioned in Appendix 1.

DISCUSSION

The recent literature on open innovation provides fresh insight into changing realities influencing business and the complexity of business relationships. However, it is clear that the basic model of open innovation is insufficient, by itself, to analyze the reality of SMEs. Indeed, relations between firms in terms of joint R&D, IP or the creation of spin-off companies, or evidence of how globalization creates new market imperatives, provide very little guidance for success with open innovation. This is why it seemed appropriate to add specific organizational approaches such as dynamic, absorptive and appropriative capacities to the basic open innovation model, all the more so because it is in this direction Lichtenthaler and Lichtenthaler (2009), Teece (2007) and Van de Vrande et al. (2009) were moving as well.

In making this case study, it was expected to see how a small enterprise would interact with the enhanced OI model. One of the questions was: How exactly does the company studied make use of open innovation? The study has made three observations. Firstly, it has been noted that certain features of the basic open innovation model such as joint R&D and co-product development have been incorporated into Gaspor's business culture. This aspect is particularly interesting because it suggests that open innovation has a transcendent value. In other words, Gaspor has demonstrated its permeability to external socioeconomic influences. Its participation in open innovation need not be explicitly formalized. Referring to open innovation as a "paradigm" (Chesbrough et al., 2006) would seem to be confirmed by this observation, although, in this case, the incentive for the OI paradigm was almost certainly socio-economic.

The second observation concerns the complexity of open innovation, from a managerial point of view. Although Gaspor has adopted key methods associated with the basic model of open innovation, it encountered some difficulty in understanding the links between these approaches. In other words, when taken one by one, the components of open innovation are easily understandable, but integrating them into a coherent and synergistic program takes time and dedication. But finding time cannot be taken for granted in small enterprises such as Gaspor where entrepreneurs may be saddled with an impossible workload. In addition, since financial resources are limited for most of them (Julien, 2008), budgeting time for strengthening their open innovation would be required. With respect to Gaspor, the company is aware of these requirements as well as its own limitations and realizes that implementing open innovation on a larger scale would require more resources.

The third observation relates to the type of business. For a food processing company that works with living creatures, it seems that the open innovation model developed by Chesbrough (2003) is less well suited to this type of business. Indeed, Chesbrough (2003) describes large U.S. companies whose activities are in the field of semiconductors or computers. However, these products have no specific expiration date. They can spend months on the shelves without decay. But the context of our SME contains an additional variable: live perishable products. Indeed, milk-fed piglets reach a specific weight within a limited number of days. When this equation is disrupted, the entire chain of production and distribution suffers, especially when it comes to products which are sold fresh. The company's business model must consider first and foremost the constraints of raising live animals. Technological innovations, backed up by strong organizational capacities, can then be integrated into the model.

FUTURE RESEARCH DIRECTIONS

It seems clear that more research is needed to better understand the relationship between technological innovation and organizational capacities. Open innovation will ideally be seen as a model, rather than a paradigm. This helps make clear how flexible an open innovation model can be. The expression "one size fits all" is far from being realistic when working with SMEs. Each one has its own value and idiosyncrasies so the open innovation model should always be tailored to make a perfect fit. This will help to identify with more accuracy the factors for success or failure.

Another avenue for future research concerns the socio-economic context which can have a strong impact on the likelihood that a given business will explore using OI and ultimately, of successful implementation of OI. As it has been seen before, innovation within SMEs depends, of course, on their own efforts, but also on external efforts, on a regional or national scale. To study open innovation at this larger scale would certainly be an asset.

Lastly, another area for future study would focus on how an enhanced open innovation model could work during times of economic recession. The basic open innovation model takes for granted that the economy will continue to expand. But as the whole planet has seen since 2008, this is not a realistic model. When downsizing, one needs to rethink what innovation means and how it can serve people differently.

CONCLUSION

The research objective was to describe the process of open innovation and validate the presence of organizational capacities. This research's working hypothesis was that open innovation is often driven by technological innovation but requires strong organizational capacities in order to succeed. Research examining open innovation from an organizational point of view is still in its infancy. In making a case study with Gaspor, the study has seen how a small company that had never heard of such a form of innovation had yet practiced it empirically for several years.

Thus, the study has found that Gaspor have (a) set aside considerable resources for change and innovation, (b) conducted research and development (R&D) in conjunction with another company, (c) undertaken initiatives to co-develop products in an environment where intellectual property was based solely on a handshake, (d) had the same business model for the five years since its inception, and (e) shown a strong determination to expand on its ventures by making use of proven organizational capacities.

In addition, it has been observed that although the agrifood business company in the study was already familiar with the characteristics of open innovation, they still found it challenging to develop the organizational capacities to ensure success. This result showed, among other things, that in addition to the encouragement, which can be found in various magazine articles, for adopting open innovation as a "new" form of innovation, the company was shaped by a wide range of socio-economic trends. What proved most difficult was to integrate the various components of open innovation and to fully understand their relationships with each other. The study, therefore, concludes that more information and training would be necessary for dealing with the relationship between collaborative R&D and absorptive capacity or between dynamic capacities and co-product development, for example.

ACKNOWLEDGMENT

The author is grateful to the Fonds québécois de recherche sur la société et la culture (FQRSC) and to Gaspor for their time and financial support in carrying out this research.

REFERENCES

Allen, R. C. (1983). Collective invention. *Journal of Economic Behavior & Organization, 4*(1), 1–24. doi:10.1016/0167-2681(83)90023-9

Amable, B., Barré, R., & Boyer, R. (1997). *Les systèmes d'innovation à l'ère de la globalisation*. Paris, France: Economica.

Bergson, H. (1932). *Les deux sources de la morale et de la religion*. Paris, France: Les Presses universitaires de France, 58e édition, version électronique (Chicoutimi).

Bertrand, Y., & Valois, P. (1992). *École et sociétés*. Laval, Canada: Éditions d'Agence d'Arc, coll. Éducation.

Bonnin, G. (2006). Physical environment and service experience: An appropriation-based model. *Journal of Service Research, 6*, 45–65.

Chesbrough, H. (2003). *Open innovation. The new imperative for creating and profiting from technology*. Boston, MA: Harvard Business School Press.

Chesbrough, H. (2006). *Open business models. How to thrive in the new innovation landscape*. Boston, MA: Harvard Business School Press.

Chesbrough, H., Vanhaverbeke, W., & West, J. (2006). *Open innovation. Researching a new paradigm*. New York, NY: Oxford University Press.

Cohen, W. M., & Levinthal, D. A. (1990). Absorptive capacity: A new perspective on learning and innovation. *Administrative Science Quarterly, 35*(1), 128–152. doi:10.2307/2393553

De Bandt, J. (2002). L'émergence du nouveau système technique ou sociotechnique. *Revue d'économie Industrielle, 100*(1), 9-38.

Dewey, J. (1975). *Démocratie et éducation: Introduction à la philosophie de l'éducation*. Paris, France: A. Colin, coll. U.

Doloreux, D., & Melançon, Y. (2007). Réseaux d'innovation dans les PME en Estrie. *Revue Internationale PME, 20*(1), 69–90.

Dyer, W. G., & Wilkins, A. (1991). Better stories, not better constructs, to generate better theory: A rejoinder to Eisenhardt. *Academy of Management Review, 16*(3), 613–619.

Eisenhardt, K. M. (1989). Building theories from case study research. *Academy of Management Review, 14*(4), 532–550.

Elci, N. (2009). *An exploration of open innovation: An environmental scanning perspective*. Doctoral dissertation, University of Maryland. Ann Arbor, MI: UMI Dissertation Publishing.

Elmquist, M., Fredberg, T., & Ollila, S. (2009). Exploring the field of open innovation. *European Journal of Innovation Management, 12*(3), 326–345. doi:10.1108/14601060910974219

Evan, F. (2009). *Defining success in open innovation*. White paper. Retrieved from www.ninesigma.com.

Fosfuri, A., & Tribo, J. A. (2006). Exploring the antecedents of potential absorptive capacity and its impact on innovation performance. *OMEGA – The International Journal of Management Science, 36*(2), 173-187.

Fredberg, T., Elmquist, M., & Ollila, S. (2008). *Managing open innovation. Present findings and future directions*. Vinnova Report: Swedish Governmental Agency for Innovation Systems.

Garcia, R., & Calantone, R. (2002). A critical look at technological innovation typology and innovativeness terminology: A literature review. *Journal of Product Innovation Management, 19*(2), 110–132. doi:10.1016/S0737-6782(01)00132-1

Hastbacka, M. A. (Dec. 2004). Open innovation: What's mine is mine... What if yours could be mine, too? *Technology Management Journal*, December, 1-4.

Héraud, J.-A., 7 Nanopoulos, C. (1994). Les réseaux de l'innovation dans les PMI: Illustration sur le cas de l'Alsace. *Revue Internationale PME, 7*(3-4), 65-86.

Houssaye, J. (1994). *Quinze pédagogues: Leur influence aujourd'hui: Rousseau, Pestalozzi, Fröbel, Robin, Ferrer, Steiner, Dewey, Decroly, Montessori, Makarenko, Ferrière, Cousinet, Freinet, Neill, Rogers*. Paris, France: A. Colin, coll. Formation des enseignants.

Huang, F., & Rice, J. (2009). The role of absorptive capacity in facilitating open innovation outcomes: A study of Australian SMEs in the manufacturing sector. *International Journal of Innovation Management, 13*(2), 201–220. doi:10.1142/S1363919609002261

Innovaro (2006). *Innovation briefing 07-06: Open innovation*. London, UK: Innovaro Innovation Briefing. Retrieved June 16, 2009, from http://www.innovaro.com/inno_updates/Innovation%20 Briefing%2007-06.pdf

Julien, P.-A. (Dir.). (2003). *L'entreprise-réseau: Dix ans d'expérience de la Chaire Bombardier Produits récréatifs*. Sainte-Foy, Canada: Presses de l'Université du Québec, coll. Entrepreneuriat & PME.

Julien, P.-A. (2008). Réseautage riche et croissance. In C. Schmitt (Dir.), *Regards sur l'évolution des pratiques entrepreneuriales* (pp. 103-122). Québec, Canada: Presses de l'Université du Québec.

Kolk, A., & Püümann, K. (2008). Co-development of open innovation strategy and dynamic capabilities as a source of corporate growth. *Working Papers in Economics, 25*(168-180), 73-83.

Lazzarotti, V., Manzini, R., & Pellegrini, L. (2010). Open innovation models adopted in practice: An extensive study in Italy. *Measuring Business Excellence, 14*(4), 11–23. doi:10.1108/13683041011093721

Leadbeater, C. (2010). *Two faces of open innovation*. Retrieved February 24, 2010, from http://www.openbusiness.cc/ 2007/03/14/two-faces-of-open-innovation/

Lee, S., Park, G., Yoon, B., & Park, J. (2010). Open innovation in SMEs – An intermediated network model. *Research Policy, 39*(2), 290–300. doi:10.1016/j.respol.2009.12.009

Lichtenthaler, U. (2008). Open innovation in practice: An analysis of strategic approaches to technology transactions. *IEEE Transactions on Engineering Management, 55*(1), 148–157. doi:10.1109/TEM.2007.912932

Lichtenthaler, U., & Lichtenthaler, E. (2009). A capability-based framework for open innovation complementing absorptive capacity. *Journal of Management Studies, 46*(8), 1315–1338. doi:10.1111/j.1467-6486.2009.00854.x

Maillat, D., Quévit, M., & Senn, L. (1993). *Réseaux d'innovation et milieux innovateurs: Un pari pour le développement régional*. Neuchâtel, Switzerland: GREMI/EDES.

Maula, M., Keil, T., & Salmenkaita, J. P. (2006). *Open innovation in systemic innovation contexts*. In H. Chesbrough, W. Vanhaberbeke & West, J. (Dir.). (2006). *Open innovation: Researching a new paradigm*. New York, NY: Oxford University Press.

Miles, M., & Huberman, A. M. (1984). *Qualitative data analysis: A sourcebook of new methods*. California: Sage Publications.

OECD, Manuel d'Oslo. (2005). *Principes directeurs pour le recueil et l'interprétation des données sur l'innovation. La mesure des activités scientifiques et technologiques*. Commission européenne: Éditions OCDE, 3e éd.

Paquette, C. (1976). *Vers une pratique de la pédagogie ouverte*. Laval, Canada: Éditions N.H.P.

Popper, K. (1962). La société ouverte et ses ennemis.: *Vol. 1. L'ascendant de Platon*. Paris, France: Éditions du Seuil.

Popper, K. (1966). La société ouverte et ses ennemis.: *Vol. 2. Hegel et Marx*. Paris, France: Éditions du Seuil.

Rowell, A. (Sept. 2008). *Open innovation: The new R&D paradigm driving profitable product development*, (pp. 1-5). Aberdeen Group: Analyst Insight.

Schumpeter, J. A. (1947). The creative response in economic history. *The Journal of Economic History*, *7*(2), 149–159.

Stoyanov, S. (2008). *Linking value creation and value capture to open innovation*. Hasselt, Belgique: Université Hasselt, mémoire de maîtrise.

Teece, D., Pisano, G., & Shuen, A. (1997). Dynamic capabilities and strategic management. *Strategic Management Journal*, *18*(7), 509–533. doi:10.1002/(SICI)1097-0266(199708)18:7<509::AID-SMJ882>3.0.CO;2-Z

Teece, D. J. (1982). Towards an economic theory of the multiproduct firm. *Journal of Economic Behavior & Organization*, *3*(1), 39–63. doi:10.1016/0167-2681(82)90003-8

Teece, D. J. (2007). Explicating dynamic capabilities: The nature and microfoundations of (sustainable) enterprise performance. *Strategic Management Journal*, *28*(13), 1319–1350. doi:10.1002/smj.640

van Bertalanffy, L. (1993). *Théorie générale des systèmes*. Paris, France: Dunod.

Van de Vrande, V., de Jong, J. P. J., Vanhaverbeke, W., & de Rochemont, M. (2008). *Open innovation in SMEs: Trends, motives and management challenges*. Scientific, Analysis of Entrepreneurship and SMEs (SCALES), Working paper.

Van de Vrande, V., de Jong, J. P. J., Vanhaverbeke, W., & de Rochemont, M. (2009). Open innovation in SMEs: Trends, motives and management challenges. *Technovation*, *29*(6-7), 423–437. doi:10.1016/j.technovation.2008.10.001

Van de Vrande, V., Vanhaverbeke, W., & Gassman, O. (2010). Broadening the scope of open innovation: Past research, current state and future directions. *Technology Management*, *52*(3/4), 221–235.

Van der Meer, H. (2007). Open innovation – The Dutch treat: Challenges in thinking in business models. *Creativity and Innovation Management*, *16*(2), 192–202. doi:10.1111/j.1467-8691.2007.00433.x

Vanhaverbeke, W., & Cloodt, M. (2006). *Open innovation in value networks*. In H. Chesbrough, W. Vanhaverbeke, & J. West (Dir.), *Open innovation. Researching a new paradigm*. New York, NY: Oxford University Press.

Wang, J., & Kleiner, B. H. (2005). The evolution of R-D management. *Management Research News*, *28*(11/12), 88–95. doi:10.1108/01409170510785273

Watson, R. (2008). *Open innovation and other foolish ideas*. Retrieved July 4, 2009, from http://www.fastcompany.com/resources/innovation/watson/open-innovation/061107.html

Whittemore, R., Chase, S. K., & Mandle, C. L. (2001). Validity in qualitative research. *Qualitative Health Research*, *11*(4), 522–537. doi:10.1177/104973201129119299

ADDITIONAL READING

Ambert, A. M., Adler, P. A., Adler, P., & Detzner, D. F. (1995). Understanding and evaluating qualitative research. *Journal of Marriage and the Family*, *57*(4), 879–893. doi:10.2307/353409

Arbussà, A., & Coenders, G. (2007). Innovation activities, use of appropriation instruments and absorptive capacity: Evidence from Spanish firms. *Research Policy*, *36*(10), 1545–1558. doi:10.1016/j.respol.2007.04.013

Ayerbe, C. (2006). Innovations technologique et organisationnelle au sein de PME innovantes: complémentarité des processus, analyse comparative des mécanismes de diffusion. *Revue internationale PME, 19*(1), 9-34.

Bartunek, J. (1984). Changing interpretative schemes and organizational restructuring: The Example of a Religious Order. *Administrative Science Quarterly*, *29*(3), 355–372. doi:10.2307/2393029

Becker, M.C. and Zirpoli, F. (2007). *Organizing open innovation: The role of competences, modularity, and performance integration.* Communication présentée à l'Academy of Management Proceedings.

Blackwell, K. and Fazzina, D. (2008). *Open innovation: Facts, fiction, and future*. Tolland: A Nerac Publication, 1-15.

Blais, A. (1987). Les indicateurs. *In* B. Gauthier (dir.), *Recherche sociale: de la problématique à la collecte de données* (p. 153-173). Québec: P.U.Q.

Bughin, J. R., Chui, M., & Johnson, B. (2008). The next step in open innovation. *The McKinsey Quarterly*, (4): 112–122.

Chesbrough, H. (2007). Business model innovation: it's not just about technology anymore. *Strategy and Leadership*, *3*(6), 12–17. doi:10.1108/10878570710833714

Chesbrough, H., & Crowther, A. K. (2006). Beyond high tech: early adopters of open innovation in other industries. *R & D Management*, *36*(3), 229–236. doi:10.1111/j.1467-9310.2006.00428.x

Chesbrough, H., & Rosenbloom, R. S. (2000). *The role of the business model in capturing value from innovation: Evidence from XEROX corporation's technology spinoff companies*. Boston, Massachusetts: Harvard Business School.

Chesbrough, H. and Schwartz, K. (2007). Innovating business models with co-development partnership *Research Technology Management, 50*(1), 55-59.

Christensen, J. F., Olesen, M. H., & Kjær, J. S. (2005). The industrial dynamics of open innovation – Evidence from the transformation of consumer electronics. *Research Policy*, *34*(10), 1533–1549. doi:10.1016/j.respol.2005.07.002

De Wit, J., Dankbaar, B., & Vissers, G. (2007). Open innovation: The new way of knowledge transfer? *Journal of Business Chemistry*, *4*(1), 11–19.

Deshpande, R., & Farley, J. U. (2004). Organizational culture, market orientation, innovativeness, and firm performance: An international research odyssey. *International Journal of Research in Marketing*, *21*(1), 3–22. doi:10.1016/j.ijresmar.2003.04.002

Dittrich, K., & Duysters, G. (2007). Networking as a means to strategy change: The case of open innovation in mobile telephony. *Journal of Product Innovation Management*, *24*(6), 510–521. doi:10.1111/j.1540-5885.2007.00268.x

Dodgson, M., Gann, D., & Selter, A. (2006). The role of technology in the shift towards open innovation: The case of Proctor & Gamble. *R & D Management*, *36*(3), 333–346. doi:10.1111/j.1467-9310.2006.00429.x

Ellonen, H.-K., Wikström, P., & Jantunen, A. (2009). Linking dynamic-capability portfolios and innovation outcomes. *Technovation*, *29*(11), 753–762. doi:10.1016/j.technovation.2009.04.005

Fan, T., Jones, N., Kumaraswamy, A., Narasimhalu, D., Phan, P., & Tschang, T. (2004). Dynamic capabilities: New sources of industrial competitiveness? *International Engineering Management Conference, 2*(18-21 oct.), 666-668.

Gassmann, O. (2006). Opening up the innovation process: towards an agenda. *R & D Management, 36*(3), 223–228. doi:10.1111/j.1467-9310.2006.00437.x

Glaser, B. G., & Strauss, A. (1967). *The discovery of grounded theory: strategies for qualitative research*. New York: Aldine, Hawthorne, éd.

Hammersley, M. (1992). *What's wrong with ethnography? Methodological exploration*. London: Routledge.

Harison, E., & Koski, H. (2010). Applying open innovation in business strategies: Evidence from Finnish software firms. *Research Policy, 39*(3), 351–359. doi:10.1016/j.respol.2010.01.008

Hashai, N., & Almor, T. (2008). R&D intensity, value appropriation and integration patterns within organizational boundaries. *Research Policy, 37*(6-7), 1022–1034. doi:10.1016/j.respol.2008.03.001

Hogan, J. (2005). Open innovation or open house: how to protect your most valuable assets. *Medical Advice Technology, 16*(3), 30–31.

Hugues, B., & Wareham, J. (2010). Knowledge arbitrage in global pharma: a synthetic view of absorptive capacity and open innovation. *R & D Management, 40*(3), 324–343. doi:10.1111/j.1467-9310.2010.00594.x

Innovations (2007). The era of openness. Open innovation is now a recognized management tool. *Strategic Direction, 23*(1), 35-37.

Jacobides, M. G., Knudsen, T., & Augier, M. (2006). Benefiting from innovation: Value creation, value appropriation and the role of industry architectures. *Research Policy, 35*(8), 1200–1221. doi:10.1016/j.respol.2006.09.005

Janesick, V. J. (1994). The dance of qualitative research design: Metaphor, methodolatry and meaning. In *N.K. Denzin and Y.S. Lincoln (dir.). Handbook of qualitative research* (pp. 209–219). Thousand Oaks, CA: Sage.

Jusko, J. (2008). Kraft crafts an open innovation strategy. *Industry Week/IW, 257*(12), 60-61.

Keupp, M. M., & Gassmann, O. (2009). Determinants and archetype users of open innovation. *R & D Management, 39*(4), 331–341. doi:10.1111/j.1467-9310.2009.00563.x

Kodama, M. (2005). Knowledge creation through networked strategic communities: Case studies on new product development in Japanese companies. *Long Range Planning, 38*(1), 27–49. doi:10.1016/j.lrp.2004.11.011

Langley, A. (1999). Strategies for theorizing from process data. *Academy of Management Review, 24*(4), 691–710.

Laursen, K., & Salter, A. (2006). Open for innovation: The role of openness in explaining innovation performance among U.K. manufacturing firms. *Strategic Management Journal, 27*(2), 131–150. doi:10.1002/smj.507

Leiponen, A., & Byma, J. (2009). If you cannot block, you better run: Small firms, cooperative innovation, and appropriation strategies. *Research Policy, 38*(9), 1478–1488. doi:10.1016/j.respol.2009.06.003

Lichtenthaler, U. (2008). Integrated roadmaps for open innovation. *Industrial Research Institute, Inc.*, May-June, 45-49.

Lichtenthaler, U., & Ernst, H. (2008). Intermediary services in the markets for technology: Organizational antecedents and performance consequences. *Organization Studies, 29*(7), 1003–1035. doi:10.1177/0170840608090531

Lincoln, Y. S. (1995). Emerging criteria for quality in qualitative and interpretative research. *Qualitative Inquiry*, *1*(3), 275–289. doi:10.1177/107780049500100301

Mir, R., & Watson, A. (2000). Strategic management and the philosophy of science: The case for a constructivist methodology. *Strategic Management Journal*, *21*(9), 941–953. doi:10.1002/1097-0266(200009)21:9<941::AID-SMJ141>3.0.CO;2-D

Morgan, L., & Finnegan, P. (2010). Open innovation in secondary software firms: An exploration of managers' perceptions of open source software. *ACM SIGMIS Database*, *41*(1), 76–95.

Morrison, E. (2008). *Open innovation models of economic and workforce development*. White paper version 4.1. Cleveland: Institute for Open Economic Networks, 1-20.

Murovec, N., & Prodan, I. (2009). Absorptive capacity, its determinants, and influence on innovation output: Cross-cultural validation of the structural model. *Technovation*, *29*(12), 859–872. doi:10.1016/j.technovation.2009.05.010

Québec. Conseil de la science et de la technologie (2010). *Innovation ouverte. Défis et enjeux pour le Québec*. Résumé du Rapport de conjoncture 2009. Québec: Les Publications du Québec.

Reitzig, M., & Puranam, P. (2009). Value appropriation as an organizational capacity: The case of IP protection through patents. *Strategic Management Journal*, *30*(7), 765–789. doi:10.1002/smj.761

Schreyögg, G., & Kliesch-Eberl, M. (2007). How dynamic can organizational capabilities be? Towards a dual-process model of capability dynamization. *Strategic Management Journal*, *28*(9), 913–933. doi:10.1002/smj.613

Scott, W. R. (2004). Reflections on a half-century of organizational sociology. *Annual Review of Sociology*, *30*(1), 1–21. doi:10.1146/annurev.soc.30.012703.110644

Shafer, S. M., Smith, H. J., & Linder, J. C. (2005). The power of business models. *Business Horizons*, *48*(3), 199–207. doi:10.1016/j.bushor.2004.10.014

Sigurdson, J., & Reddy, P. (1995). National appropriation of University innovations: The failure of ink jet technologies in Sweden. *Technology Analysis and Management*, *7*(1), 41–62. doi:10.1080/09537329508524194

Spithoven, A., Clarysse, B., & Knockaert, M. (2010). Building capacity to organise inbound open innovation in traditional industries. *Technovation*, *30*(2), 130–141. doi:10.1016/j.technovation.2009.08.004

Van de Vrande, V., de Jong, J. P. J., Vanhaverbeke, W., & de Rochemont, M. (2009). Open innovation in SMEs: Trends, motives and management challenges. *Technovation*, *29*(6-7), 423–437. doi:10.1016/j.technovation.2008.10.001

Wincent, J., Anokhin, S., & Boter, H. K. (2009). Network board continuity and effectiveness of open innovation in Swedish strategic small-firm networks. *R & D Management*, *39*(1), 55–67. doi:10.1111/j.1467-9310.2008.00539.x

Winter, S. G. (2003). Understanding dynamic capabilities. *Strategic Management Journal*, *24*(10), 991–995. doi:10.1002/smj.318

Zahra, S. A., Sapienza, H. J., & Davidsson, P. (2006). Entrepreneurship and dynamic capabilities: A review, mode and research agenda. *Journal of Management Studies*, *43*(4), 917–955. doi:10.1111/j.1467-6486.2006.00616.x

KEY TERMS AND DEFINITIONS

Absorptive Capacity: Managerial strength to identify and apply new external knowledge to the firm.

Appropriative Capacity: Personal strength to support and strengthen the innovation efforts of the firm.

Case Study: A methodology or approach to better understand the characteristics and specificities of one or several firms.

Dynamic Capabilities: Managerial strength to diversify and better fit key roles throughout strategies.

Open Innovation: The improvement of products and services using resources outside to those owned by the firm.

SMEs: Small and medium sized enterprises often with a single manager, the owner.

Technological Innovation: New products or processes improved through radical or incremental conception and offered to the market.

ENDNOTES

1 These companies have started to open their innovation process while in crisis. They have started to do R&D and co-development with other companies, and buying/selling/sharing some intellectual properties.

2 The exact expression, in French, is « Éducation nouvelle » or « École nouvelle ».

3 It is well noted that Chesbrough spent a couple of years to study these companies through a multiple-cases study approach prior the publication of his book.

4 In 1992, IBM lost $5G and thousands of employees. This has been a turning point.

5 In Canada, businesses can use numbers such 347… Canada inc. rather than names to be incorporated. The complete number here is not revealed for confidentiality purpose.

APPENDIX 1: ANALYSIS STATEMENTS FOR BUSINESS MODELS

Statements Related to Open Innovation (statements from type 4 and above are associated with Open Innovation)	
• Type 1: statements 1-4: Undifferentiated • Type 2: statements 5-7: Differentiated • Type 3: statements 8-12: Segmented • Type 4: statements 13-17: Externally aware	• Type 5: statements 18-22: Integrated with business model • Type 6: statements 23-27: Platform player shapes market

(Use a check for each statement)
A check in "A" and "B" is possible for the same statement. A current reality (A) could be reinforced towards an ideal one (B)

A (Current business model)	B (Ideal business model)	Statements*
		1. Undifferentiated business model, convenience products
		2. Competitiveness based on price and availability
		3. Very little or no innovation at all
		4. No intellectual property rights (IP)
		5. Some differentiation in the business model
		6. First initiatives for innovation
		7. First intellectual property rights (IP)
		8. First attempt to segment markets, prioritizing and selecting its innovation projects according to its business model
		9. Innovation results from organizational activities (planned and not by chance)
		10. Innovation is regarded as a long term investment for a more sustainable company
		11. In addition to the functions of product development and R&D, all other business functions are put together to advance innovation
		12. Intellectual properties (IP) are managed by designated manager
		13. Technologies developed by other companies are integrated and can be used to serve similar markets.
		14. The company scans outside innovation and its own opportunities for innovation
		15. Suppliers and customers play a major role in innovation projects of the company
		16. The innovation is a multidisciplinary activity within the company
		17. Intellectual properties (IP) are considered an asset to serve similar markets and generate value
		18. The business model is looking for new markets and new business associates and is inviting customers and suppliers to follow
		19. The internal and external R&D is integrated into the business model
		20. The innovation plan (road map) is communicated to suppliers and clients and their plan is also known
		21. Innovation becomes a function of the company under the direction of designated manager
		22. Intellectual properties (IP) are considered a financial asset and are managed as profit centers
		23. The business model drives the business model of suppliers and key customers

		24. Innovating the business model is part of the innovation process of the company
		25. The technical and financial risks associated with innovation as well as the rewards are shared by the participating partners
		26. Intellectual properties (IP) are considered a strategic asset, facilitating new business ventures, relationships with customers and suppliers and can lead to the abandonment of a particular niche, if necessary
		27. The management of innovation and intellectual property (IP) is integrated into every aspect of business
		* Adapted from Chesbrough (2006)

APPENDIX 2: SEMI-STRUCTURED INTERVIEW GUIDE

Technological Innovation

- **Product/process:** Product development and process development, incremental and radical differences.
- **Co-developing products:** Familiarizing shareholders and staff with co-product development, knowledge of other companies engaged in co-development; factors of success and failure.
- **Research & Development:** Current R&D work, number of designated persons; joint R&D; reasons for joint R&D, factors of success and failure in collaborative R&D.
- **Intellectual properties (IP):** Patents, trademarks or other IP; patent operated, bought or sold; collaborative IP; knowledge of other companies doing collaborative IP.
- **New licenses:** Corporate exploitation; bidding or buying; knowledge of other companies using licenses.
- **Spin-off:** Creation of derivative businesses; motivation to do so; knowledge of other companies having created spin-offs; fears and motivations.

Organizational Capacities

- **Dynamic capabilities:** Employees with knowledge and skills; flexibility in employees work assignments; combination of employees for productivity gains, continuing education program in business; planning for change; knowledge of other companies reorganizing their resources.
- **Absorptive capacity:** The ability of employees to identify new information, assimilate and apply it to business development; speed adoption of competitors' practices; information monitoring; formal training in business.
- **Appropriative capacity:** Customization of innovation by employees' personal initiative; workers' sense of belonging; personal commitment of employees; constraints that inhibit employee engagement in innovation projects; positive reinforcement; adaptability of employees; employees' ability to reorganize the resources at their disposal; tendency for employees to emulate the practices of competitors.

Chapter 3
Building Innovative Business Ideas in Small Business Enterprises:
The Case of Bird ICT Company

Elham Metwally
Misr International University, Egypt & The American University in Cairo, Egypt

ABSTRACT

This study of "Bird ICT" is a case of a business development company in the field of information and communication technology (ICT). The chapter explores, describes, and examines the story and circumstances of building and managing innovative business ideas in an Egyptian small business enterprise. It specifically addresses specific circumstances, opportunities, and challenges with which a small enterprise pursuing innovation strategies is faced, and analyzes how effective deployment of innovative business ideas and bringing them to the market has contributed to strategic competitiveness. Findings highlight the patterns of developing new business ideas and mainly conclude the following: strategic competitiveness is accounted for by management's efficiency, commitment, cheap labor, and most importantly, fast delivery of easy hit technologies and innovative, reliable, flexible, yet easy to use innovative advertising solutions. This study used qualitative research strategy in connection with theory to gain full understanding of a natural setting, and as a comprehensive method of substantiating or un-substantiating the context of achieving strategic competitiveness through new innovations and technologies, and provides insights into the nature of management of innovations, from which lessons can be drawn for other organizations both in small business sector organizations and generally.

INTRODUCTION

With recent globalization trends, the role of small business enterprises in innovation, global expansion, and job creation, particularly for nations with growing populations, has been increasingly contributing to these countries' socio-economic development. But the breakthrough of the Internet, and speedy generation and development of innovations have changed the way firms operate and deliver their products and services to their clients. Technological advances eliminated time

DOI: 10.4018/978-1-61350-519-9.ch003

and distance barriers and allowed better access to trading markets across borders, which expedites economic development and accelerates globalization of business. According to the National Association of Software & Service Companies (NASSCOM) (2005), a study on countries that invested more in IT revealed that a 10% increase in IT investments raised GDP by 3.6%, and increased labor productivity by 4%. On the other hand, in economies with less investments in IT, an equally invested amount in IT produced only a 1.6% raise in GDP, and was of no statistical significance on productivity as it did not help in reducing labor hours, which could later, on the long-term, have a positive impact on labor productivity.

With the emergence of virtual business, information technology should be regarded as a competitive trade instrument that should allow firms to increase visibility and market access, promote image, improve customer service, increase customer satisfaction, and optimize business resources; all of which will have a positive impact on their profitability and competitiveness and on the national economy in general. Thus, innovations, investments in information technology capital, IT applications or systems are becoming of increasing significance not only to individual firms but also to enhancing nations' economies in general. At the same time, as the role of small and medium enterprises (SMEs) has been increasingly contributing towards achieving countries' socio-economic development, they are playing a crucial part in the transformation of innovation strategies to achieving value creation and strategic competitiveness. However, despite the value creating nature of innovation, and though it is regarded as a competitive weapon, competitive boon will rely more on deploying the right blend of innovative technologies and human factors with clients. According to a 2002 Gartner study, firms worldwide lost $A800 billion Australian Dollars annually on poorly managed IT investments (Thorp, 2005).

The development of a competitive SME sector is greatly determined by companies' readiness to adopt more advanced and innovative products in conducting their business, and their ability to position themselves for long-term competitive advantage, which requires vast improvements in their performance, and emphasis on core competencies when formulating strategies. In Egypt, SMEs constitute 99% of enterprises, employs 75% of the total labor force in the private sector, and are responsible for three quarters of the economy's private sector value added. A close examination of the literature and empirical research revealed that SMEs share in Egypt's manufactured exports still does not exceed 0.5%, which when compared to other countries like India (50%) indicates poor performance, marketing difficulties, and lack of competitiveness on the world market. In order to assist the development of SMEs, sustain and enhance their role as a competitive sector both in local and global markets, and become a driver of national development, particularly after Egypt's adoption of the comprehensive economic reform program in the early nineties, several initiatives have been undertaken by the government, non-governmental institutions, donors and assistance projects, as well as local and international development organizations to support the development of the IT sector and improve services in the telecommunications sector. Despite these efforts, and the value creating nature of IT, the impact on the development of the small and medium enterprise sector in Egypt is still limited, and up till now the country's competitiveness in terms of e-business readiness has been very modest. The Economist Intelligence Unit world's e-readiness ranks (of 70 countries) placed Egypt at the 57th position, and the country's overall e-readiness score was 4.33 out of 10 (The Economist Intelligence Unit, 2009). On the firm level, SMEs in Egypt face increasing competitive pressures and are further aggravated by their inability to adopt and use cutting edge technologies. In fact, as these small enterprises continue to be poorly managed and use simple outdated technologies, or not use technologies at all, it is fairly unlikely that they will be able to position themselves in the global economy. At the crux of developing powerful trading systems and attracting inflows of investments, is strengthening the functioning of business enterprises.

Since the main objective of this study is to explore, describe, and examine, the story and circumstances of building and managing innovative business ideas in an Egyptian small business enterprise, and in view of the fact that it specifically addresses opportunities and challenges with which a small enterprise pursuing innovation strategies is faced, and analyzes how effective deployment of innovative business ideas and bringing them to the market has contributed to strategic competitiveness; the subsequent task is to present the literature review, which explores the factors that contribute to the successful implementation of IT-based innovations in business organizations towards achieving strategic competitiveness. In this respect, the review of literature hinges upon issues of small and medium enterprises, open innovation, diffusion of innovation theory, and strategic competitiveness.

LITERATURE REVIEW

Small and Medium Enterprises

Previous research confirms that small firms are more likely to use new technologies to compete effectively with their rivals, and the Internet and Intranets to find new customers and communicate with them than their larger counterparts (Daniel & Grimshaw 2002: 133-147; Elsammani et al., 2001: 562-76; Hamill & Gregory, 1997: 9-28; Keeling et al., 2000: 57-70; Lynn et al., 1999: 9-18; O'Keefe et al., 1998: 629-43; Poon, 2000: 72-81; Poon & Jevons, 1997: 29-41).

Several studies have been undertaken on SMEs with regards to issues like the uses of e-commerce in advertising and brand building (Daniel & Grimshaw, 2002: 133-147; Doherty et al., 2001: 251-7; Hamil & Gregory, 1997: 9-28; Poon & Swatman, 1999: 9-18; Webb & Sayer, 1998: 815-27). Adopting e-commerce permits companies to achieve improved branding (Jacobs & Dowsland, 2000: 590-6) and to tap new as well as existing markets that could be unaffordable with traditional marketing tools due to the limited access to financial resources or lack of suitable human skills as compared to large companies (Da Costa, 2001: 45, 79; Daniel & Grimshaw, 2002: 134; Hamil & Gregory, 1997: 9-28; Poon & Jevons, 1997: 29-41). It is, therefore, that SMEs have started services, such as promoting their products in several languages, to enhance this specific marketing tool and, hence, realize optimum possible market reach (Daniel & Grimshaw, 2002: 134). But whilst SMEs financial resource might represent a significant constraint on their growth (Ballentine et al., 1998: 241-51; Binks & Ennew, 1996: 17-25; Chang & Powell, 1998: 199-215; Daniel & Grimshaw, 2002: 133-147), SMEs could be more advantageous than larger companies since their processes, structures, and systems are simpler, which allows them to be more flexible in adapting to new ways of conducting business transactions (Daniel & Grimshaw, 2002: 133-147; Lynn et al. 1999: 9-18).

There is no one exclusive definition for small and medium enterprises. Countries as well as organizations define SMEs differently based upon several criteria among which are the number of employees, economic activity, annual sales, and the level of capitalization or fixed assets as defined by some countries. For example, in defining SMEs, South Korea uses the "Labor" and "Capital" variables in its manufacturing, transportation, and mining sector, whereas it uses "Labor" and "Annual Sales" in its services and trade sectors. Japan uses the "labor" and "capital" variables in defining its small business enterprises, while it uses only the "labor" criterion in defining its micro businesses. Hong Kong and the United Kingdom, however, use the "labor" criterion only in defining

SMEs. Canada, still has two definitions for small businesses; one depends on the "labor" criterion, while the other relies on "sales revenues". Finally, in defining small businesses, the United States relies on only one criterion for each sector or sub-sector. For example it uses the "labor" variable in its manufacturing and wholesale sectors, whereas it relies on "gross receipts" for its retail, services and construction sectors (Ministry of Foreign Trade 2004: 73-86).

As regards SMEs in Egypt, the multi-sector definition of micro, small, and medium enterprises (MSMEs) depends on several variables according to the policy objectives of each entity. Egypt uses the "fixed assets" criterion rather than using "total assets" to avoid current assets' value fluctuations. Fixed assets comprise land, buildings, machinery, and equipment and are based on the book value or market value of assets, whichever is lower. As regards to the "labor" variable, different entities in the government of Egypt preferred to rely on the number of employees registered under social security since it is difficult to count those who are not registered. Therefore, the widespread criteria in defining SMEs in Egypt rely on both the number of employees, as well as on fixed assets (Ministry of Foreign Trade 2004: 27-30; SME Website 2004). According to Law number 141 of 2004, the Small Enterprise Development law, a small enterprise is a firm that is involved in an economic activity, such as delivering a product and/or a service, and its paid capital varies between L.E. 50,000 and 1,000,000, and whose employed labor force does not go beyond 50 workers. On the other hand, a micro enterprise is a firm that its paid capital is less than L.E. 50,000 (Egypt's Official Newspaper, 2004; Ministry of Foreign Trade 2004: 73-86). It is worth noting that the law used both the capital and labor force criteria to define small enterprises; whereas, it used only capital in defining a micro enterprise.

Law 141 of 2004 further opted to provide some motivation to small and medium enterprises that included, but were not limited to: formation of financing funds in governorates, assigning 10% of area offered for investment for such enterprises (at the cost price), and the commitment of ministries and other authorities to procure 10% of their production.

In businesses relying on IT-based technologies, not only has the role of IT in carrying out business functions become a necessity but also it has been expanding as more and more previously manual tasks and operations have been automated.

Open Innovation in Small Business Enterprises

Researchers have usually defined "innovation" in organizations as the development as well as adoption of new ideas or actions (Amabile, 1988: 123-67; Barringer & Ireland, 2010: 45; Damanpour & Wischnevsky, 2006: 269-91; Walker, 2008: 591-615; Zaltman, Duncan, & Holbek, 1973). Ideas or actions may concern a product, process, service, technology, technique, or practice. Firms develop ideas and/or actions for their individual utilization or for utilization in other firms. Innovation is a process that results in an outcome that is original to a firm's stakeholders (Daft, 1978: 193-210; Damanpour & Wischnevsky, 2006: 269-91). Incorporation of an innovation is a procedure that ends with the assimilation of a creation to the implementing firm (Damanpour & Wischnevsky, 2006: 269-91; Walker, 2008: 591-615). The innovation adoption practice has two main stages: initiation and implementation (Nystrom, Ramamuryth, & Wilson, 2001: 221-47; Rogers, 1995).

In recent times, and with contemporary globalization trends there has been a growing research on the importance of open innovation (Kowalski, S.P., 2009). According to Henry Chesbrough (2010), "Open Innovation" is defined as "*...the use of purposive inflows and outflows of knowledge to accelerate internal innovation and expand the markets for external use of innovation, respectively.*" As such, on one hand open innovation includes, in addition to R&D tools and facilities, dynamic set of collaborative connections with other companies, suppliers, customers, partners, governments, as well as public research institutes, universities, and exchanging of external information and know-how (Chesbrough, 2010; Hansen & Birkinshaw, 2007; Kowalski, 2009). Nonetheless, open innovation still entails disbursement fees for intellectual property. On the other hand, "*Closed Innovation*" is the conventional approach. It involves a controlled, direct and chronological process that starts with basic research, going through product development, manufacturing and sales. This is often completed inside doors, using internal resources and contractual provisions (Chesbrough, 2003; Van de Vrande, de Jong, Vanhaverbeke, & de Rochemont, 2008: 347-363). Therefore, dynamism and flexibility are major characteristics of open innovation; as compared to command and arrangements that are main features of closed innovation.

Extensive interest in small business enterprises has highlighted their contribution towards countries' socio-economic development, global expansion, job creation, and in general, improving poverty. However, away from their positive impact on numerous issues, SMEs are considered at the crux of innovation and sustainable development as they symbolize creativity (Warner, 2001). Thus, given the growing importance of globalization, and the rapid progress in technologies, the issue of "*Open Innovation*" becomes inevitable to SMEs as it can speed up their internationalization (Kowalski, 2009).

But dealing with opportunities and perils of open innovation could pose some challenges to SMEs. Small firms suffer by and large from lower absorptive competences, as they do not have the capability to sustain committed resources and staff to build arrangements to find and capture valuable exterior knowledge (Chesbrough, 2010). Consequently, SMEs are also frequently unappealing as partners to others; to be given helpful thoughts from experts, and technologies. These people, for example might prefer to work with larger, better-known, and more prestigious companies, or alternatively might have a preference to be of assistance to start up a new business, than work in partnership with an on hand SME. Frequently also, SMEs lack available resources or an institutionalized, well-structured innovation procedure, which reduces their ability to formulate long-term strategies (Chesbrough, 2010). In several cases, SMEs' operations and businesses are jeopardized as they frequently have fewer resources and inadequate expertise in, for example intellectual property issues when weighted against larger multinational corporations, which is indispensable for gaining the value of a firm's intellectual property rights (Kowalski, 2009).

Nevertheless, SMEs could have structural edges that could allow them to capitalize on opportunities of open innovation. Their small size, focus, and specialization, for example, makes niche markets more appealing and cost-effective than they are to larger companies who are loaded with large overhead costs that make it costly to exploit new opportunities in smaller markets. SMEs also appeal more to smaller entrepreneurs than larger firms, and can respond more promptly to input from customers or competitors, and develop their business models more quickly, which allows them to achieve a competitive advantage and achieve above average returns, given the speedy changes in markets (Chesbrough, 2010). Being small, focused, and specialized, invites alliances with large firms that are interested in shared innovation collaboration. Big firms generate technology programs and keenly hire SMEs to grow products for these programs. SMEs are dynamic users of several new technologies and may increase performance of these technologies, which picks up on the quality or competence of a technology. SMEs have a better capability to specialize than bigger firms, and this specialization is more advantageous in particular when markets are more open for innovative actions (Chesbrough, 2010).

But the concept of "*Open Innovation*" is broad, and includes different dimensions. According to Van de Vrande, de Jong, Vanhaverbeke, & de Rochemont, (2008: 347-363), "*....there is the inside-out movement, or technology exploitation, in which existing technological capabilities are leveraged outside the boundaries of the firm.... and, there is an outside-in movement, also referred to as technology exploration, in which external sources of innovation are used to enhance current technological developments.*" In technological exploitation, companies can apply a variety of strategies to commercialize technologies using external ways, such as creating and spinning out innovative business enterprises, and providing intellectual property rights to external parties (Chesbrough, 2003; Van de Vrande, de Jong, Vanhaverbeke, & de Rochemont, 2008: 347-363). Earlier research by Chesbrough (2003) and Lord et al. (2002: 115-121) on open innovation has examined the spin off process of large firms, and concluded that some large high-tech firms spin off new ventures as the business idea might not be aligned with the on hand business model. The possibility for these spin-off firms is huge. Moreover, companies can also gain from their own and unused intellectual property, because other companies might find them lucrative (Chesbrough, 2006: 229-236; Van de Vrande, de Jong, Vanhaverbeke, & de Rochemont, 2008: 347-363)

Nevertheless, there are as well many dimensions of technology exploration. Customers, for example, are rich sources of innovative ideas for product development and commercialization (Cooper & Kleinschmidt, 1995: 374-391; Gales & Mansour-Cole, 1995: 77-109; Leonard-Barton, 1988: 251-267; Lettl et al., 2006: 251-272; Van de Vrande, de Jong, Vanhaverbeke, & de Rochemont, 2008: 347-363). Employees, too, contribute to the companies' total innovative performance. In addition to being the sources of new ideas, employees also contribute to innovative performance through informal relationships with employees of other firms, which are central to realize how new products are formed and commercialized (Chesbrough et al., 2006: 229-236; Morgan, 1993; Van de Vrande, de Jong, Vanhaverbeke, & de Rochemont, 2008: 347-363). Inter-organizational networking is also a third way in which a company can get access to external sources of knowledge. For example, R&D alliances (Gomes- Casseres, 1997: 33-44), with non-competing firms, R&D cost sharing (Rowley et al., 2000: 369–386), research collaboration with universities (Cohen et al., 1998; Fontana et al., 2006: 309–323; George et al., 2002: 33-44), for acquiring a "*window*" on new advancements and leveraging technological capabilities (Van de Vrande, de Jong, Vanhaverbeke, & de Rochemont, 2008: 347-363). Another form of inter-organizational networking could be *participation in new or existing companies*, for example through minority holdings or corporate business enterprise capital investments (Chesbrough, 2002: 4-11; Dushnitsky and Lenox, 2005a: 947-965; Ernst et al., 2005: 233-242). Through these kinds of equity investments, firms gain a "*window*" on new technological developments. However, the researchers argued that the degree to which innovation processes in SMEs use these different forms depends on their size and the kind of business (Van de Vrande, de Jong, Vanhaverbeke, & de Rochemont, 2008: 347-363). In general, firms mingle both "technology exploitation" and "technology exploration" so as to produce utmost value from their technological capacity or other competencies.

SMEs innovation strategies differ significantly according to their size classes. In fact, there is quiet large difference in the innovation strategies of small firms and medium sized enterprises. For example, the degree to which SME firms create new projects as a component of their open innovation plan is apt to depend on company size. Outsourcing of R&D also calls for a structured and well-organized innovation course of action. Moreover, involvement in other firms is feasible for a certain amount of financial resources to engage in these types of investments. Van de Vrande, de Jong, Vanhaverbeke, & de Rochemont (2008: 347-363) examined 605 Dutch innovating SMEs, and results show that larger SMEs adopted open innovation practices more quickly than smaller firms.

There are numerous reasons among firms to decide upon an open approach to innovation. A primary reason is meeting customer demand, including gathering of new ideas and experience, improving innovative accomplishments, continuous development and economic motives (Van de Vrande, de Jong, Vanhaverbeke, & de Rochemont, 2008: 347-363). Other researchers like Jacobs and Waalkens (2001: 891–898) realized that the most important factors deciding innovation within firms can be found in better usage of internal creativity. This author found some research on the impact of organizational culture on innovation and knowledge transfer (Meschi, 1997: 211- 228; Simonin, 1999: 595-623). For example, Simonin (1999: 595-623) finds that both cultural and organizational distances are connected to uncertainty, which in turn harmfully affects knowledge transmission.

Diffusion of Innovation Theory

There is abundant literature in the area of diffusion and making efficient use of technology resources, and in this aspect Rogers (1983) addressed how products, processes, ideas, or new practices, spread within an organization, or society. According to Rogers (1983), "*Diffusion is the process by which an innovation is communicated through certain channels over time among the members of a social system*" (Rogers, 1983:5). Rogers revealed that in developing an innovative entrepreneurial culture, organizations should possess several traits such as a favorable attitude with regard to change, openness, decentralized decision making, informal structure, as well as a large size (Wheelen & Hunger, 2004:290). In State Street Boston Corporation, essentially a commercial bank, management team viewed information technology as one of the bank's strategic resources. In October 1994, David Sexton, executive vice president, information technology at the bank, revealed the contributions that information technology had made to State Street's earlier achievement and considered potential challenges. Information technology was observed as a catalyst to allow the accomplishment of planned objectives. The mission of State Street is to combine information technology with banking, trust, investment, and securities processing capabilities to sustain the investment strategies of its customers worldwide. The bank brings concurrently advanced information technology and experiences of worldwide financial markets to allow its customers to perform their wide-reaching investment plans and achieve their objectives (Stoddard & Francalanci, 1994).

Literature on the strategic role of small and medium enterprises in economies, with recent globalization trends and rapid technological changes praised the flexibility of SMEs in adopting innovations. Contrary to larger companies, SMEs may be more nimble, and innovative due to their organic oriented structures which are flat, decentralized, and depend on mutual adjustment, and which, as a result, ultimately promote flexibility and speeding adjustment to changing environments (Rao 2003: 11-32). And, literature on characteristics of e-commerce handled opportunities which adopters could capitalize on to create more value through lower costs and easier use, and hence increases adoption rates of the new technology (Daniel & Grimshaw, 2002: 134), and improve interconnectivity and exchange of information among different stakeholders (Daniel & Grimshaw, 2002: 134).

Rogers identified five different ideal attributes of those who acquire and use technologies. First, there are "innovators" or "venturesome" who are always eager to undertake new ideas. Second, there are "early adopters" or "respectable" that relies on opinions of innovators in trying something new and serves as a role model in reducing uncertainty about this new thing by adopting it and disseminating their estimation of it via interpersonal networks.

Third there is "early majority" or the "deliberate" that also rely on peer opinions, and usually adopt the innovation before average people but rarely take the lead communicating with others to reduce uncertainty about the innovation. Fourth is "late majority" or the "skeptical" who adopt the innovation after average people do. Finally, those who are called "laggards" or "traditional" are the last ones to use the innovation (Rogers, 1983: 247-251).

Rogers also determined five attributes of innovation that predict rate of adoption, which he defines as "*the relative speed with which an innovation is adopted by members of a social system measured as the number of individuals who adopt a new idea in a specified period*". These characteristics are: cost, relative advantage, compatibility, complexity, trial-ability, and observability or visibility to others (Rogers, 1983: 210-232; Rogers, 1995; Tornatzky & Klein, 1982). Rogers also emphasized that organizational innovativeness, adoption and implementation are dependent upon several structural characteristics such as leaders' traits that comprise attitudes towards change, etc., internal characteristics of organizational structure like complexity, size of the firm, and so forth, as well as other external factors like system openness and the like (Rogers, 1983: 347-361; Wheelen & Hunger, 2004: 290).

Even with the speedy progress in information technology and recognition by large businesses of such technologies, similar levels of acceptance is not apparent among small and medium enterprises (Bode & Burn 2002).

Strategic Competitiveness

Literature within the strategic management discipline has approached the concept of competitiveness from different angles. According to Hitt, Ireland, and Hoskisson (2001) "Strategic competitiveness" is achieved when a firm successfully formulates and implements a value-creating strategy. When a firm implements such strategy that other companies are unable to duplicate or find too costly to imitate, this firm has a "*sustained, or sustainable, competitive advantage*." The range of literature discussing the source of competitiveness is vast. The Resource-Based View of the Firm (RBV), builds on a paradigm found in the writings of Schumpeter (1934), Penrose (1959), Wernerfelt and Montgomery (1988), Amit and Shoemaker (1993), Peteraf (1993), Dierickx and Cool (1989), Rumelt (1991), and others who were microeconomics oriented and emphasized the role of the firm's incomparable resources and capabilities in value creation and sustained competitive advantage. The topic of gaining competitive advantage has been extensively covered by Michael Porter in many of his books and articles (Porter, 1998b, 1985, 1979), and maximizing corporate value creation and building a relative competitive advantage has been explained by his Competitive Advantage and Value Chain models. At the core of the topic is the concept of "*Margin*." A firm creates margin when the value of products or services it produces exceed the cost of production. The Competitive Advantage model argues that a firm can build a relative competitive advantage and achieve above average returns through three types of business-level strategies: cost leadership strategy where a firm is the lowest cost producer of goods or services in its industry; a differentiation strategy in which a firm delivers goods or services that are unique relative to that of competitors; and/or a focus strategy in which a firm uses its core competences to be the best in a certain competitive segment or segments (Hitt, Ireland, & Hoskisson, 2001: 153-170). In Porter's Value Chain model framework, reaching above-average performance within an industry rests upon the way a firm performs relative to its competitors in regards its two main activities: primary activities and support activities. A firm's primary activities include both inbound and outbound logistics, operations, marketing and sales, as well as services activities. Support activities, on the other hand, comprise the firm's infrastructure, human resource management, technology development, as well as inputs' procurement activities (Porter, 2001, 1998b, 1985, 1979).

Therefore, extensive research on competitiveness have identified sources or factors like firms' resources, innovative capabilities, learning capabilities, organizational culture, and firm strategies that affect organizational competitiveness and responsiveness to the market.

As mentioned earlier, the main objective of this study is to explore, describe, and examine, the story and circumstances of building and managing innovative business ideas in an Egyptian small business enterprise. Specifically, the case looks at the application and management of a new product, the Bluetooth advertising, to mobile handsets targeting other organizations like advertising agencies and solution integrators. The Bluetooth offers a safe and sound way to link and exchange information among appliances such as mobile phones, personal computers, laptops, printers, etc. Thus, the case will examine the influence of effective adoption and implementation of innovative business ideas in an Egyptian small business enterprise on strategic competitiveness. As a result, the following research proposition is advanced:

Main Research Proposition: Effective deployment of innovative business ideas contributes to strategic competitiveness.

And the following research question is advanced:

Research Question: To what extent does effective management of innovative business ideas contribute to strategic competitiveness?

To answer the research question, the researcher draws on an Egyptian small enterprise data to further understand the link between competitiveness, and management of innovative business ideas and explore ways in which utilizing innovations contributes to strategic competitiveness.

BUILDING INNOVATIVE BUSINESS IDEAS IN SMES: THE CASE OF BIRD ICT

It was late at night on November 2nd, 2005 when Khaled Awad, recently graduated from the Medical School of Alexandria University, sat in his armchair, and watched TV, with his hands on his knee, and surfed through the channels. During those few years after graduation, he had still some doubts about what career should he be in. Dr. Khaled could only think of two possible outcomes: either he would simply carry on with his medical profession, or because this was his passion, he would establish his own business in information technology. He sighed, presuming he had already reached a decision about his line of business. His pressing worry was how to start-up the business and position it to be one of the biggest information communication technology holding companies globally. Dr. Khaled was positive about his vision. But, he thought increasingly about how will he be able to manage that business and what strategies to assume.

The twenty-seven year old entrepreneur Dr. Khaled Aly Awad, founded Bird ICT in 2006, with around 10 million Egyptian pounds venture capital investment. After completing his study journey in medicine at the University of Alexandria, he ultimately found his passion in information communication technology. "*We have a team of creative and experienced personnel who share the same vision and who are willing to put much effort to achieve the same goals. Together we were able to develop two solutions for businesses in the Information Communication Technology, ICT sector. These are: PlusXMS and iBlue Co.,*" declared Dr. Khaled. "*Strategically managing the company is our aim,*" said Dr. Khaled. This study on "Bird ICT" represents a case of a business development of small enterprise in the field of information and communication technology. The mission of Bird ICT is to generate and put together innovative business thoughts, build up their business models, and set up them into other/ new companies.

Hence, the goal of Bird ICT is to become one of the biggest information communication technology holding companies globally. And as successful exploitation of new ideas is the main business strategy that enables the company to compete and have an edge in the ever more competitive and fierce worldwide setting, Bird ICT continuously makes every effort to generate and assemble innovative ideas in the ICT turf.

However, Bird ICT faced some challenges throughout the implementation phase that included lack of information to make clear decisions and lack of the suitable human resources needed to build innovation. The limited experience of the working team, delays, bugs and bad selection of the framework were some of the issues that were faced by the management during the development phase; whereas, cash and receivables delays, and using wrong channels to advertise on were issues that challenged the marketing phase.

Open Innovation in Bird ICT

"Innovation" is the strategic business practice that permits the company to compete successfully in the ever more changing and competitive business setting. In this respect, the company always identifies key strategic markets indicators, examines different industries around the world and is always alert to new technological innovations. The main goal of Bird ICT is to become one of the biggest information communication technology holding companies globally. Towards this end, Bird ICT owned and developed a wide range of enterprise solutions, introduced them to the market, and implemented them in ways that are unique enough to satisfy specific market needs. In 2007, Bird ICT introduced PlusXMS, a value added service that targeted mobile operators and web services' providers. PlusXMS is a comprehensive messaging service that allows mobile and web users to exchange rich text messages, which could also have different colors, images, smiley faces, as well as backgrounds. PlusXMS is a platform that facilitates exchanging rich text messages among mobile users, and between mobile users and web users. It is build out of three main pieces: a mobile component that runs on mobile phones, a web component that runs on web portals, and an engine that associates any web conversation program with mobile phones. The key advantage of using PlusXMS is that it uses the same SMS protocol to send rich messages. This means that PlusXMS users will be able to better express their feelings using the rich messaging feature of the PlusXMS at the same cost of normal SMS. Also, PlusXMS enhances web traffic; thereby, helping its clients to better serve their subscribers and online community, and allowing them to achieve higher average revenues per user (ARPU).

Bird ICT succeeded in starting up iBlue™ as an independent company fully owned by Bird ICT that develops Bluetooth data communication enterprise solutions. The iBlue company is specialized in innovative mobile advertising solutions for business organizations of the ICT sector, thus enabling integrated solutions based on Bluetooth technology. The mission of iBlue company is to provide enterprises with innovative, reliable, flexible, yet easy to use solutions by utilizing Bluetooth data communication technology and its expertise in their state of the art. iBlue powerful solutions can turn any mini PC, laptop, and/or desktop to a full featured Bluetooth Advertising solution. To maximize its resellers and partners return on investments, Bird ICT permits them to assemble the devices in-house, thereby, cutting the logistics' expenses dramatically, and maximizing profits.

Bird ICT launched also "LongArm" as a unique Bluetooth marketing solution in the Middle East and North African region. Through "LongArm", marketers reach their clients on their Bluetooth-enabled electronic devices efficiently, as the solution allows broadcasting any file format to a large number of Bluetooth enabled devices simultaneously with a very high flexibility in creating and managing marketing campaigns. "*The power of LongArm is the result of its unique features. Out of all the features of LongArm the six key ones are: serial coupons, support of all file types, campaign scheduler, statistics viewer, back-up utility, and high connection range,*" stated Dr. Khaled Awad, Bird ICT managing director. The serial coupons feature makes it easier for LongArm users to send a group of successive documents or coupons to their customers where each customer gets only one of those coupons. This characteristic will allow the organization control this kind of coupons in a clever way in one mouse click. Also, LongArm supports all file types; thereby allowing marketers more room to do their innovative works and launch it to their clients in the design they wish for.

Moreover, LongArm users can automatically schedule all their campaigns using LongArm's flexible campaign scheduler. The scheduler allows them to state the lively time of each campaign in terms of time and date and days of the week. A statistics viewer gives the marketers clear graphs of the status of campaigns in terms of the number of clients who have received their marketing material and who have not. With iBlue introducing their smart back-up function, marketers no longer have to worry about losing their digital marketing campaign. This characteristic helps LongArm users to support all their marketing campaigns and specific configuration in a file that can be used to restore this data as needed. The advanced hardware of LongArm, together with its embedded software, enables fast immediate files transmission and allows the users to remain linked to their clients up to a distance of 150 meters. According to Mohamed Kash, Bird ICT business development manager, there are no limits to where LongArm can be used. Wherever Bluetooth enabled devices exist, LongArm can exist. He stated that iBlue is targeting marketers in different business areas ranging from shopping malls, sporting clubs, and outlet stores to airports, hotels, events management, and advertising agencies.

Implementation Phases of Bird ICT Innovations

Introducing Bluetooth advertising to mobile handsets started with creating the idea, analyzing it, out lining the main functions and uses, and finding the resources to develop the product. The development phase started with building the wire frame of the product, then building the use case, finding the framework to work with, defining development tasks, assigning tasks to developers, and reviewing tasks, testing the product and fixing bugs, and releasing the beta version, until eventually, releasing the final version. The founders, the Chief Executive Officer and Chief Technical Officer, and Project Manager of the company, were mainly responsible for the business innovation phases. The CEO was responsible for recruitment and hiring, building the marketing strategy, and finding funds. The CTO was responsible for all technical issues. All critical decisions were made using a round table discussion. Training involved that accompanied the product's development was carried out through online tutorials and recorded courses. The new products' implementation phases influenced the organizational structure in the company. For example, during the products' development phase, technical persons played the leading role. After that, business developers take over, and then as the company starts commercializing the product, the marketing team begins playing the main role. Management depended heavily on negotiations and agreements in managing change. To market the product, Bird ICT carried out several steps that started with building the identity of the product (logos & color & product name), defining the targeted segment, building the marketing materials for the target sector, publishing the marketing materials to public, reviewing the marketing efficiency, and hiring distributors and sales team.

Diffusion of Innovation in Bird ICT

Bird ICT is a small enterprise that is blessed with a team of creative and experienced staff who share a common vision and expected future direction, and whose efforts are aligned towards common goals. Together, they were able to grow solutions for businesses in the ICT sector. "*The most important part of what we are is who we are,*" stated Dr. Khaled. Composed of a mixture of talents and interests, skills and ideas, Bird ICT staff includes project manager, HR managers, developers, design professional and more. "*We are more than our work, and by bringing our creativity and experience to bear on our projects, we all benefit,*" continued Dr. Khaled. In this respect, Bird ICT has a group of qualified management team whose experience insures the reliability of the solutions. People interact with others and perform tasks that assist them in meeting their top-priority satisfaction needs. The dominating culture emphasizes achievement, self-actualization, encouragement and affiliation. As such, Bird ICT encouraged feedback from employees throughout innovation of products. "*We shifted from using certain development framework to another one because one of our employees proved it cut the development time,*" declared Dr. Khaled. People are very committed and the evidence is that they go beyond the basic requirements of the job and might change their job to facilitate organizational functioning. Moreover, people are emotionally attached to the company, and the prevailing mindset nature and decision to remain in the organization reflects an affective desire.

Bird ICT target market includes advertising agencies and solution integrators. PlusXMS targets users who are interested in communicating their feeling of love, apology, greetings, friendship, congratulations, etc. "*PlusXMS adds life to short mobile messages and bridges the gap between mobile and web users. People can now use PlusXMS to express their feelings.*" stated Mohamed Kash, Bird ICT business development manager.

Bird ICT provides its users with forums to communicate with its management team, encourages them to recommend new features, provide their feedback, contribute to PlusXMS events, etc. Also, through these forums, users are enticed to develop their own content, images, smiley faces, and backgrounds that they prefer to see in messages they share with others. PlusXMS team is also working on making the PlusXMS platform highly adaptable to accept changes requested and desired by the community, and work on any computer and mobile program.

Innovation and Strategic Competitiveness in Bird ICT

In Dr. Khaled's opinion, there are several factors that contributed to strategic competitiveness of the company. "*First, Bird ICT's organizational commitment profile, cheap labor, and fast delivery of easy hit technologies, influenced the company's ability to create value to different stakeholders. As employees get more experience and more connected to the market key persons, shareholders see that the company is growing, and clients are happy due to more competitive products that match their needs,*" asserted Dr. Khaled. "*Moreover, ICT helps us achieving competitiveness because of its low barriers, all you want to have is to know how to write a code using new technologies. Non human resources within ICT are cheap. And leadership is the most important part, especially in small companies, to keep everything working in harmony,*" added the young entrepreneur.

In building and managing innovative business ideas, Bird ICT was able to achieve strategic competitiveness and used several measures in this respect. For example, in measuring customers' satisfaction and value creation, the company uses response rate to customers' needs and the number of customers who purchased ICT-based produced as measures of good performance. For example, the rate of response to customers' needs is set by the company before one to two months;

whereas, the number of customers who found solutions as useful, competitive, and matching their needs, and accordingly purchased them were about 700. Employees were happy getting more experience and connections to the market through key persons? Bird ICT was also able to use cheap labor, and thereby it was able to cut on salaries. Shareholders were happy too, with the company's growth.

DISCUSSION

This case study explored, described, and examined activities and management practices throughout implementation of the Bluetooth advertising, to mobile handsets targeting other organizations like advertising agencies and solution. The company achieved what it set out to accomplish as it realized desired results of building and managing business ideas and bringing them into the market. Though obtaining financial measures of performance of Bird ICT was difficult, there were other measures that relied on in keeping score with stakeholders and measuring achieved strategic competitiveness, in accordance with literature on strategic competitiveness.

Neither the company's clients nor employees showed any resistance to the new innovations and products. In fact, they were appealing to most of the clients, and created more value through lower costs and easier use, which coincides with previous literature (Daniel & Grimshaw 2002). Challenges to innovation included technical problems and work overloads. In this respect, the management adopted several methods to overcome such challenges among which were dividing works among teams according to their members' experiences, brainstorming with staff, and conducting training courses. Some of the challenges faced were lack of information and appropriate human resources that were capable of finding and capturing valuable exterior knowledge. Also, in managing innovations the company did not undertake specific strategies; on the contrary, they used the trial and error method because of the small size of the firm, and this coincides with Chesbrough's (2010).

But the small size of the company, focus, and specialization allowed it to find a niche market (advertising agencies and solution integrators) more appealing and cost-effective, and responded promptly to inputs from customers and employees, which is also in agreement with writings of Chesbrough (2010). Launching iBlue company and "LongArm" as a unique Bluetooth marketing solution in the Middle East and North Africa that targeted different business areas ranging from shopping malls, sporting clubs, and outlet stores to airports, hotels, events management, and advertising agencies, agreed with literature on dimensions of open innovation (see Van de Vrande, de Jong, Vanhaverbeke & de Rochemont, 2008). Simple structures, processes and systems allowed Bird ICT to be more flexible in adapting to new ways of conducting business (see Daniel & Grimshaw, 2002: 133-147; Lynn et al. 1999: 9-18). Further, Bird ICT generated and put together innovative business ideas, built up business models, and set up them into companies, and all the time seeking up-to-date ideas in the ICT turf, which matches literature on the concept of open innovation.

Bird ICT possessed several leadership traits, such as passion in innovations and technologies, openness, decentralized decision making, and an entrepreneurial culture that promoted achievement, encouragement, and eagerness to undertake new ideas, which matches literature on diffusion of innovation (Rogers, 1983, 1995; Tornatzky & Klein, 1982), and on the impact of organizational culture on innovation and knowledge transfer (Meschi, 1997: 211- 228; Simonin, 1999: 595-623). Bird ICT assumed several leadership actions to get people involved. Management from several departments with different educational backgrounds and fields of experience cooperated throughout the implementation phases. Implementing new systems triggered re-organization of staff activities and each staff member had his/her role defined according to his/her area of expertise. Some work adjustments and new tasks were added to existing employees' job descriptions. New systems assisted people working in the customer services department to be more sales oriented. Bird ICT invested heavily in information technology and training staff. It started small, then after testing the product, finding that it is profitable, and acceptable to users, it changed its distribution strategy and starts implementing another product.

Main Findings and Results

1. Bird ICT realized strategic competitiveness, as it was able to create more value for its stakeholders including customers, shareholders, and employees.
2. Bird ICT relied mainly on in-house resources in implementing, exploiting, and marketing new innovations.
3. Because of the small size of the firm, Bird ICT did not undertake specific strategies in managing innovations; on the contrary, they used the trial and error method.
4. The entrepreneurial culture that promoted achievement, encouragement, and eagerness to undertake new ideas, and supported employees' involvement and participation in the implementation of new technologies.
5. Leadership traits, such as passion in innovations and technologies, openness, and decentralized decision making influenced details of activities during the implementation process.

Practical and Policy Implications

1. Enhance in-house resources in contrast to other external-based support, such as outsourcing, in implementing new delivery systems, particularly in small and medium enterprises.
2. Increase employees' involvement and participation.
3. Develop a suitable environment that would induce firms' foreign investment.

Recommendations

1. Enhance identifying new product market opportunities.
2. Enhance responding to open innovation developments
3. Enhance SMEs' organizational capabilities and human resources skills to promote open innovation, particularly the exploration dimension.

FUTURE RESEARCH DIRECTIONS

Future research, therefore, might include other factors that could influence innovation in SMEs and their ability to achieve strategic competitiveness. Examples could be the influence of external environment variables like competitors, the economic, social, and political environments, might also have an impact on the dependent variable. Other interventions to enhance the competitiveness level of SMEs, which are different from innovation, could include improving profitability, expanding market share, human resources development and sharpening skills, greater customer orientation, corporate governance, complying with international standards, development of a service quality image, advantages of scale, distribution strategy, and enhancing customer relationship management issues with open innovation. Suggestions for future research also include outcomes to innovation other than strategic competitiveness. Examples could be the role of innovation in explaining organizational creativity, cultural differences, and individual variations amongst different industries.

CONCLUSION

This case described Bird ICT's activities, management practices, and circumstances throughout building and implementation of innovative business ideas in an Egyptian small enterprise. The small enterprise specialized in introducing the Bluetooth advertising, to mobile handsets targeting other organizations like advertising agencies and solution integrators. There were several factors contributed to Bird ICT's strategic competitiveness in generating and developing innovative ideas. New innovative business ideas adopted by Bird ICT created value to all the stakeholders. They increased service availability to customers, shareholders' profits, employees' experience, reduced workload stress on labor, facilitated work, and reduced costs and lead-time. Employees were also highly committed and emotionally attached to the organization. Organizational commitment further improved customer services provided with implementation of new technologies. New technologies were also indispensable to achieve strategic competitiveness. Management's orientation and openness to change and technology solutions, including passing it on to employees, contributed to achieving strategic competitiveness. And though after implementation, the company faced some problems, they were mostly technical challenges that emerged.

The Contribution of this Study

In spite of the importance of studying the topic of building innovative business ideas as being catalysts of SMEs' strategic competitiveness and development, up to the researcher's knowledge no research has been conducted on the research topic, particularly on Egyptian SMEs. As such, this research is the first to examine the topic of interest, and therefore, it has contributed to the literature on SMEs, as well as research on open innovation, where little investigation has taken place on issues concerning effective management of open innovation and strategic competitiveness. It explored and provided a theoretical understanding of some of the attributes of SMEs, open innovation, diffusion of innovation, and strategic competitiveness. It highlighted open innovation issues and challenges facing building and managing innovative business ideas in an Egyptian small business enterprise. This case explored how effective deployment of innovative business ideas and bringing them to the market has contributed to an SME. Thus in effect, going through the research proposition, whether eeffective deployment of innovative business ideas contributes to strategic competitiveness, this study validates the research question, as to what extent does effective management of innovative business ideas contribute to strategic competitiveness. As such, this research has added to the understanding of several theories, models, and frameworks of open innovation and strategic management, and the impact of building innovative business ideas on strategic competitiveness.

REFERENCES

Amabile, T. M. (1988). A model of creativity and innovation in organizations. In Cummings, L. L., & Staw, B. M. (Eds.), *Research in organizational behavior*.

Amit, R., & Shoemaker, P. J. H. (1993). Strategic assets and organizational rent. *Strategic Management Journal, 14*(1), 33–46. doi:10.1002/smj.4250140105

Ballantine, J., Levy, M., & Powell, P. (1998). Evaluating information systems in small and medium-sized enterprises: Issues and evidence. *European Journal of Information Systems, 7*(4), 241–251. doi:10.1057/palgrave.ejis.3000307

Barringer, B. R., & Ireland, D. (2010). *Entrepreneurship: Successfully launching new ventures*, global ed. New Jersey: Pearson Higher Education.

Binks, M. R., & Ennew, C. T. (1996). Growing firms and the credit constraint. *Small Business Economics, 8*(1), 17–25. doi:10.1007/BF00391972

Bird, I. C. T. (n.d.). *Website*. Retrieved June 14th, 2010, from http://www.birdict.com/

Bode, S., & Burn, J. M. (2002). Strategies for consultancy engagement for e-business development - A case analysis of Australian SMEs. In Burgess, S. (Ed.), *Managing Information Technology in small business: Challenges and solutions* (pp. 227–245). Hershey, PA: Idea Group, Inc.

Chang, L., & Powell, P. (1998). Towards a framework for business process re-engineering in small and medium-sized enterprises. *Information Systems Journal*, *8*(3), 199–215. doi:10.1046/j.1365-2575.1998.00035.x

Chesbrough, H. (2002). Making sense of corporate venture capital. *Harvard Business Review*, (March): 2002.

Chesbrough, H. (2003). *Open innovation: The new imperative for creating and profiting from technology*. Boston, MA: Harvard Business School Press.

Chesbrough, H. (2010). Open innovation: A key to achieving socioeconomic evolution. How smaller companies can benefit from open innovation. *Economy, Culture & History: Japan Spotlight Bimonthly JAPECO*, 169. Japan Economic Foundation (JEF). Retrieved January 21, 2011, from http://openinnovation.haas.berkeley.edu/papers/How_Smaller_Companies_Can_Benefit.pdf

Chesbrough, H., & Crowther, A. K. (2006). Beyond high tech: Early adopters of open innovation in other industries. *R & D Management*, *36*(3). doi:10.1111/j.1467-9310.2006.00428.x

Cohen, W. M., Florida, R., Randazzese, L., & Walsh, J. (1998). Industry and the academy: Uneasy partners in the cause of technological advances. In Noll, R. (Ed.), *Challenges to the university*. Washington, DC: Brookings Institution Press.

Cooper, R. G., & Kleinschmidt, E. J. (1995). Benchmarking the firm's critical success factors in new product development. *Journal of Product Innovation Management*, *12*(5). doi:10.1016/0737-6782(95)00059-3

Da Costa, E. (2001). *Global e-commerce strategies for small businesses*. London, UK: The MIT Press.

Daft, R. L. (1978). A dual-core model of organizational innovation. *Academy of Management Journal*, 21.

Damanpour, F., & Wischnevsky, J. D. (2006). Research on organizational innovation: Distinguishing innovation-generating from innovation-adopting organizations. *Journal of Engineering and Technology Management*, *23*(4). doi:10.1016/j.jengtecman.2006.08.002

Daniel, E. M., & Grimshaw, D. J. (2002). An exploratory comparison of electronic commerce adoption in large and small enterprises. *Journal of Information Technology*, *17*, 133–147. doi:10.1080/0268396022000018409

Dierickx, I., & Cool, K. (1989). Asset stock accumulation and sustainability of competitive advantage. *Management Science*, *35*(December), 1504–1511. doi:10.1287/mnsc.35.12.1504

Doherty, N., Hughes, F., & Ellis-Chadwick, F. (2001). An investigation into the factors affecting the level of e-commerce uptake amongst SMEs. In M. Roberts, M. Moulton, S. Hand, & C. Adams (Eds.), *Sixth Annual Conference of UKAIS*, Portsmouth, April, (pp. 251-57). Manchester, UK: Zeus Press.

Dushnitsky, G., & Lenox, M. J. (2005a). When do firms undertake R&D by investing in new ventures? *Strategic Management Journal*, 26.

Egypt's Official Newspaper. (2004). Small enterprise development law 141, June 10th, 2004. Retrieved November 21, 2010, from http://www.eedpegypt.org/download/pdf/SMEs_law_no_141-2004-AR.pdf

Elsammani, Z., Scown, P., & Hackney, R. (2001). A case study of the impact of a diffusion agent on SMEs adoption of Web presence. In M. Roberts, M. Moulton, S. Hand & C. Adams (Eds.), *Sixth Annual Conference of UKAIS*, Portsmouth, April (pp. 562-76). Manchester, UK: Zeus Press.

Ernst, H., Witt, P., & Brachtendorf, G. (2005). Corporate venture capital as a strategy for external innovation: An exploratory empirical study. *R & D Management*, 25.

Fontana, R., Geuna, A., & Matt, M. (2006). Factors affecting university–industry R&D projects: The importance of searching, screening and signaling. *Research Policy*, 35.

Gales, L., & Mansour-Cole, D. (1995). User involvement in innovation projects: Toward an information processing model. *Journal of Engineering and Technology Management*, 12.

George, G., Zahra, S. A., & Wood, D. R. (2002). The effects of business-university alliances on innovative output and financial performance: A study of publicly traded biotechnology companies. *Journal of Business Venturing*, *17*, 577–609. doi:10.1016/S0883-9026(01)00069-6

Gomes-Casseres, B. (1997). Alliance strategies of small firms. *Small Business Economics*, , 9.

Hamil, J., & Gregory, K. (1997). Internet marketing in the internationalization of UK SMEs. *Journal of Marketing Management*, *13*(1), 9–28. doi:10.1080/0267257X.1997.9964456

Hansen, M., & Birkinshaw, J. (2007). The innovation value chain. *Harvard Business Review.* Retrieved January 21, 2011, from http://hbr.org/2007/06/the-innovation-value-chain/ar/1

Hitt, M. A., Ireland, R. D., & Hoskisson, R. E. (2001). *Strategic management: Competitiveness and globalization*. United States: South-Western College Publishing.

Jacobs, D., & Waalkens, J. (2001). *Innovatie: Vernieuwingen in de innovatiefunctie van ondernemingen. AWT achtergrondstudie, 23*. Deventer, The Netherlands: Kluwer.

Jacobs, G., & Dowsland, W. (2000). The dot-com economy in Wales: The long road ahead. In P. Beynon-Davies, M. D. Williams, & I. Beeson (Eds.), *UK Academy of Information Systems Conference*, Swansea, (pp. 590-596). Maidenhead, UK: McGraw-Hill.

Keeling, K., Vassilopoulou, K., McGoldrick, P., & Macaulay, L. (2000). Market realities and innovation in small to medium enterprises: Facilitators and barriers to the use of electronic commerce. *New Product Development and Innovation Management*, *2*(1), 57–70.

Kowalski, S. P. (2009). *SMES, open innovation and IP management: Advancing global development.* Retrieved January 20, 2010, from http://www.wipo.int/edocs/mdocs/sme/en/wipo_smes_rom_09/wipo_smes_rom_09_b_theme02_2related1.pdf

Lecocq, X., & Demil, B. (2006). Strategizing industry structure: The case of open systems in low-tech industry. *Strategic Management Journal*, 27.

Leonard-Barton, D. A. (1988). Implementation as mutual adaptation of technology and organization. *Research Policy*, 17.

Lettl, C., Herstatt, C., & Gemuenden, H. G. (2006). Users' contributions to radical innovation: Evidence from four cases in the field of medical equipment technology. *R & D Management*, 36.

Lord, M. D., Mandel, S. W., & Wager, J. D. (2002). Spinning out a star. *Harvard Business Review*, 80.

Lynn, G., Maltz, A., Jurkat, P., & Hammer, M. (1999). New media in marketing redefine competitive advantage: A comparison of small and large firms. *Journal of Services Marketing*, *13*(1), 9–18. doi:10.1108/08876049910256041

Meschi, P. X. (1997). Longevity and cultural differences of international joint ventures: Toward time-based cultural management. *Human Relations*, 50.

Ministry of Foreign Trade Website. (2004). *MSME definition study (phase II)*. Retrieved.

Morgan, G. (1993). *Imaginization* (p. 43). Sage.

National Association of Software & Service Companies (NASSCOM). (2005). *Research reports: IT in the economy of India*. Retrieved December 31, 2005, from http://www.nasscom.org/artdisplay.asp?Art_id=4289

National Science Foundation. (2006). *Science resource studies*. Survey of Industrial Research Development.

November 18, 2004, from www.moft.gov.eg

Nystrom, P. C., Ramamurthy, K., & Wilson, A. L. (2002). Organizational context, climate and innovativeness: Adoption of imaging technology. *Journal of Engineering and Technology Management*, *19*, 221–247. doi:10.1016/S0923-4748(02)00019-X

O'Keefe, R. M., O'Connor, G., & Kung, H. J. (1998). Early adopters of the web as a retail medium: Small company winners and losers. *European Journal of Marketing*, *32*(7/8), 629–643. doi:10.1108/03090569810224038

Penrose, E. T. (1959). *The theory of growth of the firm*. London, UK: Basil Blackwell.

Peteraf, M. A. (1993). The cornerstones of competitive advantage: A resource-based view. *Strategic Management Journal*, *14*(March), 179–191. doi:10.1002/smj.4250140303

Poon, S. (2000). Business environment and internet commerce benefit: a small business perspective. *European Journal of Information Systems*, *9*(2), 72–81.

Poon, S., & Jevons, C. (1997). Internet-enabled international marketing: A small business network perspective. *Journal of Marketing Management*, *13*(1), 29–41. doi:10.1080/0267257X.1997.9964457

Poon, S., & Swatman, P. M. C. (1999). An exploratory study of small business internet commerce issues. *Information & Management*, *35*(1), 9–18. doi:10.1016/S0378-7206(98)00079-2

Porter, M. E. (1979). How competitive forces shape strategy. *Harvard Business Review*, *20*(8), 1–10.

Porter, M. E. (1985). *Competitive advantage* (pp. 11–15). New York, NY: The Free Press.

Porter, M. E. (1998b). *Competitive advantage: Creating and sustaining superior performance*. New York, NY: The Free Press.

Porter, M. E. (2001). Strategy and the internet. *Harvard Business Review*, *75*(3), 63–78.

Rao, S., Metts, G., & Monge, C. (2003). Electronic commerce development in small and medium sized enterprises. *Business Process Management Journal*, *9*(1), 11–32. doi:10.1108/14637150310461378

Report, G. C. (2010). *Country rankings 2010-2011*. Retrieved November 1, 2010, from http://gcr.weforum.org/gcr2010/

Rogers, E. M. (1983). *Diffusion of innovations*. New York, NY: The Free Press.

Rogers, E. M. (1995). *Diffusion of innovations*. New York, NY: The Free Press.

Rowley, T., Behrens, D., & Krackhardt, D. (2000). Redundant governance structures: An analysis of structural and relational embeddedness in the steel and semiconductor industries. *Strategic Management Journal*, 21.

Rumelt, R. P., Schendel, D., & Teece, D. J. (1991). Strategic management and economics. *Strategic Management Journal*, *12*(Winter Special Issue), 5-29.

Schumpeter, J. A. (1934). *The theory of economic development: an inquiry into profits, capital, credit, interest, and business cycle*. Cambridge, MA: Harvard University Press.

Simonin, B. L. (1999). Ambiguity and the process of knowledge transfer in strategic alliances. *Strategic Management Journal*, 20.

Stoddard, D. B., & Francalanci, C. (1994). State Street Boston Corp.: Leading with Information Technology. *Harvard Business Review*, *19*(5), 1–35.

The Economist Intelligence Unit. (2009). *E-readiness rankings 2009, the usage imperative*. Retrieved November 20, 2010, from http://graphics.eiu.com/pdf/E-readiness%20rankings.pdf

Thorp, J. (2005, October 4). Managing change the key to realizing IT value. *Fujitsu's Electronic Customer Magazine*. Retrieved May 18, 2006, from http://www.fujitsu.com/au/interaction/archieves/2005/200510-04.html

Tornatzky, L. G., & Klein, K. J. (1982). Innovation characteristics and innovation adoption-implementation: A meta-analysis of findings. *IEEE Transactions on Engineering Management*, *29*(1).

Van de Vrande, V., de Jong, J. P. J., Vanhaverbeke, W., & de Rochemont, M. (2008). *Open innovation in SMEs: Trends, motives and management challenges*. SCALES-initiative (SCientific AnaLysis of Entrepreneurship and SMEs), SMEs and Entrepreneurship Programme. Netherlands Ministry of Economic Affairs, Zoetermeer, November 2008.

Walker, R. M. (2008). An empirical evaluation of innovation types and organizational and environmental characteristics: Towards a configuration approach. *Journal of Public Administration: Research and Theory*, *18*(4).

Warner, A. M. (2001). *Small and medium sized enterprises and economic creativity*. United Nations Conference on Trade and Development: Improving the Competitiveness of SMEs in Developing Countries. The Role of Finance To Enhance Enterprise Development (UNCTAD/ITE/TEB/Misc.3). New York, NY: United Nations.

Webb, B., & Sayer, R. (1998). Benchmarking small companies on the internet. *Long Range Planning*, *31*(6), 815–827. doi:10.1016/S0024-6301(98)80018-6

Wernerfelt, B., & Montgomery, C. A. (1988). Tobin's q and the importance of focus in firm performance. *The American Economic Review*, *78*, 246–250.

Wheelen, L. W., & Hunger, J. D. (2004). *Concepts in strategic management and business policy*. New Jersey: Pearson Prentice Hall.

Zaltman, G., Duncan, R., & Holbek, J. (1973). *Innovations and organizations*. New York, NY: Wiley.

ADDITIONAL READING

Arora, A., Fosfuri, A., & Gambardella, A. (2001a). *Markets for Technology: The Economics of Innovation and Corporate Strategy*. Cambridge, Massachusetts: MIT Press.

Baldwin, C., Hienerth, C., & von Hippel, E. (2006). How user innovations become commercial products: A theoretical investigation and case study. *Research Policy*, *35*(9), 1291–1313. doi:10.1016/j.respol.2006.04.012

Betz, F. (1993). *Strategic Technology Management*. New York: McGraw Hill.

Boutellier, R., Gassmann, O., & von Zedtwitz, M. (2000). *Managing Global Innovation: Uncovering the Secrets of Future Competitiveness*. Berlin: Springer.

Caloghirou, Y., Kastelli, I., & Tsakanikas, A. (2004). Internal capabilities and external knowledge sources: complements or substitutes for innovative performance? *Technovation*, *24*(1), 29–39. doi:10.1016/S0166-4972(02)00051-2

Chesbrough, H. (2004). Managing Open Innovation: Chess and Poker. *Research & Technology Management*, *47*(1), 23–26.

Chesbrough, H., & Kusonoki, K. (2001). The Modularity Trap: Innovation, Technology Phase Shifts and the Resulting Limits of Virtual Organizations. In Nonaka, I., & Teece, D. (Eds.), *Managing Industrial Knowledge*. London: Sage Press.

Chesbrough, H., & Rosenbloom, R. S. (2002). The role of the business model in capturing value form innovation: evidence from Xerox corporation's technology spin-off companies. *Industrial and Corporate Change*, *11*(3), 529–555. doi:10.1093/icc/11.3.529

Chesbrough, H., Vanhaverbeke, W., & West, J. (Eds.). (2006). *Open Innovation: Researching a New Paradigm*. Oxford: Oxford University Press.

Chesbrough, H. W., & Teece, D. J. (1996). When Is Virtual Virtuous? Organizing for Innovation. *Harvard Business Review*, *74*(1), 65–73.

Clark, K., & Wheelwright, S. (1992). *Revolutionizing Product Development*. New York: Free Press.

Cohen, W. M., & Levinthal, D. A. (1990). Absorptive capacity: a new perspective on learning and innovation. *Administrative Science Quarterly*, *35*(1), 128–152. doi:10.2307/2393553

Correa (Eds), Environmental Strategy and Competitive Advantage, Northhampton: Edward Elgar Academic Publishing, pp. 259-285.

Day, D. (1994). Raising Radicals: Different Processes for Championing Innovative Corporate Ventures. *Organization Science*, *5*(2), 149–172. doi:10.1287/orsc.5.2.148

Eisenhardt, K. M., & Graebner, M. E. (2007). Theory Building from Cases: Opportunities and Challenges. *Academy of Management Journal*, *50*(1), 25–32. doi:10.5465/AMJ.2007.24160888

Fabrizio, K. R. (2006). The use of university research in firm innovation. In Chesbrough, H. W., Vanhaverbeke, W., & West, J. (Eds.), *Open innovation: researching a new paradigm*. Oxford: Oxford University Press.

Gallagher, Scott. and Park, Seung Ho. (2002). Innovation and competition in standard-based industries: A Historical Analysis of the U.S. Home Video Game Market. *IEEE Transactions on Engineering Management*, *49*(1), 67–82. doi:10.1109/17.985749

Gans, J. S., & Stern, S. (2003). The product market and the market for 'ideas': commercialization strategies for technology entrepreneurs. *Research Policy*, *32*(2), 333–350. doi:10.1016/S0048-7333(02)00103-8

Gassmann, O., & von Zedtwitz, M. (2002). Market versus technology drive in R&D internationalization: four different patterns of managing research and development. *Research Policy*, *31*(4), 569–588. doi:10.1016/S0048-7333(01)00125-1

Gawer, A., & Cusumano, M. A. (2002). *Platform Leadership: How Intel, Microsoft, and Cisco Drive Industry Innovation*. Boston: Harvard Business School Press.

Giarratana, M. S. (2004). The birth of a new industry: entry by start-ups and the drivers of firm growth: The case of encryption software. *Research Policy*, *33*(5), 707–806. doi:10.1016/j.respol.2004.01.001

Granstrand, O. (1982). *Technology Management and Markets*. London: Pinter.

Granstrand, O., & Sjölander, S. (1990). Managing innovation in multi-technology corporations. *Research Policy*, *19*(1), 35–60. doi:10.1016/0048-7333(90)90033-3

Hagedoorn, J., & Cloodt, M. (2003). Measuring innovative performance: is there an advantage in using multiple indicators? *Research Policy*, *32*(8), 1365–1379. doi:10.1016/S0048-7333(02)00137-3

Hauschildt, J. (2004). *Innovation Management*. Munich: Vahlen.

Hitt, M., Ireland, D., Camp, M., & Sexton, D. (2002). *Strategic entrepreneurship*. Oxford: Blackwell. doi:10.1111/b.9780631234104.2002.x

Hockerts, K. (2003) Sustainability Innovations - Ecological and Social Entrepreneurship and the Management of Antagonistic Assets, Bamberg: Difo.

Jelinek, M., & Schoonhoven, C. B. (1990). *The Innovation Marathon*. Oxford: Blackwell.

Kirschbaum, R. (2005). Open Innovation in Practice. *Research-Technology Management*, *48*(4), 24–28.

Laursen, K., & Salter, A. (2005). Open for Innovation: The Role of Openness in Explaining Innovation Performance among U.K. Manufacturing Firms. *Strategic Management Journal*, *27*(2), 131–151. doi:10.1002/smj.507

Louis, K. S., Jones, L. M., Anderson, M. S., Blumenthal, D., & Campbell, E. G. (2001). Entrepreneurship, Secrecy, and Productivity: A comparison of Clinical and Non-Clinical Life Sciences Faculty. *The Journal of Technology Transfer*, *26*(3), 233–245. doi:10.1023/A:1011106006976

Maula, M. V. J., Salmenkaita, J.-P., Keil, T., & Väisänen, K. (2003). "Dynamic complementarities of startups and incumbents in the creation of a systemic innovation," Paper presented at the 23rd Annual International Conference of the Strategic Management Society, Baltimore, Maryland.

McDermott, C. M., & O'Connor, G. C. (2002). Managing Radical Innovation: An Overview of Emergent Strategy Issues. *Journal of Product Innovation Management*, *19*(6), 424–438. doi:10.1016/S0737-6782(02)00174-1

Morone, J. G. (1993). *Winning in High Tech Markets*. Boston: Harvard Business School Press.

O'Connor, G. C., & Ayers, A. D. (2005). Building a Radical Innovation Competency: A Mid-Study Review. *Research-Technology Management*, *48*(1), 23–31.

Petersen, H. (2006) 'Ecopreneurship and Competitive Strategies – Striving for Market Leadership by Promoting Sustainability', in Schaltegger and Wagner (Eds), Managing the Business Case for Sustainability, Sheffield: Greenleaf Publishing, pp. 398 -411.

Prahalad, C. K. and Hamel, Gary. (1990). The Core Competence of the Corporation. *Harvard Business Review*, *68*(3), 79–91.

Prahalad, C. K. (2006). The Innovation Sandbox. *Strategy & Business*, *44*, 1–10.

Prencipe, A. (1997). Technological competencies and product's evolutionary dynamics: A case study from the aero-engine industry. *Research Policy*, *25*, 1261–1276. doi:10.1016/S0048-7333(96)00900-6

Rennings, K. (2000). Redefining Innovation - Eco-Innovation Research and the Contribution from Ecological Economics. *Ecological Economics*, *32*, 319–332. doi:10.1016/S0921-8009(99)00112-3

Rice, M. P., Leifer, R., & O'Connor, G. C. (2002). Commercializing discontinuous innovations: bridging the gap from discontinuous innovation project to operations. *IEEE Transactions on Engineering Management*, *49*(4), 330–340. doi:10.1109/TEM.2002.806721

Roberts, E., & Liu, W. K. (2001). Ally or Acquire? How Technology Leaders Decide. *Sloan Management Review, 43*(1), 26–34.

Roberts, E. B. (1977). Generating effective corporate innovation. *Technology Review, 80*(1), 27–33.

Rothaermel, F. T., & Deeds, D. L. (2004). Exploration and exploitation alliances in biotechnology: A system of new product development. *Strategic Management Journal, 25*(3), 201–222. doi:10.1002/smj.376

Siegel, D. S., Waldman, D. A., Atwater, L. E., & Link, A. N. (2003). Commercial knowledge transfers from universities to firms: improving the effectiveness of university-industry collaboration. *The Journal of High Technology Management Research, 14*(1), 111–133. doi:10.1016/S1047-8310(03)00007-5

Simard, C., & West, J. (2006). Knowledge networks and the geographic locus of innovation. In Chesbrough, H. W., Vanhaverbeke, W., & West, J. (Eds.), *Open innovation: researching a new paradigm*. Oxford: Oxford University Press.

Teece, D. J. (1989). Interorganizational requirements of the innovation process. *Managerial and Decision Economics, 10*(1), 35–42.

Tennenhouse, D. (2003). Innovation breeds success at Intel. *IEE Engineering Management, 13*(6), 44–47.

Tidd, J., Bessant, J., & Pavitt, K. (2005). *Managing Innovation. Integrating Technological, Market and Organizational Change* (3rd ed.). New York: Wiley.

Vanhaverbeke, W. (2006). The inter-organizational context of open innovation. In Chesbrough, H. W., Vanhaverbeke, W., & West, J. (Eds.), *Open innovation: researching a new paradigm*. Oxford: Oxford University Press.

KEY TERMS AND DEFINITIONS

Bluetooth Advertising: Data communication enterprise solutions to mobile handsets that allows link and exchange of information among appliances such as mobile phones, personal computers, laptops, printers, etc.

Closed Innovation: Is the conservative approach to innovation that entails a direct and sequential process that begins with research, going through product development, manufacturing and sales.

Diffusion of Innovation: Describes how products, processes, ideas, or new practices, proliferate within an organization, or society.

Open Innovation: The term "open innovation" was invented by Chesbrough (2003) to illustrate a innovation model based on inflows and outflows of knowledge to enhance internal innovation.

Small Enterprises: Organizations define small enterprises differently based upon several criteria among which are the number of employees, and the level of capitalization or fixed assets.

Stakeholders: Anyone who is interested in and/or benefits from the operation of a particular firm. Examples are employees, shareholders, investors, and customers.

Strategic Competitiveness: A firm achieves "*Strategic competitiveness*" when it successfully formulates and implements a value-creating strategy to its stakeholders.

Chapter 4
A Structural Model for the Design and Implementation of Open Innovation

Matthew C. Heim
NineSigma, USA

ABSTRACT

With the recent developments of open innovation as a formal management discipline, many organizations today are struggling to form effective internal competencies that can be leveraged to generate measured success. Many companies are using a trial-and-error approach that too often leads to unnecessary cost overruns and even failure. The model presented in this chapter provides readers with a simple, yet elegant structure necessary for the design and implementation of a successful open innovation program. The chapter explores the leading causes of failure in a new open innovation program, and offers guidelines and criteria that open innovation leaders and practitioners can use to avoid these pitfalls, and to establish a program that generates tangible returns, while motivating participants to achieve more desirable innovative behaviors.

INTRODUCTION

Open Innovation has become increasingly popular over the past decade, and is being considered one of the most exciting new management trends of our time for small, medium and large enterprises alike. Promises of reduced product development cycles and costs, as well as enhanced innovation capacity are increasing the momentum behind this current phenomenon. As practitioners begin their journeys in this promising new space, many harsh realities can be encountered that were otherwise not anticipated, often leading to perceptions of over-promising and under-delivering. Many of these issues can be avoided if the proper attention is given at the onset of the open innovation program. With the right level of awareness, structure and discipline, open innovation can lead to reduced innovation costs, earlier new product introductions and an inspired organizational culture that is aligned around the notion of innovation excellence across the entire value chain.

DOI: 10.4018/978-1-61350-519-9.ch004

This chapter discusses some of the key challenges that practitioners often encounter when attempting open innovation for the first time, and what they can do to establish an effective and sustainable open innovation program. The author provides a proven framework for the design and implementation of an integrated innovation process that incorporates and assimilates new knowledge from within the organization (i.e., other departments, business units, etc.), the organization's ecosystem of third-party suppliers and partners, and the expansive global innovation community.

BACKGROUND ON TODAY'S OPEN INNOVATION ENVIRONMENT

Although "Open Innovation" is a relatively new term, the concept has its roots back in the early stages of the Industrial Age. Until the early 20th century, most manufacturers relied heavily on external research laboratories and inventors to supply them with ideas for new products and product enhancements. As management theory advanced, giving way to such practices and disciplines as division of labor and the five year strategic plan, companies began to safeguard their intellectual property from the outside world, with the fears that competitors would pick up on their new ideas and copy them, or somehow use this competitive intelligence against them. The idea of the research and development department became increasingly popular through the 1900's, allowing companies to retain their ideas, knowledge and assets within the physical boundaries of the organization. With the slow pace of information dissemination and employee turnover throughout most of the 20th century, this idea of guarding intellectual property worked just fine.

As information technology began to skyrocket in the late 1900's, many things began to change, thus creating a disturbance in this once stable system. The Internet suddenly provided a means of rapid dissemination of new information, giving access to online publications, patent information and other sources of knowledge (Chesbrough, 2006).[1] Knowledge workers and technology-based markets began to become more transient, and information was exposed at an unprecedented rate. Management was no longer able to rely on the safety net that their corporate boundaries once provided. A fork in the road was eminent, and corporate leaders had to choose between one of two options: Find new ways to safeguard intellectual property, or accept the fact that information is no longer as safe as it once was, and focus instead on the *acceleration* of innovation. With the recent emergence of open innovation, the latter was obviously the choice of many.

The early adopters of open innovation had a tremendous impact on building momentum behind this new idea (Lichtenthaler, 2008).[2] Companies like Procter & Gamble touted their success stories of how open innovation helped them to achieve higher overall returns for less R&D investment (Huston & Sakkab, 2006),[3] and new case studies began to surface as more companies adopted this new practice, providing some anecdotal evidence that open innovation was a worthwhile investment. However, with the absence of a generally accepted list of best practices, companies took it on their own to implement open innovation programs, with mixed results. While some were able to realize immediate success, others struggled to demonstrate any tangible value from their new efforts. Most companies found it difficult to reach beyond their ecosystems of existing suppliers, partners and university contacts (Davis, 2009).[4] Other companies that were able to make valuable connections were unable to demonstrate their success in terms of financial gain, because they failed to implement the proper metrics and key performance indicators. With the lack of visibility and reporting capabilities, many open innovation programs were forced to shut down or drastically reduce their budgets and staff sizes at the first sign of recession.

In the last three decades, significant strides have been made in the attainment of best practices regarding process innovation. Process improvement methodologies, such as Six Sigma, have been introduced, focusing on the improvement of productivity and profits by implementing and monitoring metrics throughout key business processes for the purpose of continuous improvement (Eckes, 2001).[5]

New Challenges with Open Innovation

With the recent developments of open innovation as a formal management discipline, many organizations today remain challenged to form effective internal competencies, departments, processes and metrics that can be leveraged to generate measured success. Because of the lack of quantifiable business cases and best practices in the field of open innovation many companies are using a trial-and-error approach that too often leads to unnecessary cost overruns, or even failure. Both the academic and commercial communities are in need of a proven model that will provide guidelines and criteria for the design and implementation of a successful open innovation program.

In a 2009 NineSigma survey (Davis, 2009) among open innovation practitioners, when asked what has been their company's greatest challenge in adopting or implementing open innovation, the most common responses were as follows:

1. Existing mindset or culture
2. Current work processes and priorities
3. Resources and funding
4. The ability to find or execute on external ideas[6]

Beginning with the most common response, "mindset or culture," it is known that the reward systems under the old R&D paradigms of the industrial age are still commonplace in today's corporate environment. These reward systems create incentives for corporate researchers to be the *first to invent*, or the *first to patent*, and tend to work directly against the drivers behind open innovation. The drivers behind open innovation are aligned with those of the business unit, which are oriented towards the acceleration of innovation, or a *first to market* outcome. If the intention behind open innovation is to help accelerate product development and time-to-market cycles, then it becomes clear how the old reward systems can be inhibiting.

The second constraint, "current work processes and priorities," does not necessarily point to a *flawed* process, but rather, one that is absent. Open innovation does not change the way that internal R&D is performed. It does, however, necessitate a new set of processes and priorities that are to be followed, based on a key question that must now be embedded throughout the innovation cycle: Should we make it, or source it? Without a comprehensive and inclusive process for open innovation, roadblocks will be encountered on a regular basis as innovators try to find, acquire and assimilate new knowledge into the organization. Once they finally get through the hurdles within the R&D department, other hurdles begin to emerge as unanticipated requirements and restrictions are presented by the legal and purchasing departments. The absence of comprehensive and inclusive processes is one of the leading causes of delay in deal making with external solution providers (see Slowinski, 2005)[7], and excessive delay is one of the leading causes of overall deal failure after a viable solution provider has been found.

The third most common constraint in open innovation is "resources and funding." This constraint goes hand-in-hand with the previous constraint of processes and priorities. Without the proper budgets in place, innovators are left helpless when new potential breakthrough technologies are introduced to the company, and there are no funds available to acquire or license them. Not having the right resources in place to drive the success of the new program is another major issue. As with the absence of process, these resource issues can lead to significant delays in the acquisition of new knowledge, eventually leading to failure in the deal making process.

The fourth most common constraint, "the ability to find or execute on external ideas," reflects the organization's ability to extend its reach beyond the physical boundaries of the department, business unit or company to inspire and source new ideas and technologies. While the previous three constraints are primary indicators of how well *enabled* a company is to adopt new knowledge, this issue deals with the company's ability to *engage* with resources outside the organization's four walls. Figure 1 shows the three tiers of innovation access typically available to an organization. External resources do not necessarily refer to third party suppliers and innovation partners.

In fact, many companies have trouble locating existing knowledge across different departments within their own organizations (see Tier 1 in Figure 1) This lack of transparency can cause unnecessary duplications in work effort. Furthermore, the inability to identify and assess parallel technology developments outside the company (Tiers 2 and 3 in Figure 1) can lead to new products being launched with soon-to-be outdated technologies.

A Framework for Open Innovation

As outlined in the previous section of this chapter, achieving success in open innovation requires a holistic approach. Figure 2 provides a conceptual framework through which open innovation excellence can be achieved.

This model highlights the importance of developing and maintaining a balanced approach between the *Engage* and *Enable* axes of the model. *Engage* represents the extent to which an organization makes connections across all three tiers as shown in Figure 1, to find and inspire new innovative ideas. *Enable* represents an organization's ability to assimilate new knowledge into its internal innovation processes. In a complete open innovation model, competencies in both areas are of equal importance. Without the ability to adopt new knowledge from the outside, having broad access brings little value. And, of course, having a robust open innovation organiza-

Figure 1. The three tiers of open innovation access

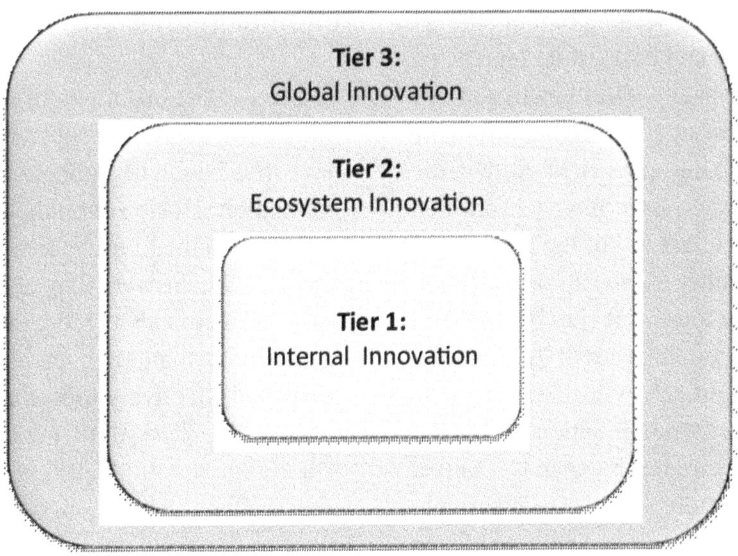

Figure 2. Open innovation development framework

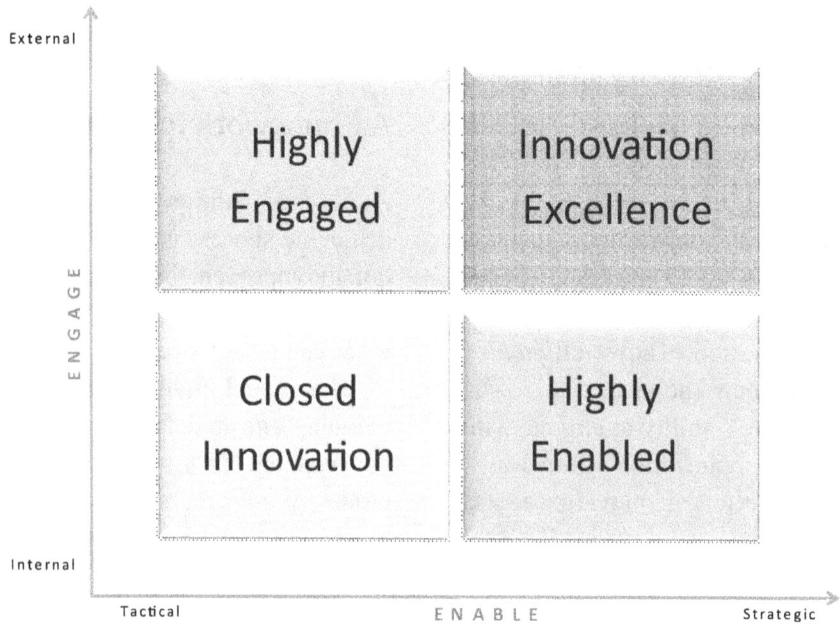

tion with integrated processes, mobilized personnel and strong sponsorship is worth little without having the access to outside knowledge. While some smaller and medium sized companies have significantly less complexity, the same principles of the *Engage* and *Enable* model apply, albeit in a more simplified and streamlined manner.

This model has been considered by many early adopters as a framework for new open innovation programs that have since yielded remarkable reductions in both time and cost, while increasing overall innovation capacity. The objective of this model is not to have an organization *leap* from the lower left quadrant to the upper right. Rather, the model is based on a progressive approach in which movement from the lower left to the upper right is foundational. In other words, an organization must first master the lower left (i.e. internal and tactical capabilities) before successfully migrating upwards and to the right, achieving higher degrees of external engagement and strategic enablement. As a company's outward reach expands, so must its tools and infrastructure, in order to effectively handle the additional sources and volume of new knowledge. By combining Figure 1 and Figure 2, the author effectively shows this progression, as seen in Figure 3. The remainder of this section discusses the steps necessary to create such a foundation, ultimately leading to open innovation excellence, where the disruption of industry norms can be more readily achieved, placing new products on the cutting edge.

Internal Innovation

Leaders in organizations of all sizes realize that not all the smartest people work for them, and have thus begun to embrace the concept of open innovation (Huston & Sakkab, 2006)[8]. However, in these organizations, there is a growing realization that the answers they seek exist not just on the outside, but also on the inside. In many large organizations internal resources are not being leveraged effectively, thus causing duplicate work efforts. '*We don't know what we already know*' is a common complaint that underlies a serious issue. When large commercial organizations closely align their innovation initiatives with their

Figure 3. A foundational approach to open innovation

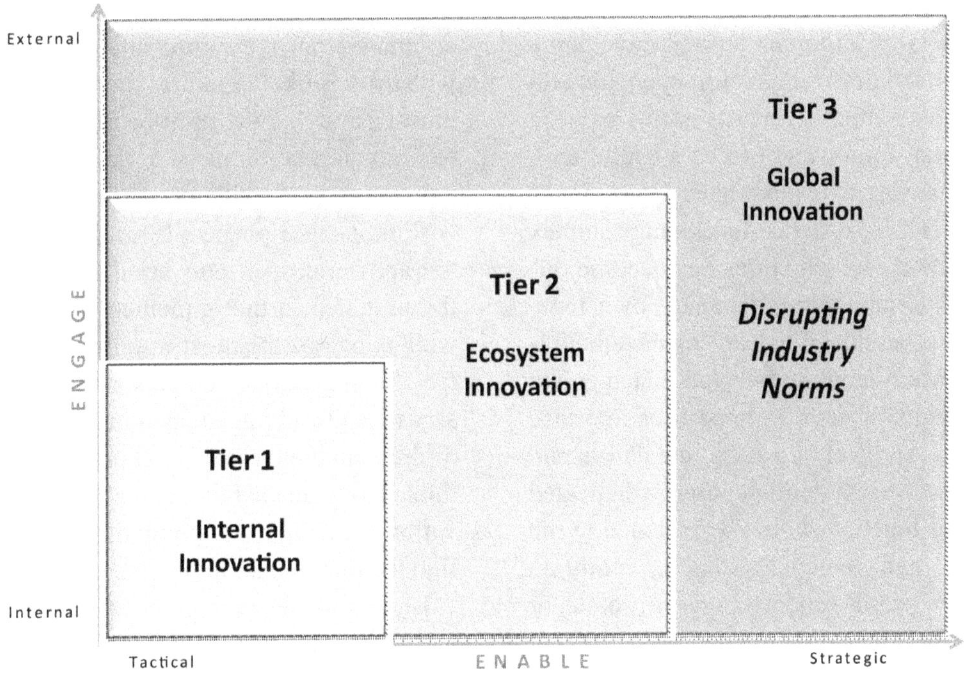

product-specific business units it often results in the business units innovating in silos, minimizing the exchange of knowledge and expertise across the company. Engaging with internal innovators from across the business is a critical first step in the open innovation process, and having the infrastructure to support this is a vital step in building a solid foundation.

Collaboration across the whole enterprise is imperative. It is important to view the entire company as a potential source of information, talent and technology, and to foster collaboration within and across all departments and business units. Interactions should be well organized, structured, and supported through a robust system of metrics and rewards to garner a high level of participation and enthusiasm. Physical and virtual rooms or spaces should be allocated, and designated as areas for cross-functional collaboration where participants from various business units and departments can create new concepts and ideas, and brainstorm necessary next steps by using proven techniques, such as mind-mapping or visioning exercises.

Soliciting new ideas for innovation from within the company should be supported by a standard process that invites participation from all employees, thus adding a more diverse perspective. Internal campaigns can be launched and managed via intranet sites, social media or crowdsourcing applications designed to allow ideas to be communicated openly across the company, and allowing others to build on these ideas through an open forum. These new ideas can then be filtered into a more structured evaluation process where ideas are strategically assessed and ranked, based on the foreseen level of effort that the initiative would require and the potential impact it would have on the company and its markets. A clear path should exist for those ideas that pass the evaluation phase, converting these ideas into newly funded projects in the innovation pipeline. Idea campaigns should generate incentives or rewards that provide recognition and status to individuals and teams contributing the best ideas.

New knowledge can manifest in many different forms, ranging from tacit knowledge to structured documentation or databases. The ability to capture and disseminate these diverse knowledge types and sources is an important step in developing a foundational infrastructure for open innovation. Knowledge management systems exist in many different forms, and can be a useful tool. However, many organizations have a tendency to over-engineer these systems, developing complex taxonomies and data structures that become difficult, or even impossible to manage over time. The key to successful knowledge management is simplicity. Many new knowledge management systems provide a more organic user interface with the look and feel of a social media system. Regardless of what system is chosen, a dedicated management function should be assigned to the knowledge management system to maintain indexes, key words, and to purge obsolete or redundant structures on a regular basis, making it easy to use with little unnecessary complexity.

Ecosystem Innovation

"Open Innovation Ecosystem" refers to a company's innovation activities involving a network of known or trusted partners. (Rohrbeck, Hoelzle & Gemuended, 2009).[9] A company's ecosystem can yield significant value when engaged properly. Whether it is with suppliers, university partners, customers, or even local government, it is essential that any relevant knowledge from the ecosystem be captured and assessed by the organization's innovation teams. Do the existing people and processes exist to extend knowledge searches to the outside? Does the organization have the right partners? Are new partners easy to add to the network? Do systems and processes exist, that capture information and solutions from this outside network? These are all important questions to consider. Relationships may already exist with these external entities, and these partners and suppliers are often quite willing to assist where they can in the innovation process, if only for the likelihood of retaining the partnership, or even the promise of a new supply agreement. However, when these relationships are expanded to include co-innovation, the nature of most interactions with them is likely to shift. The host organization must ensure that the right people, processes and technologies are in place to accommodate these new interactions without causing undue delay.

Finding the right people to manage the ecosystem and run an open innovation program is usually the first step in the implementation process, as well as one of the most important steps overall (see Lindegaard).[10] The personality types necessary to lead and run an open innovation program differ from the typical R&D profile. Big-picture thinkers are needed to connect the various innovation initiatives throughout the company, and to find the right technologies and partners to bridge existing knowledge gaps. Advanced networking and communication skills are mandatory, as relationships will need to be continuously developed and maintained. People with a natural sense of intellectual curiosity and the willingness to influence and drive change within an organization are highly beneficial in these roles. As one can see, these are not necessarily the traits of a typical R&D professional. However, a deep understanding of technology is also a necessary attribute. There are psychometric assessment tools available that can effectively measure a person's aptitude for performing in an open innovation environment. Other persons in the organization who may be impacted by the new initiative can be trained on the benefits and procedures involved in open innovation in order to increase awareness of the program and to minimize any confusion after the program is launched, and to eliminate any fears caused by the belief that open innovation will eliminate jobs.

Executive sponsorship will need to be established in order to remove any roadblocks that could threaten or slow down the process, and to make timely approvals and resource allocations, as necessary. Sponsors must ensure that the appropriate budgets are established to allow the timely acquisition of new technologies and capabilities as they arise through the open innovation process. Ongoing sponsorship may also help to alleviate unexpected issues that may arise with external partners during the search and acquisition processes. Enterprise-wide communications can be broadcast from the executive sponsor just prior to the launch of a new open innovation initiative to demonstrate the seriousness and strategic nature of the new program.

Business processes can be extended to include ecosystem partners in new knowledge searches. Various departments and business units should then be made aware of the new process of including external resources, and processes should be streamlined to ensure the timely assimilation of new knowledge sourced from the ecosystem. These processes should include both intra- and inter-departmental resources who will be involved in the due diligence, legal review and acquisition steps involved in open innovation. Incentives of R&D professionals can be modified to encourage them to go to the outside first to see if a desired technology exists before deciding to embark on a timely internal development project. New metrics and key performance indicators can be integrated into new processes to allow for monitoring, reporting and continuous improvement of the program.

Internal systems and technologies are equally as important for managing internal resources as they are for managing ecosystem resources. New knowledge search campaigns should be automatically extended to ecosystem partners by including them in knowledge management indexes and distribution lists. A company's existing Customer Relationship Management (CRM) or Supplier Relationship Management (SRM) systems can be used to include names and contact information of external partners. This database of ecosystem partner information can serve as the foundation for managed supplier and co-innovator networks that can be expanded upon with each new contact made. The adding and updating of new ecosystem partner information should be encouraged by all personnel as they meet new prospective co-innovators in their daily work routines, working with suppliers, or while visiting external events such as trade shows and conferences.

Global Innovation

Expanding a company's reach to the global innovation community can yield potentially high returns if executed properly. When an external search is performed well, solutions can be found from other industries and applications that would have otherwise never been considered or found. These cross-industry solutions can even end up being those that actually disrupt existing industry norms, creating a competitive advantage. The reason for this is that most companies in an industry struggle with the same issues and challenges, and seek resolution from the same suppliers, research labs and university departments, and even visit the same industry conferences where these same issues are resounded. This eventually leads to industry homogenization, from an innovation perspective. Breaking this pattern requires new thinking and a new set of tools. Developing a new capacity for global innovation can be done with much less effort if built upon a well designed internal and ecosystem innovation process.

When expanding an open innovation program from the ecosystem to the global innovation community, a more robust organization is typically required. This expansion requires a higher level of commitment by organizational leaders and the open innovation team. If planned properly, the previous steps taken to ensure successful internal and ecosystem innovation can serve as a foundation for a more global focus. However, the company's open innovation roles, processes and technologies must be enhanced to adequately support the casting of a broader net – a net that can source new knowledge from new sources around the globe.

When a company decides to take its innovation search capability to the global innovation community level, several new roles begin to appear out of necessity. The open innovation champion or team begins to take on a more strategic role in helping individuals, departments and business units to identify the right resources to help them find what they are looking for. The role of the Technology Scout begins to emerge, which acts as an agent for business units to help seek new external knowledge sources. The Technology Scout begins by developing or acquiring search tools and managed networks of existing and potential solution providers, based on the needs of the organization. With the introduction of technology scouting, the management of ecosystem solution provider networks may begin to migrate from individual users within the R&D department to a more centralized team of Technology Scouts and Open Innovation Champions. Individuals from across the value chain begin to seek out these new professionals to help them find the solutions they need, if they do not exist internally.

Most companies practicing open innovation engage an open innovation intermediary when they cannot find a solution either internally or within their own ecosystem or managed network. Intermediaries are companies that specialize in technology or concept search, and use a variety of tools and methodologies to help connect a solution seeker with a desired solution provider. Intermediaries can also help protect a company's strategic interest by performing the search from a neutral platform, without disclosing the name of the company seeking the new capability. The intermediary can also serve as a "buffer" between seekers and solution providers, communicating with both parties to facilitate a "win-win" outcome. Developing networks of solution providers around the world is a core competency of intermediaries, and some even offer highly developed search engines that help to identify potential solution providers, even if they are not in their existing networks. If an intermediary has a strong technical staff, they can help the seeker to *boil down* the problem statement to a basic technology description, void of any product or industry description. When launched in this fashion, the search will typically yield solutions from across many industries. It is often the case that the best solution is the one that did not come from within the seeker's own industry.

Many companies are developing external innovation websites that are designed to communicate to the outside world what types of knowledge needs or gaps the company has, and to solicit ideas and solutions from the general public. Companies such as Clorox, Kraft Foods and General Mills have launched their own websites with some degree of success. This is an effective way to solicit solutions and ideas for non protected needs (i.e., those not requiring secrecy to preserve or protect competitive advantage). However, many of the solution providers visiting these websites are already active members of that company's ecosystem. Furthermore, as more and more companies adopt this method of open innovation, it will become increasingly difficult for solution providers to choose which ones they will visit, and which ones they will not. Some type of aggregation platform will ultimately be necessary to showcase these needs to a more global community.

Industry consortia have been a common means for companies on a non-competitive or pre-competitive basis to collaborate on the resolution of a common issue. Non competing companies may form a consortium to resolve an issue such as water reduction in manufacturing, and invite companies from across various industries to compare techniques and work towards the resolution of their own unique issues. Competing companies may decide to collaborate to resolve a pre-competitive issue, or one that affects all companies in that industry (e.g. sustainable packaging), but having little or no impact on a company's competitive advantage. While consortia can be an effective way to collaborate and benchmark with other companies, most have been very slow moving, and few have actually met the goals that originally brought the companies in a consortium together in the first place. With the advancement of open innovation tools and techniques, consortia now have the option to use open innovation as a means of achieving the goals of the group, and to do so in a shorter period of time. A many-to-many model can be deployed to find intra- and inter-industry solutions in a fraction of the average consortium lifecycle duration, while allowing participating companies to share the risks, rewards and costs normally associated with such a project.

Breaking the *Not Invented Here* attitude has been an ongoing struggle with most organizations. As global innovation efforts begin to mature in an organization, a new set of *softer* metrics can be established to monitor the degree to which people within the company, or the company's culture, are accepting open innovation, and reaching to the outside to inspire and find new ways to innovate. This can help open innovation leaders to identify pockets of resistance, and to receive valuable feedback from the organization that can help improve the open innovation program. As the company's culture becomes more accepting, and open innovation takes on a more strategic role, other metrics can be shifted from a project-by-project measurement of efficiency to a single enterprise-wide ratio of percentage of R&D spend vs. corporate revenue.

Outside-In Approach

An alternative to the foundational approach mentioned above (i.e., building out the internal and ecosystem innovation capabilities, and ultimately engaging the global innovation community) is to start with a few smaller projects on a pilot basis. Many early adopters were skeptical about open innovation, and decided to try an outside-in approach by working through an open innovation intermediary to see if they could find solutions to issues that they had been struggling with for some time. Once they realized the potential of open innovation, they then decided to invest in the development of their own capabilities in a more strategic fashion.

The outside-in approach is typically accomplished by engaging an open innovation intermediary who can help to identify the current needs, develop a brief or request-for-proposal (RFP), and distribute the RFP out across their global network of solution providers. If the right intermediary is chosen, no prior experience in open innovation is necessary. A full-service intermediary will assist the seeker by identifying best practices, and helping them to identify people in the organization who will need to be involved in various review, assessment and approval steps throughout the project. An internal project *champion* should be selected, who can coordinate each step, and introduce the right resources at the right time. The project champion can also document any pitfalls, resource requirements, process steps and best practices that can be used to develop the company's future open innovation program.

If a company decides to begin with the outside-in approach, it is of utmost importance that an executive sponsor is identified, who can proactively remove potential obstacles and roadblocks that could negatively impact the pilot projects, and to ensure that some funding is made available if the projects lead to an acquisition. The most common roadblock encountered in such projects is departments not making necessary resources available for project requirements, or people not making decisions in a timely manner. An executive sponsor can minimize this, and other risks, by communicating the importance of the project or pilot, by requesting full participation of all necessary resources and departments, and by holding individual managers and department heads accountable.

As the initial pilot projects progress, new external resources will be introduced to the seeker. As these new resources are introduced, the project champion or team can begin to capture relevant information about them in a managed database.

Information regarding their key capabilities, areas of expertise and technologies can be documented, as well as their general contact information. This managed database can then be expended to include existing suppliers, university relationships and other partnerships. This database can then become the platform upon which ecosystem innovation can occur. As each new global search takes place through the intermediary, or through a technology scouting function, newly identified resources can be added, providing an ever-expanding repository of innovation partners.

After several pilot projects using an intermediary, the open innovation champion or team should begin to realize the internal resources that will be required each time an external search and acquisition takes place. With this documented experience, they can eventually begin to institute repeatable processes based on the best practices identified during the pilot projects, and establish a more formal program to develop and manage ongoing open innovation activities. Many companies take these next steps in an organic fashion, allowing the program to unfold on an as-needed basis, while others develop them on a more strategic, enterprise-wide basis. Either way, the outside-in approach can be a compelling alternative for many companies that are still skeptical about the benefits of open innovation.

FUTURE RESEARCH OPPORTUNITIES FOR OPEN INNOVATION

While open innovation itself is not necessarily a new phenomenon, open innovation as a formal management discipline is. As this field continues to grow, many new opportunities for continued research in this area will make themselves known. Best practices will continue to evolve at the corporate, industry, and cross-industry levels, and with these new best practices, research will be necessary to evaluate them and publish guidelines that give corporate managers an indication of where it is safe to invest their resources. Open innovation metrics is another area within this field where companies are struggling to find the best way to measure and communicate success. Closely associated with metrics are the reward systems that guide the behaviors of individuals participating in the open innovation program. Research in either of these areas could contribute significantly to the continued improvement of open innovation.

Research in other scientific disciplines, such as systems and network science could also contribute to the practice of open innovation. Systems research can be performed to identify existing systems paradigms, such as cybernetics, complexity theory and autopoiesis (the study of self-producing systems) that can be used to explain, and more effectively manage the activities that occur within open innovation processes at all levels. Network theory can be extended to explain the existing and potentially more optimal uses of the networks in this new field that depends so heavily on them. The study of how nodes, super nodes and meta-networks in open innovation can be most efficiently applied in the context of open innovation will help practitioners to more effectively map out their future efforts.

The more research that is directed towards improving and advancing open innovation, the sooner we will be likely to see full industry-wide adoption. However, this research will have to be done with the practitioner in mind if it is to be of any use in advancing this field. The possibilities are limitless, and the time was never better to take advantage of a newly blossoming discipline, and to make a major contribution through practical research.

CONCLUSION

There is certainly a lot that we can still learn about open innovation to make it more effective within our organizations. However, this is no excuse to sit back and not participate. Awareness of the most common pitfalls that prevent organizations from successfully implementing open innovation is paramount to a successful launch. With this foreknowledge, open innovation leaders and practitioners can craft an approach that manages and mitigates these risk factors more effectively. By following the simple structure set forth with the *Engage* and *Enable* model presented in this chapter, we can begin to implement processes that produce significant benefits without having to go through the turmoil met by other early adopters.

Existing mindsets and culture within the organization can be influenced through proper sponsorship, development and reward systems. Existing work processes and priorities can be augmented with new ones that provide proven pathways for when going to the outside for new knowledge is necessary. Resources and funding can be made available proactively, so that new incoming knowledge is not delayed, and can be immediately assimilated into the organization. The ability to find or execute on ideas from the outside is made easier through a more supportive organization and network structure.

Finally, a practice of continuous improvement should be adopted by the open innovation team to ensure that new methods and tools can be integrated as they unfold from new research and application from within the field. With the emergence of new open innovation conferences, online forums and user groups around the world, continued development of the practitioner should not be an issue. The open innovation practitioner will ultimately lead the way, modeling the new "open" behaviors that will help to transform the company from a closed and guarded institution to a thriving interconnected network where boundaries are expanded, allowing it to more effectively find and assimilate new knowledge from the environment.

REFERENCES

Chesbrough, H. (2006). *Open innovation: The new imperative for creating and profiting from technology*. Boston, MA: Harvard Business School Press.

Davis, S. (2009). Open innovation practices & outcomes benchmark survey. *NineSigma Report*. Retrieved September 22, 2010, from http://www.ninesigma.com/Register.aspx?link=/Uploads/Resources/NineSigma OI Benchmark Survey Report Spring 2009.pdf

Eckes, G. (2001). *The Six Sigma revolution: How General Electric and others turned process into profit*. New York, NY: John Wiley & Sons.

Huston, L., & Sakkab, N. (2006). Connect and develop: Inside Procter & Gamble's new model for innovation. *Harvard Business Review*, (March): 2006.

Lichtenthaler, U. (2008). Integrated roadmaps for open innovation. *Research Technology Management*, May/June 2008.

Lindegaard, S. (2010). *The open innovation revolution: Essentials, roadblocks and leadership skills*. Hoboken, NJ: John Wiley & Sons.

Rohrbeck, R., Hoelzle, K., & Gemuended, H. G. (2009). Opening up for competitive advantage – How Deutsche Telekom creates an open innovation ecosystem. *R&D Management, 39*, April 2009.

Slowinski, G. (2006). *Reinventing corporate growth*. Gladstone, NJ: Alliance Management Group, Inc.

ADDITIONAL READING

Almirall, E., & Casadesus-Masanell, R. (2010). Open vs. Closed Innovation: A Model of Discovery and Divergence. *Academy of Management Review, 35*(1), 27–47.

Boudreau, K. J., & Lakhani, K. R. (2009, Summer). How to Manage Outside Innovation, MIT. *Sloan Management Review*, *50*(Issue 4), 69–76.

Chiaroni, D., Chiesa, V., & Frattini, F. (2010). The Open Innovation Journey: How firms dynamically implement the emerging innovation management paradigm. *Technovation*, *31*(1), 34–43.

Elmquist, M., Fredberg, T., & Ollila, S. (2009). Exploring the field of open innovation. *European Journal of Innovation Management*, *12*(3), 326–345.

Gassmann, O., Enkel, E., & Chesbrough, H. (2010). The future of open innovation. *R & D Management*, *40*(3), 213–221.

Heim, M. C. (2007). *Breaking the Musashi Code: Transcending Competition Through Visionary Strategy*. Sonoma, CA: Visionary Partnership Press.

Lakhani, K. R., & Jeppesen, L. B. (2007, May). Getting Unusual Suspects to Solve R&D Puzzles. Forethought. *Harvard Business Review*, *85*(5).

Lichtenthaler, U. (2009). Absorptive capacity, environmental turbulence, and the complementarity of organizational learning processes. *Academy of Management Journal*, *52*(4), 822–846.

Mortara, L., Slacik, I., Napp, J., & Minshall, T. (2010). Implementing open innovation: cultural issues. *International Journal of Entrepreneurship and Innovation Management*, *11*(4), 369–397.

Nambisan, S., & Sawhney, M. (2008). *The Global Brain: Your Roadmap for Innovating Faster and Smarter in a Networked World*. Upper Saddle River, NJ: Wharton School Publishing.

Spithoven, A., Clarysse, B., & Knockaert, M. (2010). Building absorptive capacity to organise inbound open innovation in traditional industries. *Technovation*, *31*(1), 10–21.

Van de Vrande. (2009). *Vareska, De Jong, Jeroen, Vanhaverbeke, Wim and De Rochemont*. Maurice.

(n.d.). Open innovation in SMEs: Trends, motives and management challenges. *Technovation*, *29*(6-7), 423–437.

KEY TERMS AND DEFINITIONS

Crowdsourcing: The outsourcing of tasks or initiatives that are normally performed by employees, by posting a public challenge.

Customer Relationship Management (CRM): A system, typically supported by software, to manage the sales function in an organization.

Ecosystem: A network of collaborative innovation partners.

Engage and Enable Model: A model used to convey the importance of balancing an organization's ability to engage with the global innovation community with its ability to adopt new knowledge or solutions identified (i.e., enablement).

Global Innovation Community: Any person or organization in the world who can potentially collaborate with another person or organization to help solve innovation problems.

Innovation Capacity: The extent to which an organization is capable of innovation.

Internal Innovation: The act of an organization innovating without the help of outside collaborators.

Knowledge Management: An organization's ability to store and retrieve knowledge to support the innovation process.

Open Innovation: The act of soliciting and acquiring knowledge from outside the boundaries of the organization.

Open Innovation Champion: A designated person responsible for the coordination of internal innovation needs, and soliciting solutions for them from outside the organization's boundaries.

Open Innovation Intermediary: A services firm that helps organizations to find solutions to their innovation needs from outside the organization's boundaries.

Six Sigma: A quality-inspired process design and management methodology that focuses on the full-lifecycle SIPOC model (Supplier, Inputs, Process, Outputs, Customer).

Social Media: An Internet-based system designed to connect individuals and groups for the purpose of information sharing and collaboration.

Supplier Resource Management (SRM): A system, typically supported by software, to manage the supply chain function in an organization.

Technology Scouting: An organizational function designed to search for external collaboration partners and technologies to fulfill internal innovation requirements.

ENDNOTES

1 Chesbrough, Henry (2006). *Open Innovation: The New Imperative for Creating and Profiting from Technology*. Boston, MA: Harvard Business School Press

2 Lichtenthaler, Ulrich (2008). Integrated Roadmaps for Open Innovation. *Research Technology Management*, May/June 2008.

3 Huston, L. & Sakkab, N. (2006). Connect and Develop: Inside Procter & Gamble's New Model for Innovation. *Harvard Business Review*, March, 2006.

4 Davis, Skip (2009). Open Innovation Practices & Outcomes Benchmark Survey. *NineSigma Report*. Retrieved September 22, 2010, from http://www.ninesigma.com/Register.aspx?link=/Uploads/Resources/NineSigmaOIBenchmarkSurveyReport-Spring2009.pdf

5 Eckes, George (2001). *The Six Sigma Revolution: How General Electric and Others Turned Process Into Profit*. New York, NY: John Wiley & Sons

6 See Davis, Skip (2009).

7 Slowinski, Gene (2006). *Reinventing Corporate Growth*. Gladstone, NJ: Alliance Management Group, Inc.

8 See Huston, L. & Sakkab, N., 2006.

9 Rohrbeck, R., Hoelzle, K., & Gemuended, H.G. (2009). Opening up for Competitive Advantage – How Deutsche Telekom Creates an Open Innovation Ecosystem. *R&D Management*, 39, April 2009.

10 Lindegaard, Stefan (2010). *The Open Innovation Revolution: Essentials, Roadblocks and Leadership Skills*. Hoboken, NJ: John Wiley & Sons

Chapter 5
A Research Model for Open Innovation:
Synthesizing Opportunities and Challenges Surrounding SMEs

Hakikur Rahman
University of Minho, Portugal

Isabel Ramos
University of Minho, Portugal

ABSTRACT

With the advent of innovative technologies and long the setting of entrepreneurship development philosophy, context and business handling have been transformed from being traditional to innovative. Depending on diversity and nature of the transformation, innovation has been shifted from closed peripheries to open dimension. Looking at the immense benefits, ranging from small entrepreneurs to corporate to multinational, business houses are adopting various innovation techniques. However, the road to innovation, despite being a paradigm shift is not smooth and ready for many, particularly the small and medium enterprises (SMEs) who mainly deal with the clients at the outer peripheries of the social circle. Moreover, there is a research gap within the context of SME development through open innovation strategies. This chapter intends to initiate a research model for carrying out research on the development of SMEs through utilization of open innovation strategies. Along these contexts it has tried to synthesize aspects of opportunities and challenges surrounding SMEs development utilizing open and collaborative approaches. To set the research model, the paper has developed an innovation opportunity framework, both at the policy level and at the entrepreneur level. To validate the research hypothesis, the chapter hints at carrying out a three phase survey comprising selected SMEs.

DOI: 10.4018/978-1-61350-519-9.ch005

INTRODUCTION

Business establishments have come to realize the importance of innovation for survival in a world of global competition (Hage, 1999; Sullivan, 1998). Furthermore, it has long been recognized that small and micro businesses have a vital role to play in the economy (Stanworth & Gray, 1991) with small business accounting for 99.6% of all businesses in the UK and an estimated 3.7 million active businesses in 1998. Collectively, small and medium enterprises (SMEs) are responsible for 65% of employment and 57% of Gross Domestic Product within the UK (Madsing, 1997; Tonge, 2001). (There is no single agreed definition of a SME. A variety of definitions are applied among OECD and APEC economies, and employee number is not the sole defined criterion. SMEs are considered to be non-subsidiary, independent firms which employ less than a given number of employees.) The scenario is the same in other European countries. SMEs account for over 99.8% (the EU-26 average, first 26 countries under the European Union) of all firms in Cyprus, Greece, Italy, Portugal, and Spain (European Commission, 2008; Keupp, Lhuillery, Garcia-Torres, & Raffo, 2009).

SMEs in Europe accumulates to 23M€ (23 million Euros) investment market that account for 99% of all businesses and represent 2/3 of the total employment (Renaud, 2008). But, in spite of being key contributor to the global economy accounting for approximately 50% of local and national GDP (Gross Domestic product), 30% of export, and 10% of FDI (Foreign Direct Investment) most of the SME communities are lagging behind promoting their products at the national level and at the global level (OECD, 2006). A number of publications have highlighted these situations (De Jong, Vanhaverbeke, Kalvet, Chesbrough, 2008; European Commission, 2008; MacGregor, Bianchi, Hernandez, & Mendibil, 2007; Maes, 2009; Moore, 2006; Van de Vrande, de Jong, Vanhaverbeke, & de Rochemont, 2009) and numerous cases can be cited (Landabaso, 2010; Lemola, & Lievonen, 2008; Lindermann, Valcárcel, Schaarschmidt, & von Kortzfleisch, 2009; OECD, 2008a, 2008c) to support them.

The burgeoning nature of global competition and the rising trend in research and development (R&D) expenditures lead entrepreneurs, especially SMEs, to embrace open innovation strategies collaborating with external partners, such as suppliers, customers, consumers, supply chain partners, universities, researchers, other companies, and individuals. In a complex and highly competitive global market, entrepreneurs have to innovate to survive and grow. To innovate and develop commercially viable products and services, SMEs need to explore issues of opportunities and challenges in this open paradigm of innovation (Herstad, Bloch, Ebersberger, & van de Velde, 2008; OECD, 2008a). This paper looks into the opportunities and challenges that SMEs are facing for their growth, operation, and development surrounding open innovation strategies.

To set the theory, before going to the background, this section discusses on concepts of innovation and open innovation, where relationships of SMEs with these two parameters are being focused. As the main thrust of this chapter, later on, it looks into various opportunities and challenges in relation to open innovation and they have been categorized through a vertical literature review. Intention is there to emphasize on the most influential and original works in this field.

Conceptualizing Innovation

The term innovation means a novel way of doing something (McKeown, 2008). It may refer to incremental, radical, and revolutionary changes in thinking, products, processes, or organizations (Schumpeter, 1934). Innovation then could be defined as the creation, development, and implementation of a new product, process, or service, aiming at improving efficiency, effectiveness, or competitive advantage (Chesbrough, 2003a; 2006;

EIRMA, 2004). The term may apply to products, services, manufacturing processes, managerial processes, or the design of an organization. It is most often viewed at a product or process level, where product innovation satisfies a customer's needs and process innovation improves efficiency and effectiveness of the organization. Innovation links to creativity and the creation of new ideas, and involves taking those new ideas and turning them into reality through invention, research and new product development (Govt. of New Zealand, 2007). Furthermore, innovation can be seen as the process that translates knowledge into economic growth and social well-being. It encompasses a series of scientific, technological, organizational, economic and commercial activities. Research in the context of innovation is targeted towards one of these activities and may be carried out at different phases of the innovative process (Chesbrough, Vanhaverbeke, & West, 2006; Govt. of Australia, 2010).

Innovation relates to transformation of processes, products, or services (BlessingWhite, 2006; Chason, 2008) to create additional value and wealth (Barker, 2008; Beacham, 2006; Vaitheeswaran, 2007) leading to the benefit of community (Cox, 2008). In this aspect, the real growth of the importance of innovation activities has been linked with several simultaneously affecting societal phenomena, such as easier familiarization with the globalization, obtaining benefits from deregulation and liberalization of markets, utilizing the benefits from the ICT revolution, and adopting themselves with the dynamic changes in the demand patterns (Schienstock, & Hämäläinen, 2001).

Conceptualizing Open Innovation

Open Innovation, a term recently evolved and being used by industries and organizations to promote open ended (flexible and collaborative) ideas, thoughts, processes, and researches to improve the product development, provide better services to the clients, increase efficiency and enhance value-addition. In terms of process dynamics, it incorporates accumulation of ideas, knowledge, licenses, intellectual properties, patents, and inventions (through licensing, joint ventures, spin-offs); and in terms of concept dynamics, it incorporates user innovation, market innovation, cumulative innovation, collaborative innovation and distributed innovation. While developing new ideas for better product, process and service development, open innovation emerges, in the diverse world of widely distributed knowledge, and solitary enterprises cannot afford (economically and organizationally) to rely entirely on their own research and resources, but may instead collaborate, buy, lease-out or license processes or inventions (patents, intellectual properties) from other companies, organizations or institutions (Chesbrough 2003b; 2006; Chesbrough, Vanhaverbeke, & West, 2006). Currently, many companies are promoting open innovation and among them IBM, Nokia, ORACLE and Procter & Gamble are widely renowned, though hardly they can be categorized as SMEs. Among SMEs, to name a few, Modelon in Germany, Helphone in Spain, Peira BVBA in Belgium, Nowatec Technology in Poland, and Delta R&S in Italy has been highlighted as success cases in a study conducted by European Union (Landabaso, 2010).

Along the road of open innovation, inclusion of open ended (flexible, demand driven and collaborative) contexts are becoming popular day by day. Success stories such as those of Google, Skype and Microsoft teach us at least one lesson – no one could ever know where the next big idea will be coming from! Experience shows that innovative ideas thrive when different disciplines or mindsets come together (Oxford Business School, 2008). Moreover, the contribution of the innovation systems approach with the open ended perspectives are emphasizing the importance of external linkages for innovative success and the importance of innovation diffusion together with innovation creation. Studies on service innovations have confirmed the versatility of innovation sources. These studies have also brought up the significance of new patterns of thought and models of operation, in addition to technological innovations (Toivonen, 2004; Van der Walt, Buitendag, Zaaiman, & van Vuuren, 2009).

This chapter limits the scope of research within the context of opportunities and challenges of innovation for SMEs for providing better products and services through better processes. The paper is the outcome of a progressive research on influence of open innovation for SMEs development based on extensive literature review. This paper makes a qualitative synthesis of the parameters related to opportunities aspects of open innovation in terms of SMEs advancement. Next, it carries out a qualitative synthesis on parameters related to challenges. This research has constructed an innovation opportunities framework synthesizing the opportunities and challenges, which lead the survey work to be carried out in three phases (generic development of the benchmark, specific entrepreneurships under the first phase, and focused entrepreneurships under controlled environment) to validate the research hypotheses and continue future research.

Next, the theoretical arguments are being discussed in relation to open innovation and particularly, open innovation within the context of SMEs development focusing opportunities and challenges. One may argue further elaboration on theories. This chapter to be specifically focused to the research theme, such as the synthesis of opportunities and challenges related to SMEs, has tried to set a boundary to the literature review on generic concepts and theoretical aspects of innovation and open innovation.

BACKGROUND

SMEs and Innovation

Learning from various school of thoughts innovation has many effects on SMEs in terms of economics. It leads the SMEs to reach out for evolutionary economics, institutional economics, new regional economics, the economics of learning and knowledge, and the economics of innovation (Lundvall, 1999). Successful innovation is crucial for business accomplishment, and growth through innovation underpins any region's long term economic prosperity (Roper & Hewitt-Dundas, 2004). Innovation in SMEs leads to product innovation, as well as process innovation including employment generation (Lindermann, Valcárcel, Schaarschmidt, & von Kortzfleisch, 2009). However, this window of opening is relatively challenging, since the importance of innovations is not only due to the impact they have on employment, but also depend on other variables. Moreover, innovations have effects on other variables, such as turn-over, economic growth, etc. and firms may also have made more than one innovation where the innovation-induced employment changes are counteractive (Nählinder, 2005).

The situation complicates further. An earlier survey reveals that, when innovation is reported by SMEs themselves, and defined more widely in terms of service as well as manufacturing innovations, including process as well as product innovations, the frequency of product innovations seem lowest among peripheral SMEs (SMEs who work at the lowest level of the marketing chain), either in aggregate form or even for both manufacturing and services treated separately (Keeble, 1997). But, there is another window of opening is that despite being in a continuing debate about the role of SMEs in introducing fundamentally novel innovation than the large firms (Storey & Sykes, 1996), they do have a greater ability to take incremental role in open innovation to niche themselves in better ambiance (Storey, 1994). Furthermore, SMEs can be a starting place of important innovations, which would then be commercialized by large firms, rendering to increased employment generation (Smallbone, North & Vickers, 2003).

In recent years, the contribution of SMEs to economic growth, job creation, innovation and promotion of enterprise has been widely recognized. While SMEs are important in terms of contributing their overall share on GDP, but at the same time it is also believed that many smaller firms lack both managerial and technical skills, which inhibit their business effectiveness (Newton, 2001; Papulová & Mokroš, 2007; Tilley & Tonge, 2003). Therefore, improving the competitive advantage of SMEs is imperative to individual firms and at the same time essential for improvement of the national economy as a whole (Tilley & Tonge, 2003).

SMEs and Open Innovation

The most important benefit of open innovation to companies or business entities is that it provides an extended base of ideas and technologies. Companies look at open innovation as a close collaboration with external partners, such as customers, consumers, researchers or other people that may have an input to the future development of their company. The main motives for joining forces is to seize new business opportunities, to share risks, to pool complementary resources and to realize synergies. Companies recognize open innovation as a strategic tool to explore new growth opportunities at a lower risk. In this aspect, open technology sourcing offers companies higher flexibility and responsiveness without necessarily incurring huge costs (OECD, 2008b; Sousa, 2008). Referring to the importance of innovation in SMEs, Davenport (2006:3) stated that, "Innovation is important for SMEs - a massive 60% of innovations come from the small and medium enterprise sector, and it is crucial that this total must not diminish against competing pressures in the modern market".

The open innovation approach assumes that innovating enterprise is no longer the sole locus of innovation, nor it is the only means for reaping the benefits of research, development and innovation (RDI). In comparison to a closed innovation model, where external actors are viewed with suspicion, who could take away useful knowledge to other competitors; an open innovation environment sets a common platform to its users, customers, suppliers, public knowledge institutions, individual inventors and even competitors and each and every partners are being regarded as potential contributors of crucial pieces of information. In addition to this, through open ended collaboration, innovating partners can enhance their own technologies or reduce RDI costs, which are essentially important drivers of open innovation (Lemola & Lievonen, 2008).

In SMEs innovation collaboration or activities of open innovation is a long lasting tradition, despite their lack in resources in terms of finance and human skills, which restricts them not to maintain large innovation portfolios (De Jong, Vanhaverbeke, Kalvet, & Chesbrough, 2008), but larger enterprises are more comfortable in this case to make themselves available for external collaboration or other strategies of open innovation (Chesbrough, 2003a; De Jong, 2006; De Jong, Vanhaverbeke, Kalvet, & Chesbrough, 2008; OECD, 2008b). For sake of this chapter, literature review will be limited to opportunities and challenges that are being faced by SMEs on the way of establishing an innovation opportunity framework utilizing open innovation strategies.

Opportunities

Open innovation targeting the SMEs development revolves around various issues and perspectives, in terms of product, process and service innovation (De Jong & von Hippel, 2005; 2009; IBM, 2007; Hass & Hochrinner, 2008; Maes, 2009; OECD, 2000a; Van de Vrande, de Jong, Vanhaverbeke, & de Rochemont, 2009) leading to increased competition, demanding customers, increased knowledge acquisition, and better positioning in the market (De Jong, 2006; Lemola & Lievonen, 2008; NOUS, 2007; Storey, 1994). It has been observed that open innovation has created various opportunities for SMEs in terms of their overall business development, and a few are being categorized in Table 1. This chapter is trying to make a qualitative synthesis of those parameters. The opportunity parameters were selected as per the research emphasis by contemporary researchers through an extensive literature review, which are mostly affecting SMEs development in the open innovation paradigm.

As methodology of search, search keys has been applied to various search engines, like Scopus, ScienceDirect, Elsevier, and others with search string, for example, ['open innovation'+SMEs], [opportunities+'open innovation'+SMEs], [challenges+'open innovation'+SMEs].

As mentioned earlier, Table 1 shows the focused categories of opportunities that this research has apprehended mostly from the search and then after several iterations, each of them has been discussed further based on the searched literatures, inclusive of these researchers' understanding.

Competition

In Joseph Schumpeter' s (Tidd, Bessant & Pavitt, 2005:7) view competition in the domain of innovation is, "competition from the new commodity, the new technology, the new source of supply, the new type of organization. .. competition which . .. strikes not at the margins of the profits and the outputs of the existing firms but at their foundations and their very lives". Grama & Fotache (2007) emphasized on utilization of ICT to enhance the competition within the innovation processes. IFAD (2007), however, pointed out that SMEs often compete with protected public enterprises, which is a fact in many countries. But, in the realm

Table 1. Opportunities associated to open innovation affecting SMEs development

Opportunities	Literatures reviewed
Intense competition	Barnes, et al., 2001; Kaplinsky, et al., 2003; De Jong, 2006; NOUS, 2007; Grama & Fotache, 2007; IFAD, 2007; Tidd, Bessant & Pavitt, 2005; BERR, 2008
More demanding customers	Von Hippel, 1998; 2005; Olson & Bakke, 2001; Lilien et al., 2002; Brockhoff, 2003; Bonner & Walker, 2004; Enkel, et al., 2005; FFG, 2006; Gassmann, 2006; De Jong, 2006; BERR, 2007
Abundance of knowledge, opportunities and interactions	European Commission, 2003a; 2003b; Napier et al., 2004; Cosh, Fu & Hughes, 2005; Lemola & Lievonen, 2008; Commonwealth of Australia, 2009
Easily adopt OI strategies such as inward and outward licensing and joint ventures	Fischer, 1978; Gold, 1987; Lazzarotti et al., 2008; Sautter & Clar, 2008; Cervantes, 2009
Product innovation	Crawford, 1983; Hiebing Jr. & Cooper, 2003; OECD, 2000a; Nählinder, 2005; Salavou, 2006; Maria & Micelli, 2008; De Jong et al., 2008; Lee et al., 2010; Maes, 2009; Van de Vrande et al., 2009
Process innovation	Zuboff, 1988; Walton, 1989; Davenport, 1993; OECD, 2000a; Nählinder, 2005; Maes, 2009
Service innovation	Van Ark et al., 2003; Miles, 2005; IfM & IBM, 2008
Technological innovations	Evangelista, 2000; Telemetica Institute, 2008; Conceição & Heitor, 2000; Tilley & Tonge, 2003; Toivonen, 2004
Better positioning in the market	Storey, 1994; Grant, Laney & Pickett, 2002; Van de Vrande et al., 2008
Act as starter or catalyst	Mueller & Goic, 2003; Smallbone, North & Vickers, 2003; Telemetica Institute, 2008; Hagen, 2008; Brouwers et al., 2009
Enabler of employment generation	Sengenberger et al., 1990; Smallbone, North & Vickers, 2003; EIM, 1997; European Communities, 2000
Ability to adopt quickly with new products	Stiglitz, 1998; Cooke, 1996; Cooke & Wills, 1999; Design Council, 2005; Parrilli, 2006

of open innovation, the basis of competition has been shifted from price of product to price and quality, or price, quality, choice, and others. The field has now become mature with much of the competition shifting to marginal issues like relative interest rates (Tidd, Bessant & Pavitt, 2005). Furthermore, free trade/globalization has resulted to adapt the ways of thinking to dismantle protective tariff and other barriers and, therefore, new basis of competition emerges (Barnes, Bessant, Dunne, & Morris, 2001; Kaplinsky, Morris & Readman, 2003).

Increased Customer Demand

Researchers on open innovation recognize increased customer involvement as an essential element to expedite internal innovation process (Gassmann, 2006). Apart from Von Hippel´s (2005) initiating work, this has been supported by many other researchers (Olson & Bakke, 2001; Lilien, Morrison, Searls, Sonnack, & von Hippel, 2002; Bonner & Walker, 2004). Moreover, emphasis has been given to increased involvement of the customers at the beginning of the innovation process (Brockhoff, 2003; Von Hippel, 1998; 2005; Enkel, Kausch, & Gassmann, 2005). By observing positive sales and R&D picture in various services, which reflects robust demand from key customers in diverse forms, and to cope with the situation it is essential to build a global, flexible, cost effective network and service infrastructure that can support the dynamic needs of the future customers (BERR, 2007). Furthermore, the innovation opportunity framework should have a representation through clearly defined and focused services, in a cost-optimized structure and with a customer-oriented approach (FFG, 2006).

Knowledge Acquisition

Innovation involves generating, disseminating and applying knowledge. It is not a linear process. It involves an originator at one end and a receiver at the other. In this process, many players are involved, and they interact and influence each other in multifaceted ways. Ideas can come from anywhere and may lead to unexpected directions (Commonwealth of Australia, 2009). In this aspect, external linkages, both public (including higher education institutions) and private, benefit SME innovation. These linkages can be important sources of knowledge that directly strengthen the technological competences of the SMEs and their competitive advantage. Furthermore, collaboration with customers, suppliers, higher education institutions, even competitors, allows entrepreneurs to expand their range of expertise, develop specialize products, and achieve various other corporate advantages (Cosh, Fu & Hughes, 2005). In addition to these, the growing weight of, and policy emphasis on, innovation and knowledge as drivers of competitiveness and growth, which brings major opportunities for countries that are generally not considered to be at the forefront of knowledge creation or innovative capacity (Napier, Serger & Hansson, 2004).

Adoption of Open Innovation Strategies

The diffusion of open innovation strategies has been largely recognized within SMEs (Lazzarotti, Manzini, & Pizzurno, 2008) through collaborative and open business model (Sautter and Clar, 2008). SMEs are more adaptable to embrace open innovation strategies due to their flexibility in accelerating innovation, especially in terms of radical innovation and manage themselves within the innovation process and encourage others (Edwards, Delbridge, & Munday, 2005). Furthermore, Van de Vrande et al. (2009) argued that, even with the lack of financial resources, scant opportunities to recruit specialized personnel and small innovation portfolio, open innovation practices will be increasingly adopted in SMEs. Batterlink (2009) supported the same through a study of the decade 1994-2004 and mentions that

SMEs are catching up in recent years in adopting open innovation strategies.

Product Innovation

Product innovation is linked to the analysis of changes and innovations within the product category. Product innovation allows one to map out changes in one's enterprise's product. It compels to determine how a product should evolve to meet needs of the client and be competitive in the future (Crawford, 1983; Hiebing Jr. & Cooper, 2003). Despite some researchers have made a tiny boundary between 'major product innovation' and 'minor product innovation' (Lee, Park, Yoon, & Park, 2010), product innovation in an open innovation paradigm has opportunity factors like, cost reduction (Maria & Micelli, 2008), improved performance against an existing product (Hiebing Jr. & Cooper, 2003), or value addition (De Jong, Vanhaverbeke, Kalvet & Chesbrough, 2008).

Process Innovation

Process innovation combines the adoption of a new view of the business process with the application of innovation into the key processes. The novel and distinctive feature of this combination is its enormous potential to assist an organization in achieving major reductions in process cost or process time, or major improvements in quality, flexibility, and service levels. In this perspective, the business must be viewed not only in terms of functions, divisions, or products, but also of key processes. These processes may include redesigning them from the beginning to the end and employing whatever innovative technologies and organizational resources are available at hand (Davenport, 1993; Walton, 1989; Zuboff, 1988). However, a major challenge in process innovation is to make a successful transition to a continuous improvement in the environment. Referring to Davenport (1993), this research purview that, if a company that does not institute continuous improvement after implementing process innovation is likely to revert to the old ways of doing business.

Service Innovation

Service innovation systems are dynamic configurations of people, technologies, organizations and shared information that create and deliver value to customers, providers and partners through services. They are forming a growing proportion of the world economy and are becoming central activity for the businesses, governments, families and individuals. Nowadays, firms do not consider themselves to be 'services' or 'manufacturing', but providing solutions for customers that involve a combination of products and services. Preferably, service innovation can happen across all service sectors and one should look at all possible service activities rather than looking at any specific service sectors (IfM & IBM, 2008; Miles, 2005; Van Ark, Broersma & Den Hertog, 2003).

Technological Innovation

This manuscript has touched issues like product, process and service innovation during the earlier portion, and therefore, would like to emphasize on technological innovation in terms of SME development that enclave all the product, process and service innovations. A report of Telemetica Institute (2008) supports that, technological innovations trigger new services, better care for clients, new ways of working, and new means of exchange of ideas. Hence, understanding technological trends allows one to anticipate better on near-future possibilities for tangible problems for consumers and organizations, and in this aspect SMEs act as a catalyst of innovation promotion (Telemetica Institute, 2008). However, Conceição and Heitor (2000) have indicated that, the direct implication for innovation policies is important, but sometimes, limited role of demand at the firm level in assessing the amount of incentives for firms to introduce technological innovations hinders development of SMEs.

Better Positioning in the Market

Companies consider innovation as a key component to enhance their performance and to strengthen their competitive position in the market. As SMEs grow and reach at a critical size, they start organizing the company in a more formal way, hire specialists for a broad range of specialized job functions and formalize the firm's strategy in order to ensure better market positions against large(r) and international competitors. In many SMEs when they reach that critical size, formal R&D and innovation practices start to play crucial roles in developing and sustaining competitive advantages (Van de Vrande, de Jong, Vanhaverbeke, & de Rochemont, 2008) for better positioning in the market.

Acting as Starter or Catalyst

If entrepreneurs can genuinely be developed through skilled human resources, or at least be developed, at university (or a research house, or an entrepreneur), they will ultimately act as starters of a market economy, including social enterprises (Mueller & Goic, 2003). The ventures they create will serve as catalysts for technological progress (Hagen, 2008; Brouwers, Van Duivenboden, & Thaens, 2009).

Enabling Employment Generation

In the 1970s the World saw the reversal trend towards increasing size of enterprises and business establishments, and the share of small enterprises started to grow, especially in terms of employment (Sengenberger, Loveman, & Piore, 1990). During the 1980s, at a time when large firms were cutting down their labor, the apparent ability of small firms to create jobs attracted the attention of policy-makers in many countries (Smallbone, North, & Vickers, 2003). A survey conducted by EIM (1997) found that in UK and the Netherlands, SMEs represent a smaller share of total employment than the European Union average, which is 66% (see Renaud, 2008), while in Italy, Spain, Norway and Denmark SMEs make an above-average contribution. In Spain and Italy, about 80% of total employment is represented by small enterprises.

Earlier Adapter of New Products

Stiglitz (1998) argued that SMEs are not able to produce radical innovations due to their limited size and thus fail to create innovative market (there are counter arguments about this), but Parrilli (2006) mentioned that, due to their small size, they may rather benefit from being part of an innovation system. Cooke (1996) and Cooke and Wills (1999) supported the idea of incorporating SMEs in an innovation system comprising public and private institutions and other firms to promote innovation via collective effort. Design Council's survey (2005) found the evidence of this fact that, SMEs could be easily adapted to new products due to their flexible nature.

Challenges

Antti Peltomäki, the Deputy Director General of European Commission (European Communities, 2008) states that, innovation is a costly process. He reiterated further, that only one out of 3000 product ideas makes it to the market, which means that there are hundreds of unsuccessful products beyond every success. Moreover, even successful products may be far from being user friendly. Surveys show that 75% of all users find their technology-based tools more stressing than relaxing. In such a context, user-centric validation can play an important role in speeding up effectively the innovation process through addressing the actual user needs. OECD (2000b) finds that, most of the obstacles to growth and innovation in services are the same as in manufacturing or other technologies. Insufficient access to finance and risk capital, lack of internal capacity to innovate, insufficient expertise in applying technologies and high risks are characteristically the main barriers to innovation in all sectors.

As a newly emerged field of research, open innovation for SME development deserves a prolonged and strategically developed qualitative and quantitative synthesis on literature review, including various forms of validation. However, by taking account of the contents those are available during the early process of this research; this chapter carries out qualitative discussion based on some emphasized challenge parameters which are being depicted in Table 2. Table 2 shows the categorized challenge parameters and then each of them are being discussed further.

Organizational and Cultural Differences

Forcing of technology is one of the main reasons behind the failure of the SMEs in attempting to simply use ICTs. Initially, awareness building activities should be introduced before taking any short-term initiatives, followed by more general and long-term programs. Furthermore, by keeping in mind that the introduction of open innovation strategies in SMEs can bring a real modification in their working environment, the introduction of open innovation strategy based processes should take into account the specific culture of the company, especially the background of the entrepreneur and/or the managers, as well as their openness to innovation orientation (Wei, & Morgan, 2004; Barba-Sánchez, Martínez-Ruiz, & Jiménez-Zarco, 2007). However, doing things collectively, collaboratively and jointly has been seen as a good culture in many countries who are forerunner in open innovation for SMEs development (Callegati & Grandi, 2005; Govt. of UK, 2008a; TIEKE, 2009).

Inconsistent Turn-Over

Innovations have effects on other variables as well, such as turn-over, economic growth etc1, though they seem minor from outside, but these parameters affects innovation channels in the long run (Nählinder, 2005). Firms may have more than one innovation channel, which is being described next and in those situations innovation-induced employment changes are found to be counteractive. This in turn effects the turn-over of the company. Turn over data effects critical benchmarking indices of SMEs (European Commission, 2002), which reflects input on development contexts, as

Table 2. Challenges associated to open innovation affecting SMEs development

Challenges	Literatures reviewed
Differences in organization and culture between the individual partners	Wei & Morgan, 2004; Calegati & Grandi, 2005; De Jong, 2006; Barba-Sánchez et al., 2007; Govt. of UK, 2008a; TIEKE, 2009
Inconsistency in turn-over and pattern of economic growth	O'Sullivan, 2000; European Commission, 2002; Nählinder, 2005; OECD, 2007; NOUS, 2007; Govt. of UK, 2008b
Multiple innovation channels are counterproductive	Carlsson & Eliasson, 2002; Chesbrough, 2003b; Interact, 2004; Nählinder, 2005; Parrilli, 2006; Lazonick, 2007; De Jong et al., 2008
Naïve in fundamental innovation	Storey & Sykes, 1996; Napier, Serger & Hansson, 2004; Ahrweiler et al., 2005; Hölzl, 2006a; b; Telemetica Institute, 2008
Lack in managerial and technical skills (Small companies with innovative ideas have often found themselves without the capacity to develop or exploit their innovation)	Hadjimanolis, 1999; Van Hemel & Cramer, 2002; Tilley & Tonge, 2003; Del Brío & Junquera, 2003

such human development (OECD, 2007), and thus excessive external turn-over of personnel should be avoided (Govt. of UK, 2008b; Herstad, Bloch, Ebersberger & van de Velde, 2008).

Multiple Innovation Channels

The Open Innovation model implies that enterprises can use both internal and external ideas, technologies and knowledge to advance their innovation processes, and internal ideas can be taken to the market through external channels (e.g. spin-offs, external licensing of intellectual property) to generate additional value. This new paradigm inspires enterprises to find the most appropriate business model to commercialize new products or services, regardless of any model that exists within the enterprise or must be sought externally (Chesbrough, 2003b; De Jong, Vanhaverbeke, Kalvet & Chesbrough, 2008). In this context, utilization of new innovation channels for business cooperation are imperative (Interact, 2004), and smaller equity companies experimenting with alternative technologies creates multiple channels of innovation (Carlsson and Eliasson 2002). Often, it eliminates the need for commitment to specific technologies or R&D projects. However, problems may arise at the point of acquisition, specifically, it is critical for the acquiring firm that it gains control over not just for a specific issue but the competencies embedded in the acquired organization and its personnel (Lazonick 2007). Moreover, multiple innovation channels in SMEs are found to be counterproductive (Interact, 2004).

Fundamental Technological Innovation

Simple as it seems, open innovation for SMEs development requires a paradigm shift from traditional processes to innovative processes, integration of skill development, knowledge development, product development to service development, system development and technology development. It requires general awareness, diagnostics, pro-active counseling, management and monitoring, empowerment of people involved, new business models and new collaborations among public, private, academic and service providers (Telemetica Institute, 2008). However, some researches argue that, in order to foster new technological sectors a fundamental change of financial systems towards more market-based finance is necessary. Furthermore, there is evidence that financial systems are not always rigid, which effect institutional configurations over time (Hölzl 2006a). Research by Ahrweiler et al. (2005) also suggests that, while institutional heterogeneity persists at the level of institutional frameworks, the differences are of decreasing relevance in knowledge-intensive industries (Holzl, 2006a; b), creating a space for other agents to act as intermediaries. Therefore, firms who are aggressively engaging themselves in fundamental technological innovation are found to be in challenging situations.

Lack of Human Skills

When one talks about competences for open innovation, it often means the so called "personal competences", which are skills, experience and capacity that are embodied in an individual person. These competences may appear in terms of professional, methodological, social and individual skills essential to execute a specific role in the innovation process (iNNOWiSE, 2010). In an earlier study, Hadjimanolis (1999) found that in addition to other barriers in the open innovation process, the lack of skilled labor is an important one, which is also common for developed and developing nations. He emphasized that perception of top managers on innovation may seem as a barrier and that extends further due to lack of motivation. In another study, Van Hemel and Cramer (2002) observe that lack of appropriate knowledge is a barrier to specific fields of innovation, while Gerstenfeld and Roberts (2000) added that those could be due to lack of training and awareness. Moreover, despite SMEs importance in terms of their overall share of GDP, many smaller firms lack both managerial and technical skills (focused training or short-term orientation) that inhibit their effectiveness in innovation (Del Brío & Junquera, 2003; Tilley, & Tonge, 2003). Henceforth, effort has been given to synthesize the discussed parameters above.

Synthesis and the Innovation Opportunities Framework

This research is now constructing an innovation opportunity framework by synthesizing the illustrated parameters of challenges and opportunities. By keeping the opportunities and challenges of SMEs on open innovation in mind, a four dimensional opportunity framework has been deducted. The innovation opportunity framework can be applied in such a way, so that SMEs can find themselves comfortable within the open innovation development chain:

- By acting as a starter or catalyst of newly floated ideas from open innovation paradigms and adopting them quickly to fit into the new platform leading to improved business environment,
- By narrowing down organizational and cultural differences, removing inconsistencies in turn-over or other financial barriers, providing necessary skills development leading to better access to market, finance, business skills and information;
- By adopting open innovation strategies in product, process, service, and technology innovation leading to creation of new opportunities; and
- By enabling to compete, catering more clients and triggering employment generation leading to increased competition and revenue generation.

At the entrepreneurship level, the framework would look like Figure 1 and in more simplified format, perhaps when glimpsed from the policy level it will look like Figure 2. This adaptable and repetitive framework can easily be applied to SMEs who are exploring open innovation strategies for their competencies, and business and economic strength.

Research Recommendations

Innovation has become a key factor for the economic success of many countries and a prerequisite for sustainable development. Open innovation models have become an integral part of the innovation strategies and business models of companies in recent years. Innovation is increasingly based on knowledge assets beyond the boundaries of

Figure 1. Innovation opportunity framework at the entrepreneurship level

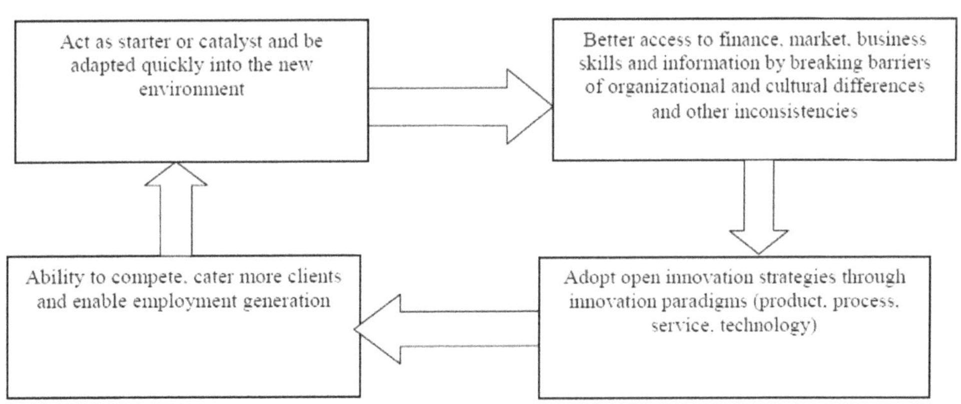

Figure 2. Innovation opportunity framework at the policy level

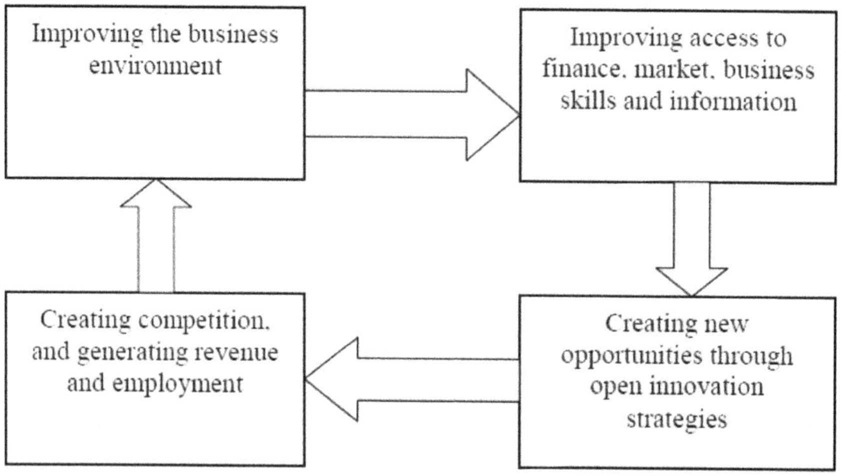

the company and co-operation has become an important way of tapping into knowledge resources from outside in order to generate new ideas and bring them quickly to the market (the "outside-in" approach). At the same time companies may spin out technologies and intellectual properties that they have developed internally. These could be the components that fall outside of their core business and thus in this way, better be developed and commercialized by outsiders (the "inside-out" approach) (OECD, 2008a;b).

Governments are paying more attention to the contribution of science and innovation to economic growth and have introduced a variety of new initiatives and reforms in many countries. Several countries, including Australia, Canada, Hungary, Ireland, South Korea and Spain have introduced comprehensive policy frameworks to guide developments in science, technology and innovation. In a number of countries, government institutions and agencies have been restructured to improve the governance of innovation systems, and at the same time policy evaluation has become more widespread. In addition to these, public research systems are being reformed to better contribute to economic and social needs by creating opportunities for further researches (OECD, 2002).

Moreover, the 'SME pact' foresees the positive mobilization of larger entrepreneurs or organizations (private or public) to promote innovative SME's development. Within this SME pact, programmes foster R&D-collaboration between innovative SMEs and large enterprises. If large enterprises are interested by SME's innovative products or services while still needing further development, these programmes support R&D projects that facilitate the testing and adaptation of the product, process and service innovation for the specific needs of the large enterprise (Cervantes, 2009; OECD, 2007; 2008b).

This research has highlighted twelve opportunities and five challenges parameters on open innovation that are being emphasized by contemporary researches/researchers for the overall development of SMEs. They are not exhaustive, even if the research intention was to make them so. Open innovation, though not a very new field of research anymore, future research can conduct extended literature review focusing wider area of interpretation and may suggest further improvement on these parameters leading to a more improved and sustained business model. Also, it is expected that the survey work that will be carried out in several phases will provide some more deliberations in this aspect.

CONCLUSION

Innovation is a prime factor for economic success in many countries. In a complex and highly competitive global market, entrepreneurs have to innovate and develop commercially viable products and services faster than ever for their survival. Moreover, open innovation is a new paradigm, which assumes that enterprises can and should use both external and internal ideas and paths to the market, while enterprises are looking for discovering and realizing innovative opportunities. The open innovation paradigm assumes that internal ideas can also be taken to markets through external channels, outside the current businesses of the enterprise, to generate value. However, the Open Innovation model does not completely upset traditional policymaking to legitimize policy interventions, including spillovers, system failures and market failures. Still these processes are being applied equally to SMEs despite the newly evolved challenges (De Jong, Vanhaverbeke, Kalvet & Chesbrough, 2008).

To meet these new challenges, companies are adopting new approaches to their innovation strategies and processes (OECD, 2008a). All businesses need to innovate, though it may take any number of forms, from the steady refinement of established products to the leap in the unknown when an untried idea is launched. Whether introducing new technologies and getting people to work in new ways or creating new products, one must innovate to survive, and these applies to SMEs equally. In a world of transformation and competition, innovating is not a luxury anymore, it is essential (De Jong, 2006; ACCA, 2008; Melkas, Uotila & Kallio, 2010).

This research, has tried to categorize aspects of opportunities and challenges surrounding the development of small sectors enterprises adopting open innovation. In doing so, it has clustered them, argued in favor and in places argued against for further deliberations. It argues that by knowing the scopes, opportunities and challenges small enterprises will be benefited the most. Furthermore, instigating extended research around some of the discussed parameters will be beneficial to the research communities.

Keeping these in mind, if one would like to initiate an entrepreneurship, the finding of this research will assist in better positioning in terms of opportunities and at the same time will make the journey safe by familiarizing the challenges in ahead. Finally, future literature review, validation of hypotheses through the intended survey and analysis in these contexts would establish extended platform of research and devise necessary tools to be utilized by the entrepreneurs.

REFERENCES

ACCA. (2008). *Director's briefing: Strategy*. UK: BHP Information Solutions Ltd.

Ahrweiler, P., Gilbert, N., & Pyka, A. (2005). *Institutions matter, but ...: Organizational alignment in knowledge-based industries, the case of German and British bio-pharmaceuticals*. Paper presented at the EMAEE Conference in Utrecht 2005.

Barba-Sánchez, V., Martínez-Ruiz, M., Del, P., & Jiménez-Zarco, A. I. (2007). Drivers, benefits and challenges of ICT adoption by small and medium sized enterprises (SMEs): A literature review. *Problems and Perspectives in Management*, *5*(1), 103–114.

Barker, J. E. (2008). *The foundations of innovation, transformation and change*. AussieInvention WiKi book.

Barnes, J., Bessant, J., Dunne, N., & Morris, M. (2001). Developing manufacturing competitiveness within South African industry: The role of middle management. *Technovation*, *21*(5), 293–309. doi:10.1016/S0166-4972(00)00047-X

Batterlink, M. (2009). *Profiting from external knowledge: How firms use different knowledge acquisition strategies to improve their innovation performance*. PhD Thesis, Wageningen University, Netherlands.

Beacham, J. (2006). *60 minute guide to innovation: Turning ideas into profit, succeeding through innovation*. UK: Department of Trade and Industry, Published by TSO.

BERR. (2007). *The 2007 R&D scoreboard: The top 850 UK and 1250 global companies by R&D investment.* Department for Business Enterprise and Regulatory Reform (BERR), UK.

BERR. (2008). *The 2008 value added scoreboard: The top 800 UK and 750 European companies by value added: Commentary & analysis*. Department for Business Enterprise and Regulatory Reform (BERR), UK.

BlessingWhite. (2006). *Leading technical professionals 2006 report*. Princeton, NJ: BlessingWhite, Inc.

Bonner, J., & Walker, O. (2004). Selecting influential business-to-business customers in new product development: Relational embeddedness and knowledge heterogeneity considerations. *Journal of Product Innovation Management, 21*, 155–169. doi:10.1111/j.0737-6782.2004.00067.x

Brockhoff, K. (2003). Customers' perspectives of involvement in new product development. *International Journal of Technology Management, 26*(5/6), 464–481. doi:10.1504/IJTM.2003.003418

Brouwers, J., Van Duivenboden, H., & Thaens, M. (2009). The triple helix triangle: Stimulating ICT-driven innovation at regional level. A Paper from the 2009 Annual Conference of EGPA, Malta, September 2009.

Callegati, E., & Grandi, S. (2005). Cluster dynamics and innovation in SMEs: The role of culture. Working paper No. 03/2005, International Centre for Research on the Economics of Culture, Institutions, and Creativity (EBLA). Turin, Italy: Università di Torino.

Carlsson, B., & Eliasson, G. (2002). Industrial dynamics and endogenous growth. *Industry and Innovation, 10*(4), 435–455. doi:10.1080/1366271032000163676

Cervantes, M. (2009). *Open innovation in global networks: Recent OECD work, policy issues and open questions*. Trade and Development Board Investment, Enterprise and Development Commission Multi-Year Expert Meeting on Enterprise Development Policies and Capacity-Building in Science, Technology and Innovation, Geneva, 20–22 January 2009.

Chason, M. (2008). *The Innovator, 1(1)*. Boston, MA: IXL Center.

Chesbrough, H. W. (2003a). *Open innovation: The new imperative for creating and profiting from technology*. Boston, MA: Harvard Business School Press.

Chesbrough, H. W. (2003b). The era of open innovation. *MIT Sloan Management Review, 44*(3), 35–41.

Chesbrough, H. W. (2006). *Open business models: How to thrive in a new innovation landscape*. Boston, MA: Harvard Business School Press.

Chesbrough, H. W., Vanhaverbeke, W., & West, J. (Eds.). (2006). *Open innovation: Researching a new paradigm*. London, UK: Oxford University Press.

Commonwealth of Australia. (2009). *Powering ideas: An innovation agenda for the 21st century*. Commonwealth of Australia 2009.

Conceição, P., & Heitor, M. (2000). *Engineering and technology for innovation in Portugal: A study on the dynamics of technological change.* A Preliminary Summary, Center for Innovation, Technology and Policy Research, Instituto Superior Tecnico, Technical University of Lisbon. Retrieved from http://in3.dem.ist.utl.pt/et2000/

Cooke, P. (1996). Building a twenty-first century regional economy in Emilia-Romagna. *European Planning Studies, 4*(1). Oxfordshire, UK: Carfax Publishing.

Cooke, P., & Wills, D. (1999). Small firms, social capital and the enhancement of business performance through innovation programs. [Kluwer.]. *Small Business Economics*, 13.

Cosh, A., Fu, X., & Hughes, A. (2005). *Management characteristics, collaboration and innovative efficiency: Evidence from UK survey data*, (p. 8). Working Paper no. 311. Cambridge, UK: Centre for Business Research.

Cox, G. (2008). Defining innovation: What counts in the University of Cape Town landscape? *Proceedings of Ascilite Melbourne*, 2008.

Crawford, C. M. (1983). *New products management.* Homewood, IL: Richard D. Irwin, Inc.

Davenport, C. (2006). Innovation is possible in every company. In J. Beacham (Eds.), *60 minute guide to innovation: Turning ideas into profit, succeeding through innovation*, (p. 3). Department of Trade and Industry, Published by TSO, UK.

Davenport, T. H. (1993). *Process innovation: Reengineering work through information technology. Ernst & Young.* Harvard Business School Press.

De Jong, J. P. J. (2006). *Open innovation: Practice, trends, motives and bottlenecks in the SMEs. (Meer Open Innovatie: Praktijk, Ontwikkelingen, Motieven en Knelpunten in het MKB).* Zoetermeer: EIM.

De Jong, J. P. J., Vanhaverbeke, W., Kalvet, T., & Chesbrough, H. (2008). *Policies for open innovation: Theory, framework and cases.* Research project funded by VISION Era-Net. Helsinki, Finland.

De Jong, J. P. J., & von Hippel, E. (2005). *User innovation in SMEs: Incidence and transfer to producers.* SCALES-initiative (SCientific AnaLysis of Entrepreneurship and SMEs), the Netherlands Ministry of Economic Affairs.

De Jong, J. P. J., & von Hippel, E. (2009). *Transfers of user process innovations to process equipment producers: A study of Dutch high-tech firms.* (MIT Sloan School of Management Research, Paper No. 4724-09).

Del Brío, J. Á., & Junquera, B. (2003). A review of the literature on environmental innovation management in SMEs: Implications for public policies. *Technovation, 23*(12), 939–948. doi:10.1016/S0166-4972(02)00036-6

Design Council. (2005). *National survey of firms.* London, UK: Design Council.

Edwards, T., Delbridge, J., & Munday, M. (2005). Understanding innovation in small and medium-sized enterprises: A process manifest. *Technovation, 25*, 1119–1120. doi:10.1016/j.technovation.2004.04.005

EIM. (1997). *The European Observatory for SMEs: Fifth annual report. Report submitted to the Enterprise Directorate-General of the Commission of the European Communities.* Zoetermeer: EIM Small Business Research and Consultancy.

EIRMA. (2004). *Technology access for open innovation.* Working Group Report WG63. Paris, France: Author.

Enkel, E., Kausch, C., & Gassmann, O. (2005). Managing the risk of customer integration. *European Management Journal, 23*(2), 203–213. doi:10.1016/j.emj.2005.02.005

European Commission. (2002). *Observatory of European SMEs: SMEs in Europe, including a first glance at EU candidate countries.* Report submitted to the Enterprise Directorate-General of the European Commission by: KPMG Special Services and EIM Business & Policy Research in the Netherlands in co-operation with: European Network for SME Research (ENSR), and Intomart.

European Commission. (2003a). *Innovation policy in seven candidate countries: The challenges,* vol. 1. Brussels.

European Commission. (2003b). *Innovation policy in seven candidate countries: The challenges, vol. 2.7: Innovation policy profile: Turkey.* Brussels.

European Commission. (2008). *SBA fact sheet Portugal- Enterprise and industry.*

European Communities. (2000). *The European Observatory for SMEs: Fifth annual report. Report submitted to the Enterprise Directorate-General of the Commission of the European Communities.* KPMG Consulting and EIM Small Business Research and Consultancy, European Communities.

European Communities. (2008). *Antti Peltomäki.* Luxembourg: Office for Official Publications of the European Communities.

Evangelista, R. (2000). Sectoral patterns of technological change in services. *Economics of Innovation and New Technology, 9,* 183–221. doi:10.1080/10438590000000008

FFG. (2006). *SME associations in Austria: An overview.* Austria: FFG Austrian Research Promotion Agency.

Fischer, W. A. (1978). Follow-up strategies for technological growth. *California Management Review, 21*(1), 11–20.

Gassmann, O. (2006). Opening up the innovation process: towards an agenda. [Blackwell Publishing Ltd.]. *R & D Management, 36*(3), 223–228. doi:10.1111/j.1467-9310.2006.00437.x

Gerstenfeld, A., & Roberts, H. (2000). Size matters: Barriers and prospects for environmental management in small and medium-sized enterprises. In Hillary, R. (Ed.), *Small and medium-sized enterprises and the environment: Business imperatives* (pp. 106–118). Sheffield, UK: Greenleaf Publishing.

Gold, B. (1987). Approaches to accelerating new product development. *Journal of Product Innovation Management, 4,* 81–88. doi:10.1016/0737-6782(87)90054-3

Govt. of Australia. (2010). *Glossary of key terms.* Australian Research Council, Government of Australia. Retrieved December 20, 2010, from http://www.arc.gov.au/general/glossary.htm

Govt. of New Zealand. (2007). *Digital strategy 2.0.* Ministry of Economic Development, Government of New Zealand.

Govt. of UK. (2008a). *Accelerating the SME economic engine: Through transparent, simple and strategic procurement.* Crown Copyright.

Govt. of UK. (2008b). *Innovation nation: Department for innovation, universities & skills.* Presented to Parliament by the Secretary of State for Innovation, Universities & Skills, the Chancellor of the Exchequer and the Secretary of State for Business Enterprise and Regulatory Reform by Command of Her Majesty, March 2008, Crown Copyright.

Grama, A., & Fotache, D. (2007). ICT and ERP applications challenges in Romanian SMEs. *Tomul LIV Ştiinţe Economice, 2007,* 23–28.

Grant, K., Laney, R., & Pickett, B. (2002). A process of innovation in marketing by SME's. *ANZMAC 2002 Conference Proceedings,* Deakin University, Australia, (pp. 2531-2538).

Hadjimanolis, A. (1999). Barriers to innovation for SMEs in a small less developed country (Cyprus). *Technovation, 19*(9), 561–570. doi:10.1016/S0166-4972(99)00034-6

Hage, J. T. (1999). Organizational innovation and organizational change. *Annual Review of Sociology, 25*, 597–622. doi:10.1146/annurev.soc.25.1.597

Hagen, S. (2008). From tech transfer to knowledge exchange: European universities in the marketplace. In Engwall, L., & Weaire, D. (Eds.), *The university in the market* (*Vol. 84*, pp. 103–117). Wenner-Gren International Series, Portland Press.

Hass, J., & Hochrinner, H. (2008). *Enhancing product- and process innovation in SMEs through cooperative education.* ASET Annual Conference 2008, The Placement and Employability Professionals' Conference, 2-4 September 2008, Robbins Conference Centre, Plymouth, UK.

Herstad, S. J., Bloch, C., Ebersberger, B., & van de Velde, E. (2008). *Open innovation and globalisation: Theory, evidence and implications.* Vision Era.Net.

Hiebing, R. G. Jr, & Cooper, S. W. (2003). *The successful marketing plan: A disciplined and comprehensive approach.* The McGraw-Hill Companies, Inc.

Hölzl, W. (2006a). *Scoping paper: Gazelles.* Europe Innova.

Hölzl, W. (2006b). Convergence of financial systems: Towards an evolutionary perspective. *Journal of Institutional Economics, 3.*

IBM. (2007). *Service science: The next frontier in service innovation.* IBM Singapore Pte Ltd.

IFAD. (2007). *Country strategic opportunities programme.* The Republic of Moldova, International Fund for Agricultural Development (IFAD), Executive Board — Ninety-second Session, Rome, 11-13 December 2007.

IfM & IBM. (2008). *Succeeding through service innovation: A service perspective for education, research, business and government.* Cambridge, UK: University of Cambridge Institute for Manufacturing. iNNOWiSE. (2010). Organizational competences for open innovation. Retrieved November 22, 2010, from http://www.think-innowise.de/?p=418

Interact. (2004). *Project "Embrace" preparation and implementation: Lessons learnt and recommendations to potential applicants.* INNOSTART National Business and Innovation Centre, a project paper submitted to European Regional Development Fund, European Union and presented at Translational INTERREG IIIB CADSES Seminar, Katowice, Poland.

Kaplinsky, R., Morris, M., & Readman, J. (2003). The globalisation of product markets and immiserising growth: Lessons from the South African furniture industry. *World Development, 30*(7), 1159–1178. doi:10.1016/S0305-750X(02)00029-3

Keeble, D. (1997). Small firms, innovation and regional development in Britain in the 1990s. *Regional Studies, 31*(3), 281–293. doi:10.1080/00343409750134692

Keupp, M. M., Lhuillery, S., Garcia-Torres, M. A., & Raffo, J. (2009). *Economic focus study on SMEs and intellectual property in Switzerland, SME-IP: 2nd report.* Bern, Switzerland: Swiss Federal Institute of Intellectual Property.

Landabaso, M. (Ed.). (2010). *The smart guide to innovation-based incubators (IBI).* Luxembourg: European Commission.

Lazonick, W. (2007). The stock market and innovative enterprise. *Industrial and Corporate Change, 16*(6), 983–1035. doi:10.1093/icc/dtm030

Lazzarotti, V., Manzini, R., & Pizzurno, E. (2008). Managing innovation networks of SMEs: A case study. *Proceedings of the International Engineering Management Conference: Managing Engineering, Technology and Innovation for Growth* (IEMC Europe 2008), Estoril, Portugal, (pp. 521-525).

Lee, S., Park, G., Yoon, B., & Park, J. (2010). Open innovation in SMEs- An intermediated network model. *Research Policy, 39*(2), 290–300. doi:10.1016/j.respol.2009.12.009

Lemola, T., & Lievonen, J. (2008). *The role of innovation policy in fostering open innovation activities among companies*. A report from Vision Era.net, Finland.

Lilien, G. L., Morrison, P. D., Searls, K., Sonnack, M., & von Hippel, E. (2002). Performance assessment of the lead user idea-generation process for new product development. *Management Science, 48*(8), 1042–1059. doi:10.1287/mnsc.48.8.1042.171

Lindermann, N., Valcárcel, S., Schaarschmidt, M., & von Kortzfleisch, H. (2009). SME 2.0: Roadmap towards Web 2.0-based open innovation in SME-networks – A case study based research framework. In *IFIP Advances in Information and Communication Technology, 301*, 28-41. Boston, MA: Springer.

Lundvall, B.-Å. (1999). Innovation policy and economic theory. In Schienstock, G., & Kuusi, O. (Eds.), *Transformation towards a learning economy, report no. 213*. Helsinki: The Finnish National Fund for Research and Development - Sitra.

MacGregor, S., Bianchi, M., Hernandez, J. L., & Mendibil, K. (2007). *Towards the tipping point for social innovation*. 12th International Conference on Towards Sustainable Product Design (Sustainable Innovation 07), Farnham, Surrey, UK.

Madsing, D. (1997). *NCC presentation on the information technology initiative*. Manchester Metropolitan University.

Maes, J. (2009). *SMEs' radical product innovation: The role of the internal and external absorptive capacity spheres. Job Market Paper July 2009*. Belgium: Katholieke Universiteit Leuven – Faculty of Business and Economics.

Maria, E. D., & Micelli, S. (2008). *SMEs and competitive advantage: A mix of innovation, marketing and ICT. The case of "Made in Italy"*. Working paper #70, University Degli Studi Di Padova, Italy.

McKeown, M. (2008). *The truth about innovation*. London, UK: Prentice Hall.

Melkas, H., Uotila, T., & Kallio, A. (2010). Information quality and absorptive capacity in service and product innovation processes. *Interdisciplinary Journal of Information, Knowledge, and Management, 5*, 357–374.

Miles, I. (2005). Innovation in services. In Fagerberg, J., Mowery, D. C., & Nelson, R. R. (Eds.), *The Oxford handbook of innovation* (pp. 433–458). Oxford University Press.

Moore, G. A. (2006). *Dealing with Darwin: How great companies innovate at every phase of their evolution*. Chichester, UK: Capstone Publishing.

Mueller, S. L., & Goic, S. (2003). East-west differences in entrepreneurial self-efficacy: implications for entrepreneurship education in transition economies. *International Journal of Entrepreneurship Education, 1*, 613–632.

Nählinder, J. (2005). *Innovation and employment in services: The case of knowledge intensive business services in Sweden*. PhD Thesis, Department of Technology and Social Change, Linköping University, Sweden, Unitryck Linköping.

Napier, G., Serger, S. S., & Hansson, E. W. (2004). *Strengthening innovation and technology policies for SME development in Turkey– Opportunities for private sector involvement. International Organisation for Knowledge Economy and Enterprise Development*. IKED.

Newton, K. (2001). (submitted to). Management skills for small business. [Small Business Policy Branch, Industry Canada.]. *Report*.

NOUS. (2007). *Today's knowledge intensive SME: Where, why and how?* Discussion Paper, Department of Innovation, Industry and Regional Development, The NOUS Group, Australia.

O'Sullivan, M. (2000). *Contests for corporate control*. Oxford University Press.

OECD. (2000a). *Enhancing the competitiveness of SMEs in the global economy: Strategies and policies enhancing the competitiveness of SMEs through innovation*. Conference for Ministers responsible for SMEs and Industry Ministers, Bologna, Italy, 14-15 June 2000.

OECD. (2000b). *OECD science technology and industry outlook 2000*. France: OECD.

OECD. (2002). *OECD science technology and industry outlook 2002*. France: OECD.

OECD. (2006). *The Athens action plans for removing barriers to SME access to international markets*. OECD Report, France.

OECD. (2007). *SME policy index 2007*. Report on the Implementation of the European Charter for Small Enterprises in the Western Balkans, prepared by: the European Commission Directorate General for Enterprise and Industry and the OECD Investment Compact for South East Europe. In consultation with: the European Bank for Reconstruction and Development and the European Training Foundation Published with the financial support of the European Commission and the OECD Investment Compact for South East Europe.

OECD. (2008a). *Open innovation in global networks: Policy brief. OECD Observer*. Paris, France: OECD.

OECD. (2008b). *Globalization and open innovation*. Paris, France: OECD.

OECD. (2008c). *Summary record*. Business Symposium on Open Innovation in Global Networks, Copenhagen, Denmark.

Olson, E., & Bakke, G. (2001). Implementing the lead user method in a high technology firm: A longitudinal study of intentions versus actions. *Journal of Product Innovation Management, 18*(2), 388–395. doi:10.1016/S0737-6782(01)00111-4

Oxford Business School. (2008). *Introducing the Oxford Centre for Entrepreneurship and Innovation retail policy issues. The Retail Digest* (pp. 30–31). Oxford SAID Business School.

Papulová, Z., & Mokroš, M. (2007). *Importance of managerial skills and knowledge in management for small entrepreneurs*. E-Leader 2007, Prague.

Parrilli, M. D. (2006). *A tripartite innovation framework for small firms in developing regions: Key issues for analysis and policy*. Paper Number 2007-01, Institute for Economic Development Policy Discussion Paper Series, University of Birmingham, UK.

Renaud, P. (2008). *Open Innovation at Oseo Innovation: Example of the Passerelle Programme*. Presentation at the OECD Business Symposium on Open Innovation in Global Networks, Copenhagen, 25-26 February, 2008.

Roper, S., & Hewitt-Dundas, N. (2004). *Innovation, networks and the diffusion of manufacturing best practice: A comparison of Northern Ireland and the Republic of Ireland*. A report from Northern Ireland Economic Research Centre Queen's University of Belfast, UK.

Salavou, H. E. (2006). *Exploring product innovativeness determinants in SMEs*. PhD Thesis, Athens University of Economics & Business, 2006.

Sautter, B., & Clar, G. (2008). *Strategic capacity building in clusters to enhance future-oriented open innovation processes*. Foresight Brief No. 150, The European Foresight Monitoring Network. Retrieved from www.efmn.info

Schienstock, G., & Hämäläinen, T. (2001). *Transformation of the Finnish innovation system: A network approach*. The Finnish National Fund for Research and Development - Sitra. Report series No. 7. Helsinki.

Schumpeter, J. A. (1934). *The theory of economic development*. Cambridge, MA: Harvard University Press.

Sengenberger, W., Loveman, G., & Piore, M. J. (Eds.). (1990). *The re-emergence of small enterprises: Industrial restructuring in industrialized countries*. Geneva, Switzerland: International Institute of Labour Studies.

Smallbone, D., North, D., & Vickers, I. (2003). The role and characteristics of SMEs in innovation. In Asheim, B. T., Isaksen, A., Nauwelaers, C., & Todtling, F. (Eds.), *Regional innovation policy for small-medium enterprises* (pp. 3–20). Edward Elgar Publishing.

Sousa, M. (2008). *Open innovation models and the role of knowledge brokers* (pp. 18–22). Inside Knowledge Magazine.

Stanworth, J., & Gray, C. (1991). *Bolton 20 years on: The small firm in the 1990s*. London, UK: Paul Chapman.

Stiglitz, J. (1998). *Towards a new paradigm for development. 9th Raul Prebisch Lecture*. Geneva: UNCTAD.

Storey, D. J. (1994). *Understanding the small business sector*. London, UK: International Thomson Business Press.

Storey, D. J., & Sykes, N. (1996). Uncertainty, innovation and management. In Burns, P., & Dewhurst, J. (Eds.), *Small business and entrepreneurship* (pp. 73–93). London, UK: Macmillan.

Sullivan A. (1998, July 5). Research outlays: The real bottom line: Expenditures color a company's outlook. *Int. Herald Tribune*, (pp. 15-17).

Telemetica Institute. (2008). *Service innovation and ICT: Vision and ambition*. The Netherlands: Telemetica Institute.

Tidd, J., Bessant, J., & Pavitt, K. (2005). *Managing innovation: Integrating technological, market and organizational change* (3rd ed.). John Wiley and Sons.

TIEKE. (2009). *ICT cluster Finland review 2009*. Finland: TIEKE Finnish Information Society Development Centre.

Tilley, F., & Tonge, J. (2003). Introduction. In Jones, O., & Tilley, F. (Eds.), *Competitive advantage in SMEs: Organising for innovation and change* (pp. 1–12). John Wiley & Sons.

Toivonen, M. (2004). *Expertise as business: Long-term development and future prospects of knowledge-intensive business services (KIBS)*. Laboratory of Industrial Management, Helsinki University of Technology, Doctoral dissertation series 2004/2.

Tonge, J. (2001). *A review of small business literature part 1: Defining the small business*. Manchester Metropolitan University Business School Working Paper Series: WPS025. Retrieved from http://www.business.mmu.ac.uk/wps/

Vaitheeswaran. V. (2007) Special report: Innovation. *The Economist*. Retrieved from http://www.economist.com/specialreports/displaystory.cfm?story_id=9934754

Van Ark, B., Broersma, L., & Den Hertog, P. (2003). *Service innovation, performance and policy: A review*. Research Series no. 6, Synthesis report in the framework of the project on Structural Information Provision on Innovation in Services SIID for the Ministry of Economic Affairs of the Netherlands.

Van de Vrande, V., de Jong, J. P. J., Vanhaverbeke, W., & de Rochemont, M. (2008). *Open innovation in SMEs: Trends, motives and management challenges*. SCALES-initiative (SCientific AnaLysis of Entrepreneurship and SMEs), SMEs and Entrepreneurship Programme. Netherlands Ministry of Economic Affairs, Zoetermeer, November 2008.

Van de Vrande, V., de Jong, J. P. J., Vanhaverbeke, W., & de Rochemont, M. (2009). Open innovation in SMEs: Trends, motives and management challenges. *Technovation*, *29*(6-7), 423–437. doi:10.1016/j.technovation.2008.10.001

Van der Walt, J. S., Buitendag, A. A. K., Zaaiman, J. J., & van Vuuren, J. C. J. (2009). Community living lab as a collaborative innovation environment. *Issues in Informing Science and Information Technology*, *6*, 421–436.

Van Hemel, C., & Cramer, J. (2002). Barriers and stimuli for ecodesign in SMEs. *Journal of Cleaner Production*, *10*(5), 439–453. doi:10.1016/S0959-6526(02)00013-6

Von Hippel, E. (1998). Economics of product development by users: The impact of sticky local information. *Management Science*, *44*(5), 629–644. doi:10.1287/mnsc.44.5.629

Von Hippel, E. (2005). *Democratizing innovation*. Boston, MA: MIT Press.

Walton, R. E. (1989). *Up and running: Integrating Information Technology and the organization*. Boston, MA: Harvard Business School Press.

Wei, Y., & Morgan, N. (2004). The supportiveness of organizational climate, market orientation, and new product performance in Chinese firms. *Journal of Product Innovation Management*, *21*, 375–388. doi:10.1111/j.0737-6782.2004.00092.x

Zuboff, S. (1988). *In the age of the smart machine: The future of work and power*. New York, NY: Basic Books.

ADDITIONAL READING

Beije, P. (2005). Open Innovation: the European Challenge, A report submitted to Rotterdam School of Management / Erasmus University

Christensen, J. F. (2007). *Innovation concepts, processes and strategies*. Presentation in PhD Course, Copenhagen Business School.

Clark, T. R., & Gottfredson, C. A. (2008). In Search of Learning Agility: Assessing Progress from 1957 to 2008, A report submitted to TRCLARK LLC

De Backer, K. (2008). *Open Innovation in Global Networks*. OECD Publishing.

EEDA. (2008). *ERDF Regional Workshops – Autumn 2008 Review Report*. UK: East of England Development Agency.

Europe, C. S. R. (2008a). *The European Alliance for CSR Progress Review 2007: Making Europe a Pole of Excellence on CSR*, European Commission, Brussels, 4 March 2008

CSR Europe (2008b). *R&D Open Innovation: Networks with SME*, Open Innovation Network, 6 November 2008

Hanna, V., & Walsh, K. (2002). Small firm networks: a successful approach to innovation? *R & D Management*, *32*(3), 201–207. doi:10.1111/1467-9310.00253

Höllmüller, C. (2008). Chances for Small and Medium-sized Enterprises (SME) in International Innovation Networks, a presentation at the Innovation Day Swiss Texnet, Dübendorf, 28 August 2008

Huston, L., & Sakkab, N. (2006, March). *P&G's New Innovation Model*, An excerpt from "Connect and Develop: Inside Procter & Gamble's New Model for Innovation,". *Harvard Business Review*, *84*(3),

Jaruzelski, B., Dehoff, K. & Bordia, R. (2006). Smart Spenders: The Global Innovation 1000, Special feature at Smart Spenders: The Global Innovation, Strategy+Business Issue 45: 46-61

Ketels, C., & Sölvell, Ö. (2006). *The Baltic Sea Region – Top of Europe in Global Competition*. Denmark: State of the Region Report, Baltic Development Forum.

Michael, S. C., & Palandjian, T. P. (2004). Organizational learning in new product introductions. *Journal of Product Innovation Management*, *21*, 268–276. doi:10.1111/j.0737-6782.2004.00078.x

Piatier, A. (1984). *Barriers to Innovation*. London: Frances Pinter.

Pourdehnad, J. (2007). Idealized Design - An "Open Innovation" Process, a presentation from the annual W. Edwards Deming Annual Conference, Purdue University, West Lafayette, Indiana, October 2007

Pullen, A., De Weerd-Nederhof, P., Groen, A., & Fisscher, O. (2008). Configurations of External SME Characteristics to Explain Differences in Innovation Performance, a proceedings of the *High Technology Small Firms Conference 2008*: Twente University, Netherlands.

Rush, H., & Bessant, J. (1992). Revolution in three-quarter time: lessons from the diffusion of advanced manufacturing technologies. *Technology Analysis and Strategic Management*, *4*(1), 3–19. doi:10.1080/09537329208524075

Shorthouse, S. (2008). Innovation and Technology Transfer, a presentation from the International Conference DISTRICT 2008. International Conference Centre, Dresden, 29 May 2008.

West, J. & Gallagher, S. (2006). Open Innovation: The Paradox of Firm Investment in Open Source Software, R&D Management Special Issue on "*Opening-up the innovation process*", 36, 3 (June 2006).

West, J., Vanhaverbeke, W., & Chesbrough, H. (2006). Open Innovation: A Research Agenda. In: Henry Chesbrough, Wim Vanhaverbeke and Joel West (Eds.) (2006). Open Innovation: Researching a New Paradigm, Oxford University Press.

Wolf, A. (2005). smE MPOWER Outlook: Empowering SMEs for long-term research interest and increased participation in EU RTD activities! Retrieved March 15, 2009 from http://www.sme-mpower.net/Presentations/11%20Andreas%20Wolf%20on%20smE-MPOWER-Community.pdf

KEY TERMS AND DEFINITIONS

Innovation: Process of doing something new (never existed before, or drastically modified from the original, or modified to be accepted as new) for creating value addition.

Open Innovation: An innovation process, but accompanying efforts from stakeholders involved in the process, and doing it in a collaborative approach.

Open Innovation Strategies;: Processes or activities that are taken during the innovation adopting open innovation mode of development.

Open Innovation Opportunity Framework: A defined trajectory of processes leveraging the opportunities that have been created by open innovation.

SMEs: Firms belonging to small and medium enterprises sector of the business community, and varying in size or turn-over from country to country or region to region, they forms the largest part of the business community in any country.

Section 2
Innovation Marketing, Communication and Growth Management

Chapter 6
Global Innovators:
How Open Innovation Serves Humanity

Robert Girling
Sonoma State University, USA

ABSTRACT

Innovation is vital to sustain and advance current activities or it can be vital to growing new businesses. The challenge for organizations operating in a global environment is to meet the evolution of the marketplace, social needs, and the needs of society. Open innovation allows organizations to draw from the global pool of knowledge to design products and business modes that provide value while meeting social needs. This chapter looks at several SMEs that are social innovators drawing on a range of technologies to create products and services or to commercialize existing products in a new way in order to meet pressing social needs round the world.

INTRODUCTION

Traditionally international business has been considered business transactions between parties from more than one country. The focus is on cross border activities and an important aspect of international business is the collaboration which occurs between domestic and foreign partners. One text states "Global business is fundamentally about not limiting yourself to your home country but treating the entire globe as your potential playground (or battlefield)" (Peng, 2010: 8). This concept is true to many aspects of international commerce and non perhaps more than innovation, research and development.

William H. Chesbrough (2003: 1) in his book defines open innovation as "the use of purposive inflows and outflows of knowledge to accelerate internal innovation, and expand the markets for the external use of innovation, respectively." Open innovation involves the international search for new ideas and collaboration in utilizing that knowledge to meet human needs. Few would disagree that innovation is vital to sustain and advance current activities or that it can be vital to growing new businesses. "Open innovation means that companies should make much greater use of

DOI: 10.4018/978-1-61350-519-9.ch006

external ideas and technologies" (Chesbrough, 2006a: 1).

The challenge for organizations operating in a global environment is to meet the evolution of the marketplace, social needs and the needs of customers in the best possible way. Open innovation allows organizations to draw from the global pool of knowledge to design products and business modes that provide value.

Twenty-first century innovation is different from innovation in past eras. This chapter looks at several SME's that are organizational innovators: some are creating new products or services, some are commercializing existing products in a new and relevant way, while others are opening new markets. The reader can explore the nature and characteristics of the emerging organizational forms that are being built through global networks and frameworks. The chapter identifies the implications of these changing business constructs for our collective future.

Open innovation requires shedding a number of assumptions and traditional ways of doing business. Companies that embrace open innovation believe that smart people work inside and outside their companies and rather than attempting to discover technologies themselves, open innovators believe they need not originate ideas and research to profit from them (Chesborough, 2006b).

The methodology imposed in this chapter is a series of case studies. Case studies are employed in order to be able to provide a more complete assessment, one which is holistic and takes into consideration the contextual environment. The aim is to provide an empirical understanding from which an improved understanding may emerge. Despite the inherent limitation of generalizing from a limited number of cases, it contributes to an understanding of how a new breed of global innovators operate.

BACKGROUND

Globalization has been accompanied by increasing consumer concern about the health consequences and working conditions under which goods are produced. Consumers want to know "are com-

Figure 1. Why fair trade is important

Fair Trade is a growing, international movement which ensures that producers in poor countries get a fair price for their goods, one that covers the cost of production and guarantees a living income.

Fair Trade, an idea which began over 40 years ago in Europe seeks to change that equation. Fair trade is the idea that when we engage in trade, we must also think about the people behind the products we are buying. The central idea is that trade should create sustainable incomes for low income producers in Latin America, Africa, Asia and around the world by paying a fair wage in the local context, ensuring safe working conditions, providing business development support and creating stable access to outside markets. As Paul Rice, founder of TransFair, an Oakland, California non-profit devoted to extending the reach of fair trade, puts it "Business can play an enormous role in ensuring that producers are compensated so that their families can live productive and sustainable lives."

For a product to be considered fair trade it needs to adhere to several fair trade principles including (a) paying a fair, living wage in the local context, (b) providing health and safe working conditions, (c) providing education to workers about the environment. In addition they often pay a premium to workers to improve the community. And in many cases fair trade involves transparent partnerships with farmers and artisans have a say in how the business is run.

For example, each day nearly 2.5 billion cups of coffee are consumed. It is the 5th most widely traded commodity in the world and 20 to 25 million families produce and sell coffee for their livelihood and most of them are *small-scale farmers* with limited financial resources and scope to diversify out of coffee production.

A recent Oxfam research report showed that Ugandan coffee farmers only get about 2.5 percent of the final retail price of their coffee in the UK market.

Sources:www.maketradefair.com; *www.tutor2u.net/economics/revision-notes/as-markets-coffee.html*

panies treating their workers fairly, are they being paid fairly, do the products contain harmful chemicals, is the workplace safe?" A growing number of citizens are coming to realize that the decisions of companies where they work or shop has an impact on others all around the globe. And there is genuine cause for concern: Consider that only 12 percent of the price of a pair of jeans goes to the workers who made them in Honduras or Cambodia. UNITE (the Union of Needletrades, Industrial and Textile Employees) found that in El Salvador, the minimum wage for workers is 60 cents an hour, but an employee must earn at least $1.73 to pull themselves out of poverty. Consider that only 10 per cent of the price of your espresso or cappuccino went to the grower. On a world scale 850 million people earn less than a living wage and there are some 250 million child laborers between 4 and 14 years of age. These issues have led to global efforts to establish standards of fair and ethical trade (Young and Welford, 2002).[1]

MAIN THRUST

Global innovators are companies which combine their values as an integral part of their business model to earn revenue while addressing essential social and environmental issues. As such they are adding important dimensions to global business. This article explores how some enterprises are building innovative business models which extend a hand to improve working conditions and raise living standards around the globe. For example, Indigenous Designs, has a positive influence on the lives of indigenous women of the Andes. The company drew on the financial technology provided by Root Capital, a global financial intermediary which seeks to protect the environment by promoting organic production for export. Vestergaard-Frandsen is a global company that uses commerce to improve the health of the world's poorest and neediest. Meanwhile KIVA has introduced a game changing business model of partnership by sharing its technology with hundreds of micro-finance intermediaries around the world so that global citizens can provide resources to help improve the lives of small scale entrepreneurs around the world.

Indigenous Designs and Root Capital

In 1993 Joe Flood walked into Scott Leonard's surf shop in Santa Cruz, California wearing a multicolored sweater. Scott was amazed by the sweater, "You know I could sell something like that in my shop!" When Joe explained that the sweaters were made by women eking out a living high in the Andes of Ecuador, Scott decided to join Joe on a trip to Ecuador to visit small knitting cooperatives.

The weavers' use of primitive tools kept most of them in a poverty stricken state. Often trading their textiles for food, these people could not afford basic necessities. They used poor quality wool and old bicycle spokes as knitting needles and primitive tools for carding and spinning the wool. The resulting products were of poor quality, lacking uniformity in size and even sleeve length much less style. Lacking access to distribution, they walked hours to the nearest village where they would sell their sweaters for a fraction of the worth. Returning home, the women's poverty gnawed at him; he had a vision to make a difference in the world. "We decided if we could develop viable designs, control the colors, control the sizing, and put some quality controls in place, we could market to the outdoor industry and in turn pay these women a sustainable wage."

But for Leonard, getting the company going was a steep challenge and he made numerous trips back and forth to Ecuador. At this point the company was operating on Leonard's credit cards, his wife was pregnant, and he was stretched to the limit. He scrounged money to pay for a ten by ten booth at Magic, the fashion and clothing industry's annual trade show.

The show was not going particularly well when Leonard happened to see the Nature Company crew walking by. He set up an appointment to visit them and ended up with a purchase order for over $450,000. The company had a commitment to deliver 4000 sweaters and 2000 jackets; but there were problems. Could they meet the Nature Company's exacting quality standards? Could they deliver the order on time? The weavers were loosely organized into cooperatives and he was having difficulty getting the coops to produce the promised items on time.

Then at the wedding of a mutual friend, Leonard ran into Matt Reynolds, a high school friend who had lived in Ecuador and who had just returned from several years working with a large retailer in Europe. In 1995 Matt joined the partnership and immediately set off for Ecuador to get production on track. How could they standardize production among a group of cooperatives? Working with NGOs, the company created a set of prototype samples. The cooperatives selected their best knitters who worked together to train all the other knitters. The company provided 10 day free knitting workshops which trained the women in:

- The importance of sizing and color
- How to follow patterns and designs
- How to work with a wider variety of yarns
- Quality control and the benefits of reducing returns and rework

Reynolds contacted the director of each of the three coops and said, 'We're going to give you two thousand sweaters to knit.' He wrote into the agreement that if one of the three failed to produce all two thousand units, one of the others had to cover the shortfall.

This first order was a breakthrough in many ways. Until this point, the small producer cooperatives were competing among themselves. With this large an order, no single group could fulfill it alone. Three coops had to work together. As Reynolds put it "When they completed the work and received payment, a light bulb went on. It showed the women that they could do better by cooperating among themselves."

As they completed the training and began receiving increased payments for their products, the women began to realize that their products had value on a world market. Then they began to see opportunities for advancement that had never before occurred to them. Some women saw an opportunity to become trainers in the expanding network of weavers. They began to see the returns to their communities and to their families as more money flowed in to pay for education of their children.

But the relationship was riddled with challenges. Leonard and Reynolds were savvy enough to know that their company had to be fiscally sound and prosperous in order to help others. Without a company that was economically sustainable, their other goals would be short-lived.

They decided that the company would be a "quadruple bottom line" company. What that means is that since they were attempting to partner with organizations around the world to help improve economic and social conditions, instead of simply maximizing profitability or a single bottom line without considering the company's impact on the environment or the community, they would add ecological stewardship, social justice as well as the company's global impact.

Leonard and Reynolds realized that in order to get a quality product from the cooperatives they had to climb a big mountain. They had to grow or find a sizable market for the product and they would need at least $300,000 just to get their venture of the ground. They were able to piece together start-up funding from personal savings, loans from family and friends, and a $200,000 loan from the US Small Business Association which then allowed them to begin partnering with the artisans through local cooperatives to produce, inspect, transport, and sell small orders to retailers.

But when the company began contacting the cooperatives to place orders, they discovered yet another problem—the coops lacked the financial resources for anything but tiny orders; they could not purchase the materials or sustain themselves. Some of the cooperatives were working with NGOs [non-governmental organizations] which provided small micro-loans to the coops; but these were very limited. So in order to meet the growing demand Indigenous Designs needed to provide adequate working capital to the cooperatives.

By 1999 the company had exhausted its own start-up capital by extending credit to the coops. According to Reynolds "We were maxed out. Interest costs were killing us. Traditional banks and financial institutions wouldn't lend to us" The hand knit cooperatives were still viewed as high risk by traditional banks because they could not offer collateral for a traditional loan. As a result sales flattened out and the company was stretched to the breaking point.

Enter Root Capital

The solution came from Leonard and Reynold's network with another global innovator, *Root-Capital,* founded to promote conservation of natural resources in rural areas of Latin America.[2] Willie Fulbright Foote established *Root Capital* as a vehicle to provide loans to small and medium firms dedicated to exporting "green" products like organic coffee, fair trade, or other sustainably produced products. *Root Capital* sees the key to transforming rural poverty in promoting sustainable livelihoods that affirm local cultures and by strengthening community participation in natural resource management. Lending money to *Indigenous Designs* would help the women to define new economic opportunities while promoting local stewardship of regionally and globally important resources. By providing affordable loans and financial services to social businesses, *Root Capital* was able to promote its mission by supporting an environmentally sustainable company; one which would otherwise have difficulty obtaining traditional loans.

Yet there was still another problem. The loan from Root Capital required that the cooperatives pay interest on the loans before they received payment. This concept was foreign to the women. How could Indigenous Designs get the women to understand they needed to pay interest on the loans they received until the customer paid for the product, often as much as 220 days later? Working closely with *Root Capital,* Reynolds developed a series of workshops for the cooperatives showing how their earnings from increased production would pay the interest and still be leave them with more money. Crossing this barrier moved the company to another plateau and as a result, business grew by a whopping 93 per cent over the next year.

Indigenous Designs invested over seven million dollars in Peru, Ecuador, Guatemala and India to train and educate the artisans, build cooperative work-centers, provide design expertise and set up logistics. Scaling has allowed the company to supply larger mainstream companies such as *Dillard's* department stores and *Eileen Fisher.*[3]

Indigenous Designs developed ties with over 275 cooperatives each with 6 to 60 members. This added capacity allowed *Indigenous* to produce a high volume of hand knit garments.[4] The time, money, and resources that *Indigenous Designs* invested towards infrastructure paid dividends as the unique relationship with local people allowed them to build up trust and acceptance which translated into a competitive advantage.

By promoting Fair Trade practices and paying a livable wage, Indigenous Designs helps thousands of artisans and their families rise out of poverty. They also partner with non-profits and government agencies to provide family educational assistance, flexible work hours, and opportunities for promotion. By investing in training, equipment, and knitting workshops, they help promote safe and clean working conditions. The cooperatives and financial partners also ensure the workplaces are equipped with proper airflow and lighting, and do not allow child labor

The coop work-centers allowed the artisans to knit together and avoid long walks to the nearest large city in order to sell their individual products. Indigenous provided free knitting needles and assistance in financing new equipment such as looms. The company uses natural fibers and organic cotton to protect the workers' health.[5] By paying workers 200% to 300% more than if they worked independently the families and communities have become more self sufficient and prosperous.

Indigenous Designs has profited from its model. Through 2008 the company experienced double digit annual growth as the value of its initial investment grew by 20 times. Then with the onset of the worldwide economic downturn, the US clothing industry sales plummeted by 25 percent. Some of the company's competitors went bankrupt or suffered losses. Yet because of the goodwill built up over its years of work and the diversity of its supplier network, the Indigenous Designs was able to maintain its operations without a decline in sales.

Vestergaard Frandsen SA

Vestergaard Frandsen SA has built its core business around an enlightened sense of humanitarian responsibility. Vestergaard Frandsen is a Danish company that, for 40 years, made uniforms for hotel workers and retailers which has evolved through adapting a series of innovations borrowed from all quarters to make textile-based, life-saving products, including ZeroFly, a durable plastic sheeting for sheltering refugees that also kills disease-spreading insects, and LifeStraw, a water filtration tool the size of paper-towel cylinder that serves thousands who are deprived of clean drinking water.

The company got its start in Denmark in 1957 as a manufacturer of uniforms for hotel workers and retailers; it has since evolved into a global leader with headquarters in Switzerland and branch offices in Denmark, Ghana, India, Indonesia, Kenya, Nigeria, South Africa, UAE and USA. In addition, there are laboratories in Ivory Coast and Vietnam and production in Belgium, China, India, Korea, Thailand and Vietnam.

The company's president, Mikkel Vestergaard Frandsen, the grandson of the company's founder, had no interest in the growing market for women's clothing. Upon graduating high school in 1991, he fled his native Denmark to go backpacking through India and Africa where he experienced heartbreaking poverty first hand. "I started out as an entrepreneur in Africa, where at the age of 19, I opened my first company in Lagos. My core business at the time was selling trucks and truck engines, which had nothing to do with public health or doing good. But this experience ignited my passion for Africa and guided a strategic overhaul of our family textile business into a dynamic house, driving positive change within malaria, guinea worm, diarrhea, HIV and more."[6]

Facing increased competition from Asia in their core business, Mikkel and his father seized a new business opportunity, exporting textiles for distribution in the third world. On returning to Denmark, his father Torben struck a deal to buy a million yards of old wool from Sweden's civil defense stockpiles and asked Mikkel to work on having it cut and sold to the Red Cross to provide refugees in Rwanda and Kurdistan with blankets. The challenge was to produce clothes from fabric that did not fade in sunlight and did not weaken when permeated with insecticide due to the problem of sleeping sickness in Africa. Years of fabric and textile expertise positioned Vestergaard Frandsen as the frontrunner in producing durable fabrics that repelled the deadly horsefly but did not break down in extreme conditions. But by the mid 1990s, the little operation started in the back of the warehouse had outgrown the family business.

Frandsen then moved to Nairobi to build his Africa operation moving into mosquito nets in 1994, at first sourcing them out of Thailand. But soon Indian, Pakistani and Chinese manufacturers started to bid directly on NGO tenders, making it much tougher for the company to compete.

As competition grew, two basic choices emerged: make the product cheaper, or better. "Everybody else was picking the cost leadership route," Frandsen says, so he decided to differentiate using open innovation to acquire the needed technologies. "The company invested heavily in research and development and pursued tighter integration with its licensed production facilities for disease control nets in Vietnam, where its nets are still made. 'The ideas we came up with were basically the two featured at MoMA: a branded product portfolio for malaria control in development settings [Permanet]; and a branded product for malaria control in the acute phase of a complex emergency [Zerofly plastic sheeting].' His decision was spurred by a number of developments. In the late 1990s, the "Roll Back Malaria" campaign began pushing its way on to the international agenda, and a number of NGOs and UN health officials were looking for partnerships with the private sector. The WHO held a seminar bringing private and public sector players together. Mr Frandsen recalls there were a lot of people there, allowing him to handpick an expert team. "We were the people with the in-depth knowledge of the textile platform to use this on, we had the money and the logistical background, and we had the production set-up in Vietnam," he says (www.nextbillionnet/news/a-responsible-profit-from -the -war -on-malaria) [7].

Today the company's factories in Vietnam and Thailand churn out two million PermaNets each month which are then sold to privately funded companies which distribute them free or at highly subsidized rates mainly for use in Africa (Friedman, 2005)[8].

The majority of the company's products are sold in mass quantities to nonprofit, philanthropic, governmental and religious organizations

More recently the company developed a low-cost water filtering system that packs a water filter into a short plastic pipe. That pipe, LifeStraw transformed Vestergaard Frandsen's product line and expanded their global products. The LifeStraw uses hollow-fiber technology that eliminates 99.9999 percent of all bacteria, parasites and some viruses. The company now makes bigger versions that can filter five gallons an hour and can last a family three years.

As Vestergaard Frandsen evolved from a high quality, low cost manufacturer of textiles to a world leader in the fight against malaria and poverty, along with it Mikkel's image transformed from a rebellious youth to an official representative of the business world, with enough clout to influence global policy. One of his socially conscious initiatives involves motivating rural Africans to take an HIV test by offering them a free LifePack if they did so. Because the stigma of even asking for an HIV test is so strong, many Africans are reluctant to do so. Due to the promotional efforts of the free "LifePack, " a small plastic pack containing a water purifier, a sleeping net, 60 condoms and health education pamphlets, over 50,000 people were tested for HIV at a cost to the company of over $3 million. (McNeil, 2009)[9] The company is committed to a triple bottom line approach. Mikkel Vestergaard Frandsen states

We have seen how private public partnerships have matured over the last decade to a point where today there is neither controversy about, nor conflict between doing business and doing good. At Vestergaard, we have turned corporate social responsibility into our core business by dedicating our entire innovative platform to developing lifesaving technologies and concepts for people in the developing world. Our environmental agenda in relation to bed net production includes continuous optimization to reduce waste and increase recycling. This work is assisted by an environmental audit program. We have taken a major step in packing all nets in biodegradable bags, which will help reduce the environmental impact in the countries that receive our nets. We have been hugely inspired by our partners over the years. The integration of bed-net distribution into immunization campaigns for measles in 2002 and 2003 was one such inspiring idea....I would like to ... acknowledge the power of partnerships in health, which help cast a bridge between an innovation and its final users.[10]

The firm has come a long way, evolving from its humble beginnings as a regional clothing manufacturer into a global humanitarian leader with a staff of 200, headquarters in Lausanne, Switzerland, and branch offices in Ghana, India, Indonesia, Kenya, Nigeria, South Africa, the UAE and the USA. Vestergaard Frandsen also operates laboratories in Ghana and Vietnam, and production facilities in Belgium, China, India, Korea, Thailand and Vietnam.[11]

We know we're helping when we see thousands of people cueing up outside a clinic to get what we offer," says Mikkel. "We're focusing our business skills on developing a culture of trade for lifesaving products; we're focusing our investments on building factories and scaling up access to lifesaving products. This might not be capitalism in it purist form but it's business, it's doing good and it's fun. [12]

The company has been consistently profitable, [though as it is privately owned and not publicly traded, sales and profitability data is not published]. According to Sanne Fournier-Wendes "like any other private company we generate profits by the sale of our products to our customers, which in our case are mostly public agencies. The difference between us and our competitors is that we reinvest our profits in a range of social products related to our core mission. And this model has allowed us to grow dramatically."[13]

Vestergaard-Frandsen's mission is to improve the health of the poor. Together its efforts over the years and its focus on "humanitarian entrepreneurship," which the firm defines as the ability to do business and do good at the same time, has earned Vestergaard Frandsen a solid place as a Good Company. With everything from research to design to the material and distribution, Vestergaard-Frandsen keeps the end-users needs paramount. Kevin Starace, malaria advisor for the United Nations said this "Vestergaard is just different from other companies we work with. They think of the end-user as a consumer rather than a patient or a victim."[14]

KIVA[15]

Kiva's mission is to connect people, through lending, for the sake of alleviating poverty by empowering individuals to lend to an entrepreneur across the globe.

Kiva's innovation is to have developed the world's first person-to-person micro-lending website,thereby empowering individuals to lend directly to entrepreneurs around the globe. By combining the technologies of microfinance with the internet, Kiva connects people directly through lending.

Kiva was born of the following beliefs:

- People are by nature generous, and will help others if given the opportunity to do so in a transparent, accountable way.
- The poor are highly motivated and can be very successful when given an opportunity.
- By connecting people we can create relationships beyond financial transactions, and build a global community by expressing support and encouragement of one another (www.kiva.org).

Named as one of the top ideas in 2006 by the New York Times Magazine, Kiva itself is a front, a sort of backing bank for agencies that offer microloans by enabling microfinance institutions (MFIs) to raise debt capital directly from individual and group social investors via the Internet. Kiva provides partner institutions with interest-free US dollar loans to pass on to clients requiring only that they comply with transparency on the internet. MFIs on-lend this capital at prevailing in country interest rates and keep the interest income. Losses arising from client default are borne by Kiva's social investor.

Kiva shares its technology with its partners, existing microfinance institutions such as World Relief in Cambodia and Pro Mujer in Bolivia. This allows Kiva to gain access to potential entrepreneurs from impoverished communities without having to put staff in remote locations around the world. Kiva can rely on its MFI partners, who are usually short on funds to lend, to choose qualified entrepreneurs. Kiva's partners upload their entrepreneur profiles directly to the site so that lenders in California or London or Zurich can make their decision on which investments to make. Each partner organization is given a "risk rating" on the website and information about the delinquency and default rate for the micro-lender is provided as well,

Kiva's innovation is its data-rich, transparent lending platform which shows the social investors or lenders just how their money flows throughout the entire cycle, and what effect it has on the people and institutions lending it, borrowing it, and managing it along the way using the power of the internet to facilitate one-to-one connections that were previously prohibitively expensive. In contrast with high overhead child sponsorship, Kiva creates a similar interpersonal connection at much lower costs due to the instant, inexpensive cost of internet delivery.

Kiva.org was born in 2004 while Matt Flannery and Jessica Jackley rode on a bus down a bumpy road in Africa. Flannery was working as a computer programmer for TiVo while Jackley was working at the Stanford Business School where she heard Dr. Mohammed Yunus speak on the topic of microfinance. The two got to work creating a business plan. They met with Geoff Davis the CEO of Unitus, a "microfinance accelerator," and pitched their plan; Davis pointed out two challenges in the plan: (1) the difficulty of creating a scalable business, and (2) the high costs of connecting one lender to one borrower so they might see where their money was going.

Despite this they pressed ahead and were faced with yet another challenge -- whether they would present the company as a non-profit or a for profit business.[16] This led to questions about tax implications for the company and the lenders, tax write offs, and if this could possibly become a source of capital to get the idea going. But without a source of income to get the ball rolling they were stuck. It was at that point that they decided to "just start" and see where things went.

The couple spent the first few months of 2005 developing a beta version of their website and they decided on the name Kiva, which appropriately means "unity." They were almost ready to go, but did not have any entrepreneurs to loan to, so they asked Moses Onyango (a friend of Jessica's from her stay in Africa) to help them out. He found and posted seven local entrepreneurs on the site while Matt and Jessica emailed 300 friends from their wedding invitation list asking for help. That first weekend alone raised $3,500 and Kiva.org had finally taken its first breath.

The couple soon started getting calls from friends who wanted to know how their loans were doing. "Our first real success story was a fish seller in Uganda named Elizabeth Omalla. 'Hey, did you hear?' one friend called to tell me, "Elizabeth sold twice as much fish this week with the money we loaned to her!" Since Kiva began providing her loans, Elizabeth is now able to take her children to school, buy two cows and five goats and open a savings account"(Welch 2009, pp. 90-91).

Inspired by this early success, the couple prepared a 40 page business plan and started talking with microfinance professionals, UN experts and international lawyers. The advice they received from the 'experts' was that their concept just wouldn't work. "They said our idea would not scale. They said the overhead would cripple the cash flow margin. And most of all, they made false pattern-matching comparisons to preexisting models, saying that Kiva was too much like other child sponsorship organizations. It was all very disheartening" (Welch, 2009, pp.90-91).

But then they had several big breaks. First Premal Shah joined their team. He worked at EBay and when he decided to come and join the Kiva team permanently, he brought with him EBay's support which provided a key to Kiva's success because of its decision to donate free payment processing which they continue to do. Then they were featured in a blog on Daily Kos home page. The blog brought in $10,000 and over 1,000 emails as over a million people read about Kiva.

In 2006 Kiva's application for 501(c) (3) status was granted, but the firm was not generating enough funds to stay afloat. Then, on Halloween night, a show aired on Frontline that answered their prayers. It was a 15-minute piece on Kiva that overnight sent their volume skyrocketing from $3,000 per day to $30,000 per day. Previous to the airing of the show Kiva had processed a total of $500,000 in loans but seven months later that number jumped to over $5.5 million.

The next step was to establish a group of microfinance institution (MFI) partners. Once potential partners realized that the Kiva website would provide them access to a global source of capital they no longer had difficulty acquiring partners (Flannery, 2007).

Individuals from anywhere in the world who are interested in making a loan to another individual need only follow a few simple steps. First, the lender can browse a list of profiles of possible entrepreneurs in need. The list can be sorted by gender, sector (agriculture, education, health), region, or terms of the loan. This allows the lender to decide very specifically where their money is going. This differs greatly from charitable giving to other large organizations that decide specifically how to disperse the funds. Once the lender has found a person of interest, they can read their profile to get more information. For example,

Semmatimba Florence is a 45 year old widow who knows her back would be against the wall had she not taken the initiative to blaze a business trail 7 years ago. She began as a worker in a stone quarry. Prior to that she was still trying to come to grips with the death of her husband who was by then the sole bread winner for the family. In the following years she was living by her wits. Sensing that Florence was having difficulty providing for her family, a close friend of hers persuaded her to bear a hand in any business activity that could extricate her out of her predicament. It is then that this cheerful mother of 8 tried stone quarrying. With time she changed her priorities and began a restaurant and hotel business in Kilangira-Mukono. Her stone quarry business however is something she still has reins over. She has never looked back. In a week, she makes profits that sum up to 100,000 shillings and with this loan, she is looking at buying more utensils, i.e., cutlery, for her restaurant and equipment for her stone quarrying business. She needs $1100 and as of August 20, 2010 had raised $825 from 28 lenders scattered around the globe.[17] (www.kiva.org Retrieved Aug 20,2010)

The lender can choose to make a loan to the entrepreneur of their choice, making loans in $25 denominations up to the full loan amount requested. On average, it takes only one and a half days for a client to be fully funded (Kiva.org). Because of EBay's involvement with Kiva, PayPal or credit cards are accepted and Kiva assures that privacy and security are taken very seriously (Figure 2).

Once the funds have been lent, Kiva wires them on to one of their affiliated microfinance partners in the country that the lender has selected. These partners are the ones who actually administer the money to the entrepreneur. Kiva field partners are microfinance institutions (MFI) such as Hattha Kaksekar Limited (Cambodia), Emprender (Bolivia), Salone Microfinance Trust (Sierra Leone), and many others. Each Field Partner must meet Kiva's minimum standards of performance in order to get involved. For ex-

Figure 2. How Kiva works

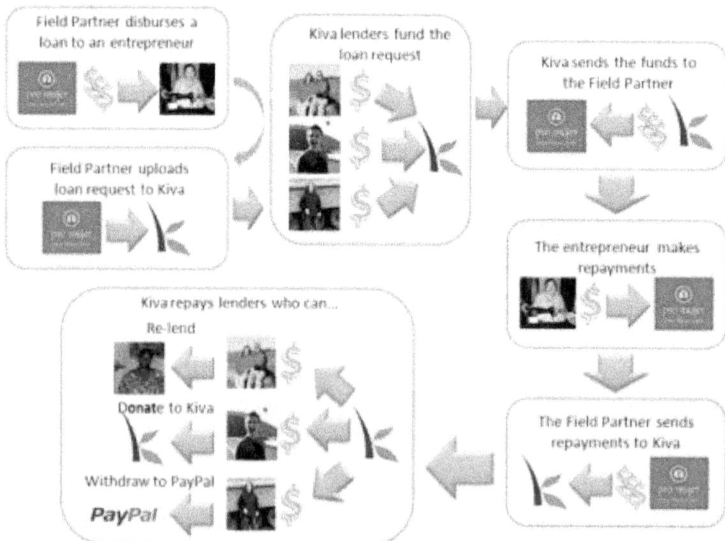

ample, each institution must "have a history (at least 2-3 years) of lending to poor, excluded, and/or vulnerable people for the purpose of alleviating poverty or reducing vulnerability; be registered as a legal entity in its country of operation; [and] have at least 1 year of financial audits". [18] (www.kiva.org)

These field partners are responsible for managing foreign exchange risks and repatriating funds to the local currency from the United States dollar and vice versa upon repayment, but their duties do not end there. They often take a very active role in providing training to increase the success chances of the entrepreneur. They are responsible for photographing the borrower, writing their business descriptions, keeping updates, and maintaining communication via Internet. Transparency is very important for Kiva and for its social investors. Because of this, field partners are required to report on the social impact of clients at least once per loan term and post the report on the Kiva website. The MFI itself can also recognize benefits from working in conjunction with Kiva, such as positive exposure to a worldwide audience of over 500,000 social investors, low costs including those for staff, administration, and foreign hedging, and 0% interest on debt capital.

When the lender is logged onto the website, they can see which Field Partner is associated with their chosen entrepreneur. For example, Christian Rural Aid Network's mission "is to work towards improvement in the quality of life for the rural poor and disadvantaged populations and communities in a holistic manner (physically, socially, economically and spiritually) with emphasis on the economic and social empowerment of women" (CRAN via Kiva.org). They have been with Kiva for 16 months, have worked with 1,624 entrepreneurs raising a total of $763,950, and have delinquency and default rates of 0%. Kiva has awarded this organization a 4-star (out of 5) rating, indicating that they have shown significant reliability in repayment and that Kiva has confidence that the Field Partner will dutifully administer and collect the loan. When it comes time for a borrower to repay her loan in monthly installments, she will give the money to her Field Partner who will then send the money back to the country from which the funds were lent. The loan is set up so that the borrower can repay the loan in monthly installments.

One borrower wrote about the impact Kiva has had on his life:

It's always unbelievable that a poor man's state can change in Africa. Here, people are too poor such that a man like me is considered to be among the poorest of the poor. I thank God for the founders of Kiva who made it possible for me to meet people with big heart for the suffering people of the continent of Africa. I received loan money worth $500 from Kiva almost a year ago. I decided to be very focused on the business that I was running through the support of Village Enterprise Fund grant money worth $100. Before the loan money, I used to buy only one cow and walk with it to the cattle market. This would cost me an amount of 230-250,000 thousand Ugandan shillings. Then sell it at around 270-280,000 thousand Ugandan shillings. Now the profit of 20-30,000 thousand shillings was not enough to meet all our basic needs and then keep the business going well. The whole situation changed when I got a Kiva loan of $500. The first step is that the stock increased from buying only one cow to buying six cows at same time. This caused a great change in our income due to a good amount of money coming in from the profit. Today, I can comfortably buy up to 12 cows and then drive them to cattle market. Our business has grown by 80% ever since we got the loan money. This is the present achievement that I've got from our business.

1. Renovate our house that was falling a part[sic]
2. Buy beddings for myself and my seven Children. At least now each Child has a pair of bed sheets (unbelievable).
3. I can afford a good nutritious food for the family. Our family is able to eat meat twice in a week, fish twice in a week, Take tea with sugar for a whole week, (this was not there before).
4. Cater for family treatment, basic essential needs.
5. Now we have our own one cow, two (2) goats, twelve (12) chicken, three (3) Turkeys and four (4) Piglets.
6. All my Children who had dropped out of school has resumed their studies because now I can afford to pay their school fees and buy scholastic materials for each of them."[19](www.kiva.org retrieved Aug 20,2010)

Today Kiva is a model followed by many other imitators in just a few short years. What is unique is that Kiva empowers individuals to lend directly to entrepreneurs in the developing world that they select in accord with Kiva's mission: *to connect people through lending for the sake of alleviating poverty* (Table 1).

Solutions and Recommendations

What is it that makes companies like Vestergaard Frandsen and Indigenous Designs, Kiva and Root Capital models of a new trend in global business? The value of technology depends upon the ability to use it and adapt it. This comprises several components: First, seeking opportunities to use the technology; second; identifying potential partners to work with you or supply complementary technology and third, disseminating and/or commercializing a new or improved product or service. Each of these companies draws on the global knowledge base to innovate. But the products and services draw upon additional characteristics of global social innovation. These include the following:

A Foundation of Generosity

Each of these companies is built on a value concept of generosity. This basis attracts financial

Table 1. Kiva's Impact (as of August 2010)

Total value of all loans made through Kiva:	$155,061,500
Number of Kiva Users:	744,822
Number of Kiva Users who have funded a loan:	474,374
Number of countries represented by Kiva Lenders:	201
Number of entrepreneurs that have received a loan through Kiva:	398,155
Number of loans that have been funded through Kiva:	213,234
Percentage of Kiva loans which have been made to women entrepreneurs:	82.12%
Number of Kiva Field Partners (microfinance institutions Kiva partners with):	119
Number of countries Kiva Field Partners are located in:	53
Current repayment rate (all partners):	98.84%
Average loan size	$384.44
Average total amount loaned per Kiva Lender (includes reloaned funds):	$207.83
Average number of loans per Kiva Lender:	6.11

Source:www. kiva.org

success, allowing them to show a profit and fund their vision of helping people and the planet. Each company is driven by the value of social justice. Indigenous ensures indigenous artisans are paid fair value for their work; Vestergaard Frandsen works to promote health and wellbeing of the neediest elements of the world's population. While Kiva's financial innovation provides the resources to financially empower thousands of poor families around the globe.

Perseverance and Partnership

The companies have each persevered to overcome obstacles by sharing technology or management systems by partnering with organizations around the world to help them meet their goals. For example, Indigenous Designs partnered with Root Capital and Vestergaard Frandsen with UNESCO. Kiva has achieved its impressive record by partnering with over 100 microfinance organizations located in the countries where it operates.

Meaning

Perhaps one of the most important developments in recent years is the recognition that the quality of what organizations accomplish is contingent on how people who do the work feel about the value of the work that they do. Among the things that employees need and expect today from their work, is a sense of meaning a place that provide meaning in their lives.

These companies care for the world; they make the world's problems their business. First, they produce goods which are beneficial and life enhancing. They do not produce, sell, promote or finance goods which are harmful or toxic to human or animal life. For example they do not produce weapons or armaments, toxic chemicals, or tobacco products. They extend the values they espouse at home to their entire supply chain. They extend their environmental practices to their suppliers. They may partner with suppliers to educate them about important environmental issues such as how to reduce waste or protect their employees from hazardous wastes. These companies seek to distribute their wealth equitably among those in the value chain who contribute to its creation.

Moreover, they partner with the communities in which they operate and are rewarded for partnering with the community.[20] What we see from these examples is that is it that values can be significant in determining the success of a new breed of global innovators. These are companies that have broadened the definition of success to include social and environmental factors. And they clearly demonstrate that social and humanitarian considerations make for good business decisions.

FUTURE RESEARCH DIRECTIONS

To date the research on Open Innovation has centerd on a small number of high-technolgy companies. The thrust of the research has been in the context of R an D as a source of competitve advantaage for the firm. Open innovation is about the inflows and outflows of knowledge to accelerate the firm's innovation and markets. The book Rembrandt in the Attic (Rivette and Klein 2000) takes note of the fact that many companies have valuable intellectual property which they fail to use.

Meanwhile there are enormous social needs for technology around the world--for health, clean water and inexpensive lighting, to name but a few. There is a need to open the concept of innovation to consider its social value and not just the economic value.

To take another example, Benetech is a company that operates much like a startup company in a venture capital environment. But it focuses on identifying needs and opportunities where technology could have a big impact improving the lives of underserved people. Benetech's first product was a machine which converted printed books to audio books for the blind, using optical-recognition software which was inspired by Fruchterman's work on smart bombs. At the time, the best available technology for a blind person to read printed text was an expensive machine the size of a washing machine, an unrealistic and unaffordable option for accomplishing daily tasks like browsing a newspaper or looking over a piece of mail. Although the technology for creating an affordable, portable machine existed, the potential customer base—blind individuals and their employers—was too small to promise a traditional return on investment. As a result, technology investors were unwilling to take the risk to develop such a product. In contrast with the typical firm, Benetech determines the viability of an idea, and then the company applies research, analysis and business planning to make decisions on behalf of society. The company is known for its Arkenstone reading machine for the blind developed by accessing a wide range of technology.

There is a need for more research on how to share technology so as to stimulate the formation of more companies like Benetech and Vestergaard Frandsen.

CONCLUSION

In contrast with many traditional innovators which focus on short term profitability and place the interests of management and investors above all other stakeholders, these socially motivated global innovators take pains to serve the balanced interests of all stakeholders. And the definition of stakeholders extends broad and wide to include farmers and employees of subcontractors far away and far down the supply chain. In this way companies which include their values as an integral part of their business model to earn revenue while addressing essential social and environmental issues are adding important dimensions to global business. Open innovation allows companies like Benetech and Kiva to draw from the global pool of knowledge to design products and services which meet social needs. This chapter has examined three SME's out of many such companies that are drawing on a range of technologies to create products and services in order to meet pressing social needs round the world.

REFERENCES

Chesborough, H. (2006a). *Open innovation: Researching a new paradigm*. Oxford University Press.

Chesbrough, H. (2006b). *Open business models*. Harvard Business Review Press.

Flannery, M. (2007). Kiva and the birth of person-to-person microfinance. *Innovations, Winter/ Spring* (2007).

Freedman, M. (2005, December 26). A fine mesh. *Forbes.com*.

Kiva. (n.d.). *Website*. Retrieved August 20, 2010, from www.kiva.org

McNeil, D. (2009, February 3). A company prospers by saving poor people's lives. *The New York Times*. Retrieved from www.nytimes.com/2009/02/03/health/research/03prof.html?pagewanted=2&_r=1

Peng, M. (2010). *Global*. South-Western Cenage Learning.

Rivette, K., & Klein, D. (2000). *Rembrandts in the attic: Unlocking the hidden power of patents*. Harvard University Press.

Roll Back Malaria. (n.d.). *Global advocacy leadership interviews*. Retrieved July 15, 2010, from www.rollbackmalaria.org/globaladvocacy/leadershipinterviews6.html

Welch, W. (2009). *Tactics of hope*. Earth Aware Press.

Young, W., & Welford, R. (2002). *Ethical shopper*. Fusion Press.

ADDITIONAL READING

Arena, C. (2007). *The High Purpose Company*. Collins.

Boschee, J., Emerson, J., Sealey, K., & Sealey, W. (Eds.). (2000). *A Reader in Social Enterprise*. Boston: Pearson Custom Publishers.

Brown, Tim and Wyatt, Jocelyn(2010) "Design Thinking for Social Innovation" Stanford Social Innovation Review. v. 8. N.1. Winter

David Bornstein. (2004) How to Change the World: Social Entrepreneurs and the Power of New Ideas, Oxford 2004

Elkington, John and Pamela Hartigan. (2008) The Power of Unreasonable People. Harvard 2008

Gregory Dees;(2001) Enterprising Nonprofits .Wiley 2001

Heal, G. (2008) When Principles Pay. Columbia University Press. London,Ted and Hart, Stuart (2010). Next Genrastion Business Strategies for the Base of the Pyramid. FT Press.

Novogratz, J. (2009). *The Blue Sweater*. Rodale Press.

Prahalad, C. K. (2005). *The Fortune at the Bottom of the Pyramid. Wharton School Publishing. Chappell,Tom. (1993) The Soul of Business: Managing for Profit and the Common Good*. Bantam.

Smith, P., & Thurman, E. (2007). *A Billion Bootstraps*. McGraw Hill.

Welsch, W. (2008). *The Tactics of Hope. Earthaware Publishers 2008 John Wood*. Leaving Microsoft to Change the World.

KEY TERMS AND DEFINITIONS

Case Studies: A detailed description of a single enterprise used to provide information on the history and to aid in understanding and interpreting its development, process and practices.

Fair Trade: Trade that satisfies certain criteria on the supply chain of the goods involved, usually including fair payment for producers; often with other social and environmental considerations.

Global Innovators: Companies that draw from the wealth of global knowledge to develop a new process or apply an existing process to a new situation or problem.

Social Innovation: A new approach or process applied to a social issue or problem. For example, Kiva combines microfinance with internet technology developed by pay-pal to connect lenders in the industrialized nations with borrowers in the emerging economies.

ENDNOTES

1 This information comes from William Young and Richard Welford. Ethical Shopper. Fusion Press. London 2002.

2 In 1996 Root Capital was known as EcoLogic Finance. In 2000, EcoLogic Finance was spun off from EcoLogic Development Fund as a separate company. Their mission is to assist "small grassroots businesses—such as coffee farmer cooperatives and artisan associations—which they see as engines for sustainable development and poverty alleviation in rural communities in the developing world. Considered too small and risky for mainstream banks and too large for microfinance, these grassroots businesses are caught in the "missing middle"; they cannot access the capital they need to grow and sustain their operations." Root Capital addresses this market failure through an innovative lending model that serves the needs of these remote grassroots businesses by creating a new class of capital sitting between micro-credit and commercial lending, enabling rural communities to unlock wealth and build sustainable livelihoods.

3 Completely hand knit, Indigenous Designs emphasizes the time and care that goes into each garment they sell. The designs are also recognized throughout the industry as fashionable and stylish. Customers know they are receiving a quality product which will last them many years. The company also owns key patents for their unique clothing designs, as well as the Indigenous Designs logo and name. They are further exploring opportunities to protect proprietary fiber blends through patents and agreements with strategic milling partners. Their name brand is becoming increasingly recognizable in the organic and fair trade industry, synonymous with fair labor prices and quality products. These may provide additional advantages to Indigenous Designs over potential competitors. Nevertheless, just how important sustainability is in the market is open to some dispute..

4 This capacity continues to grow as word spreads within the indigenous people community of the opportunities to be had by working for such a fair company. By offering fair trade wages, they are able to continue to increase the number and size of the cooperatives they work with, thus growing their production capacity and speed. Large capacity allows them to take advantage of economies of scale, resulting in lower per unit costs to produce, inspect, and ship each garment. This ultimately leads to increased margins.

5 Indigenous is an all natural fiber company utilizing wool, alpaca, silk, Tencel, organic cotton and other natural fibers created by nature certified by Skal[1].[Conventional cotton is one of the most pesticide-dependent crops, making the growing and harvesting of this fiber harmful to the water table and local communities] Fibers are procured as close to the cooperatives as possible so that we can further benefit these local communities. Indigenous was the first clothing company to blend organic cotton with Tencel, a breakthrough fiber made of sustainably har-

vested tree pulp. In many cases, Indigenous utilizes the natural colors of organic cotton and fleece, eliminating the need for dyes to achieve complex color schemes within our fashion collections. Indigenous also develops low impact dyes which contain no heavy metals and are azo - dioxazine compound free. Indigenous ensures a qualified system for waste water filtration in alignment with the company's sustainability goals and to best serve the local communities .

6 www.rollbackmalaria.org/globaladvocacy/leadershipinterviews6.html retrieved 15 July 2010

7 www.nextbillion.net/news/a-responsible-profit -from-the-war-on-malaria retrieved July 10,2010

8 Freedman, Michael. "A Fine Mesh." Forbes.com. December 26, 2005. <http://www.forbes.com/forbes/2005/1226/071.html>

9 "A Company Prospers by Saving Poor People's Lives." The New York Times. February 2, 2009. <http://www.nytimes.com/2009/02/03/health/research/03prof.html?pagewanted=2&_r=1>

10 www.rollbackmalaria.org/globaladvocacy/leadershipinterviews6.html retrieved 15 July 2010

11 Vestergaard-Frandsen is still a family-owned business and does not disclose financial data, but the company has sold over 165 million nets. Forbes reported that Vestergaard Frandsen earned $5 million pretax on sales of $40 million in 2005.

12 http://www.youtube.com/watch?v=1bjBxo0mkJw

13 Personal interview. Lausanne Switzerland. June 24,2011

14 "A Company Prospers by Saving Poor People's Lives" The New York Times. Feb 2, 2009.

15 This section was prepared with the assistance of Ashley Manges and Andrea Johnson.

16 The legal implication of the decision ended up being much more complex than they had originally thought. Due to the nature of the idea they had a very difficult time convincing a lawyer to work with them because "of the murky territory of securitization" that they were entering in; there is a long list of requirements that a company must meet when dealing with securities, because "whenever money is being exchanged between two people, someone in the government somewhere will begin to take notice". This brought them to the question as to whether they could achieve a 501(c) (3) status as a charitable organization.

17 www.kiva.org retrieved Aug 20, 2010

18 www.kiva.org retrieved Aug 20,2010

19 www.kiva.org retrieved Aug 20,2010

20 Globe Scan found that eight out of ten respondents said that their respect for a company would increase if it were partnered with a non-governmental organization.

Chapter 7
Theoretical Model for a Local Economy Open Innovation Program: A User Organization

Amiram Porath
Center for Academic Studies in Or-Yehuda, Israel

ABSTRACT

This chapter presents a model of open innovation as a collaborative effort of firms, mostly SMEs that are managed under a government support with the assistive involvement of academia. While normally industry-academia cooperation is R&D focused, the model presented is a model of open innovation not involving dedicated R&D. It deals with the process of assimilating existing technologies and methodologies; the model focuses on search, identification, and implementation phases to improve competitiveness through open innovation strategies. The model claims that specifically for SMEs, innovation not related to R&D, and especially open innovation, is hard to deal with alone. Therefore the model presents a group of SMEs working together towards that end, and the role of the governing authority supporting them. The model is supported by a short case study of a national program of Israel, the Users Association for Advanced Technologies.

INTRODUCTION

Open Innovation is a multi-disciplinary area of research, as it deals with the concept that for organizations it is not possible to relay just on their own research and that they need to license or buy technologies and methodologies from external sources (Chesbrough, 2003). The focus selected here is not on an R&D process, but on the selection and acquisition process of the technologies and methodologies. There are several areas that deal with it, and within each area there can be more than one aspect that is relevant. Among them could be counted economic, organizational, legal, and network aspects.

DOI: 10.4018/978-1-61350-519-9.ch007

Economics

The model can be viewed from the economic point of view, under the following theoretical aspects:

First, as a transaction including identification, acquiring and implementing the acquired knowledge – innovation. This analysis, under the transaction costs theory regards the firms as limited resources SMEs. It also includes the role of a governing authority in reducing costs and in influencing behavior decisions in the long term.

Second, by analyzing the influence the joint acquirement of methodologies and technologies may have on the market competition of the firms involved. Why do firms engage in the acquirement not of unique knowledge that can bring the single firm an advantage, but rather a joint acquirement of a known technology / methodology?

Organization Theory

The case of Open Innovation from the organizational point of view, has several aspects such as:

First, regarding the influence such an activity would have on the interaction of the firm with its environment. Innovation which is an important capability of the firm is imported, jointly with others from an external source. How would that influence the definition of the firm boarder, the development of internal routines in response to the external routines of the consortium?

Second, regarding the influence such cooperation may have on the firm from rate of spillovers, and of human resources turnover, up to influencing a more coherent structure and internal routines in the firm.

Following the above, the role of the government and the Triple-Helix concept (see key terms below) is also discussed briefly in the model.

Legal

The model presented has two main legal aspects relevant to its analysis:

The first is the legal environment under which the operation is carried out. As the activity described in the model has strategic value for the firms, it is important that the firms entering it would be familiar with the rules, by rules or regulations relevant to it. This familiarity makes the firms feel comfortable with them, allows them to enter the process. Lack of such clarity, and understanding would lead to problems that could cause the shutdown of such programs. The legal aspects of the activity and their transparency are of major importance as a public entity and public funds are involved.

The second relevant aspect is the transfer of knowledge. As that knowledge, new or old, may be proprietary, it is important to review the legal aspects dealing with the Intellectual Property rights which change from country to country. These include the relevant agreements between the firms involved in the activity. As before, transparency and clarity beforehand, play a major rule.

Regarding the legal analysis the specific case study must be analyzed within the legal environment it is active in, conclusions derived from the case study into the model should be dealt with caution.

Networks Theory

The activity discussed and the model are operating in what may seem to be a network environment (Laredo, 1998). However, the network environment will not be dealt with in this chapter as it would seem to be of less prominence to other theories. The network effect is felt by the author (Porath, 2008) to be stronger in R&D consortia. In the specific case of the model described below, the firms come together in order to learn and implement from sources external to the group. Any intergroup learning is incidental.

The chapter begins by laying down the theoretical foundation for the model. A short literature review has been presented to illustrate the main assumptions of a few of the relevant theories, which is followed by a case study to illustrate some of the points. A practical model has been derived from the case study based on the theoretical background and the chapter ends with recommendation for further research in order to validate the model and expand it for additional application. At the end of the chapter the reader should feel that the Users Association for Advanced Technologies (UA), as a model for Open Innovation on its characteristic parameters, players and relevant points is comprehensible, and that the reader knows how the model can be applied.

BACKGROUND

Theoretical Background

The model described below has a complex background, as the theoretical background of an activity as complex as Open Innovation would tend to be. The background is presented against the setting of some of the more relevant theories. As it is related to economic development and as it deals with firms, the author has selected to start with the economic theory and its relevance to the model, including the Triple Helix concept (Etzkowitz an&d Leydersdorrf, 1997), and other less complex models; the economic background has been followed by a short discussion regarding organizational theory, and thereafter it discusses on legal theory to end this section.

Economic Theory

There are several aspects of the theory that are relevant to the model. The more obvious at first glance would be to understand the impact of cooperation in innovation between firms on the market place. A second aspect related to the incentives for the firms to cooperate in the first place would be Transaction Costs.

When dealing with models related to cooperation in innovation and competition in the market there is a whole volume of research in that topic. Some of the earlier models e.g. Katz (1986), D'Aspremont, and Jacquemin (1988) and those developed later in the 1990's were very simple (Kamien, Muller & Zang, 1992). The models later evolved into more complex and more lifelike models (Yun, Park & Ahn, 2000; Pastor & Sandons, 2002; Miyagiwa, & Ohno, 2002; Geuna & Matt, 2006; EU, 2006). It is granted, the bulk of the work focuses on cooperation in R&D or as Porath (2008) refers to it, Collaborative Research, but the basic assumption that all models share is that the cooperation in pre-competitive research does not hurt the marketplace competition (Rosenberg, 1990; Audretsch & Feldman, 1996; Katsoulacos & Ulph, 1998; Cabral, 2000). The models show (Kamien and Zang, 2000) that the cooperation increases the social benefit derived by the firm and society together, and that it is better for the firms to collaborate than to do their own single R&D (Olk, 1991; Doz, Olk & Ring, 2000; Porath, 2008; Porath, 2010). Originally Olk (1991) and later expanded by Doz, Olk and Ring (2000), referred to spontaneous collaboration funded entirely by the firms. The reference to collaboration under government support was added by Porath (2008) in his comprehensive model. The original role of government, and their support for R&D on national and supra-national programs was discussed by Luukkonen (1998, 2000), but was discussed on a wider scope, to include the interrelations in general among Government, Industry and Academia by the researchers dealing with the Triple Helix model (Etzkowitz & Leydersdorrf, 1997). The model developed in this study is not fully compatible with the Triple Helix approach as it does not deal with the influence the industry and academia have over the government.

The model as a practical model addresses the scarcity of resources and the demand for resources that the innovation process will execute from firms attempting it. These come under the Transaction Cost and the Resource Base aspects of the theory. Transaction Costs and Resource

Base aspects would direct firms to follow paths that would require them to dedicate the smallest amount of resources possible, to try and reduce the cost of the transaction so as to allow the firm to perform it. The drive to lower the transaction cost and the focus on that aspect will interest the community the most (Parkhe, 1993; Audretsch & Feldman, 1996; Kline, 2000).

Organizational Theory

The model relates to two major aspects. The first is the boundary of the firm. Ever since Penrose (1968) defined the firm and its relations to the environment, the collaboration in R&D and innovation activities have been making the definitions open to re-examination (Kirk, 1995; Spender, 1996, Laursen & Salter, 2004). The Brown and Eisenhardt (1998) reference to Co-opetittors, as a mix between competitors and cooperators have made the definitions of different players in the firms' environment more complex. The boundaries of the firm, and the relationships it has with the outside, are relevant to Open Innovation on two fronts. First, as can be deduced from the Open Innovation term, the model will deal with deriving an important capability, innovation, from external sources. That implies some dependence but also some level of exchange. Second the model deals with a legal entity formed by the firms, acting on their behalf, supported by all and supporting most, making the border definitions of "in" and "out" less clear. Is the internal cooperation between the firms in the new legal entity considered in or out?

The second aspect is Inertia. Inertia can be expressed as elements dealing with the internal routines created within the cooperation structure, as well as elements regarding the partnership in an external organization, that demand little if any internal change occur (Stern & Pozner, 2007). It is the inherent Inertia that prevents the firm from making structural changes that will allow it to adapt better to market changes, technological changes, or adopt innovation in general (Ring & Van de Ven, 1994; Audretsch & Feldman, 1996). Here again the role of the management in selecting the sources of innovation outside the firm is much dependent upon the experience of the management (Laursen & Salter, 2004). The management will tend to focus on known and tested sources, basing its trust on past experience. It will be reluctant to try new and costly sources.

As Zhou and Wu (2010) showed, high technological capabilities may hinder the firm's ability to utilize exploration for innovation. High capabilities pushed the firms towards high exploitation based on the technical capabilities. The Inertia and the absorption difficulties increase in order to absorb innovation from the outside and put it into internal use. The firm must be allowed to recruit knowledgeable management and employees (Harison & Koski, 2010). In times of increased competition and stress in the market, SMEs' lack of resources will make it difficult for the firms to acquire such personnel, leading in turn to a reduced absorption capability, and increased need for external assistance.

Network Theory

As mentioned above, the network theory while being relevant to the inter-relations among the firms (Ring and Van de Ven, 1994), and the learning process from each other will not be discussed here. This specific learning process is of lesser significance in the model presented here than in R&D consortia models (Laredo, 1998; Dyer & Nobeoka, 2000).

Legal Aspects

The legal aspects of the cooperation are interesting. The Intellectual Property, or in the more practical aspect of it, the Intellectual Property Rights (IPR) of cooperation programs is a field of research by itself (Porath, 2006). IPR play an important role in the cooperation between firms (Walter, Lechner, & Kellermanns, 2007), wherever knowledge is part of the model. The legal bonding between the firms is less important. As can be seen, it can be a free type of tie (Olk, 1991; Stern & Pozner, 2007), or it can be an agreement, a legal body or any other form (Porath, 2008). The important issue regarding the bonding is that whether the rules are clear and allow the firms to operate internally with as little disturbance as possible. The IPR arrangements can vary (EU-IPR, 2010) but as long as they are clear before the project start, the firms will know how to adjust (Argawal & Henderson, 2001;EU, 2009;EU-IPR, 2010). The IPR can have long term implications on the partners and require, therefore, serious consideration (Thursby, Thursby, & Mukherjee, 2005). It will be further discussed below.

THE USERS ASSOCIATION FOR ADVANCED TECHNOLOGIES (UA) AS A CONCEPT

Socio Economic and Legal Background of the Model

The Users Association for Advanced Technologies (UA) is a program operated by the Magnet Authority, an executive agency within the Ministry of Industry, Trade and Labor in Israel. The model named after this program the UA model, is in fact a model of collaboration in Open innovation. It depicts firms (preferably but not exclusively SMEs) coming together and collaborating in accessing innovation from sources external to each other. The characteristics of the UA model are being described next.

Open Innovation as mentioned above comes from outside the firm. It can come from without the firm when the firm acts as an independent player seeking the innovation and implementing it. Alternatively, as in this model, several firms can come together and do it jointly. This is a very specific way that enhances advantages they have and reduces the risk of seeking innovation outside.

The short case study presented in the next section describes a program that executes the model in real life, followed by an analysis seeking to define the model based on the case study. However, the UA program as depicted in the case study and analysis shown is not a solitary tool. While it is an independent program and is not linked directly to other programs (except sharing the same Authority and financing source at the government side), it is in fact complementary in its action to other tools. In 1994 the Magnet research consortia program was officially launched. It supported collaborative research of Industry and Academia (Porath, 2008). It was considered important at that time for the improvement of the competitiveness of whole sectors to work in this way and to bring Industry and Academia together to increase the innovation in an effective way. However, the creation of new knowledge was deemed insufficient. Therefore, close to the official launch of the R&D program the UA was launched. It was assumed that a tool that would focus on the spreading and implementation of existing knowledge could complement the development of sectors and the improvement of their competitiveness.

Israel has had a reputable academic infrastructure since early in the 20th century, the Hebrew University in Jerusalem, the Technion in Haifa (both with Nobel Laureate winners), were the first, followed by the Weizman Institute and others later on (Volanski, 2005). However, technology and innovation reputation were won by the Israeli economy mainly since the 1990's. The 1990's saw Israel enact several innovation encouraging programs, the Technology Incubator program, the Yozma (meaning Initiative in Hebrew) for creating new Venture Capital funds, bringing that sector from close to 0 to over 10 Billion US$ in the early 2000's. The Magnet R&D Consortia Program was one tool among several intended to help the industry gain the advantage in technological fields, and it was complemented by the UA.

The Technology Incubator's program, the Magnet Authority and others operate under the Chief

Scientist Office in the Ministry of Industry, Trade and Labor. The programs managed by the Magnet Authority deal with generic, pre-competitive or non-competitive R&D (not leading directly to a product). Therefore, the financial support is given in the form of a grant and not a loan. Even though the firms contribute part of the budget, still most of the financing is granted by the government, and the programs come under the R&D law (Porath, 2006).

The UA is according to the Magnet website (Magnet, 2010): "A Users Association for Advanced Technologies is an incorporation of firms with common interest in advance technologies in their sectors' technologies that can be purchased or that already are in use with some of the firms' and which are not a strategic asset of the firms owning them."[1]

The above definition is aided by the principals guiding the model (Magnet, 2010), which are:

Defining common technological needs; Allowing access to technologies known elsewhere, that may be updated from time to time; and Preparing an action plan to disseminate the external knowledge among the UA members.

History (Including Legal Background) of the Users Association for Advanced Technologies Program

Soon after the establishment of the Magnet Program in 1994, another program was added to the management responsibility of Magnet. That second program was the Users Association for Advanced Technologies (UA).The UA like other programs in the Chief Scientist Office in the Ministry of Industry Trade and Labor in Israel, operates under an Order of the General Manager of the ministry. The current order number 8.5 from September 2001 replaced the former order number 4.2 from 2000.

In this section the background and legal aspects as well as general information about the UA will be described so as to lay the foundation for the model presented later in this chapter. The case study presents, in support of the model a specific UA, which has been operated in recent years. It presents the reasons for establishing it; the way it was established; the way it operates; the benefits for the participants and main characteristics of the UA.

An association (non-profit) is formed under the UA program, which needs to be re-established. Every few years (5-6), the association is dissolved and a new association is being formed. In general the aims of the UA has been mentioned in the Iltam Moto (see section below), in a more detailed conservative way, and expressed in the annual report (Iltam, 2009). The UA aims to improve the competitiveness of the knowledge based industry by identifying, learning, and implementation. To improve the competitiveness means to reduce time to market; reduce development and manufacturing costs; comply with international standards; improve quality and reliability, and increase excellence.

The IPR environment is such, that the firms are required to purchase (buy, license etc.) the knowledge individually, while the UA serves as a facilitator, but not as owner. Ownership is individual but rarely if ever, exclusive.

Model Background and Work Environment

The basic concept of Open Innovation seems to be fairly easy to understand. Firms, understanding the importance of innovation seek it outside the borders of the firm. That would have been very simple and easy to understand. So, why it is difficult to accomplish? Why are so many efforts and so much money involved? These questions are not redundant. If one compares Open Innovation to Marketing, the importance of these questions becomes clear. In Marketing, while effort and research are active, the scope and level of attention are less than that regarding Innovation generally, and Open Innovation specifically. Additionally, while there are programs or plans to support and assist marketing (normally export based) in most countries, the level of government involvement and support cannot be compared to what happens when dealing with Open Innovation. The EU has recognized that 99% of the businesses in the EU are SMEs (over 20 millions) and is attempting to help them (SME website, 2010) as well as investing in them via the 7th Framework Program (1.336 Billion euro) (EU- FP7 budget, 2010). Furthermore, there are more initiatives of similar nature from other DG's in the commission.

The reason lies in two aspects, one is the maturity of the activity and its tangible results; the other lies in the complexity of the topic addressed. Marketing has been around for many years, firms that do not know how to market themselves and their products become extinct very quickly. The topic of marketing is very mature and is in the mainstream of any Business Management course. It is, therefore, well known, and as it brings tangible measurable results, and it is easier to teach and understand. Additionally it is relatively easy to evaluate the performance of firms or teams within the firm, charged with that activity.

The case for Innovation is different, although the commission officially launched the Lisbon agenda in 2000, placing Innovation as one of the targets aimed at; it is only in recent years that the realization of its global relevance to all sectors and to all regions, has become fully appreciated. Innovation is no longer the special unique domain of High-Tech sector's or the technological sectors alone, nor even of firms or business organizations alone. Innovation has been realized to be relevant to everyone and everything, from a person, to an association, NGOs, government, industry and business, and in fact any organization or individual operating in the real world. In this context, innovation is no longer singularly attached to R&D, or to technology. These are still the domains of very specific sectors. While one may claim that tourism has little to do with R&D or technology (and even that would not be universally accepted), no one today will dare claim the same regarding Innovation in tourism. So it is clear that the Innovation is relevant to all, not only to the sectors dealing with R&D.

But the singularity was broken in another aspect as well. Not only has innovation moved into the general domain disassociating the singular link with R&D, but even for sectors relevant to R&D, technological sectors, innovation is no longer limited to R&D activity. The case study (short as it may be) shows, that even in High-Tech technological sectors, such as electronics and communications, innovation can be found other than R&D. The UA above does not deal with R&D, support or finance it. The UA deals with technologies, methodologies or other supports the member firms need for their innovation. It does not support the creation of new knowledge, but supports only the implementation of the existing one.

The UA program shows that great improvement regarding the competitiveness of a firm can be achieved through the adoption of existing knowledge. When considering the size of the budget that is used to support the creation of new knowledge, and the fact that in many cases knowledge resulting from these projects is left unused (see publication such as Research Europe by the European Commission, or the specific programs in the Framework Programs that publish unused technologies from FP5 and FP6), one has to ask whether this is the optimal mix of support. Should there not be more support for the implementation within firms of existing technologies and methodologies, than additional effort dedicated to the creation of new knowledge. The support of the creation of new knowledge creates peaks of implemented new knowledge that emphasize and increase the gap between the R&D performers and the non-performers creating a sort of mountain topography in a sector (see Figure 1).

Figure 1. R&D peaks

Figure 1 is a figurative schematic non-unitary description of the Knowledge in a sector created individualy by each firm by its own efforts, or supported efforts but without collaboration. Each peak denotes the knowledge of a single individual firm. The Y-Axis denotes the level of advance of the Knowledge and the X-axis the breadth the Knowledge covers for a general topic.

However, if dedicating the effort required to focus on widespread implementation of existing proven technologies and methodologies, and the effort that can be spread on a sectorial basis, then one could get a less "pointy" and "smoother" picture (see Figure 2). A widespread implementation would have, by nature, a wider influence, and would help decrease the gaps between the R&D leaders and the stragglers, regarding adoption of leading technologies. That is of course when dealing with competitiveness in external arenas and not in local arenas. Figure 2, in contrast with Figure 1depicts a case in which the firms cooperate and therefore the Knowledge in the sector is similar in breadth of topic covered and in the advance towards the Knowledge front. It is a more unified Knowledge. The Y and X axis are similar to those in Figure 1.

In the UA case study, the effort was dedicated towards increasing the attractiveness of producing in Israel against that of the Far-East, EU and the Americas; and it was not intended to improve the competitiveness in the local market. Thus, by allowing a large enough number of firms to be involved in the effort, the average attractiveness of a sector has been raised.

Figure 2. Implementation

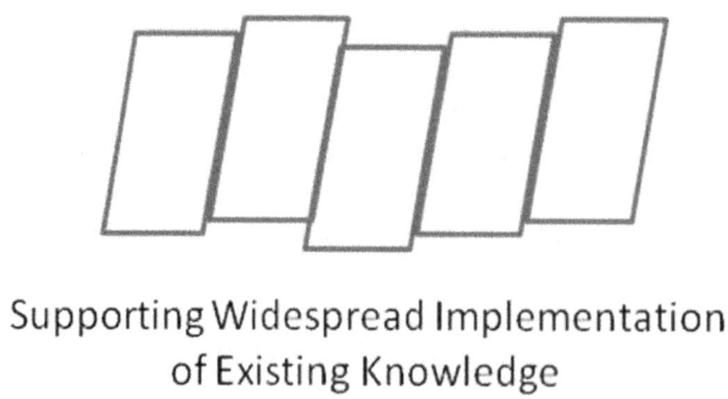

In the works of Luukkonen (1998, 2000), it was shown that the role of the support for R&D was to allow firms to take on riskier projects than they normally would, due to the reduced risk. That is to say that for the firms, the R&D support lowered the barriers towards riskier, more innovative projects. It is with similar approach that one can say that in the UA one can see a reduction of the Implementation Barrier. The Implementation Barrier is the barrier stemming from a combination of lack of resources (mind share and management time, diversified expertise) that prevents firms, from embarking on an Implementation Project. The Implementation Project is executed by following the Implementation Process, which is made of searching, identifying, selecting and implementing a needed Innovation Ingredient. The Innovation Ingredient may be a methodology, software, technology or similar tool.

The Innovation Process requires that a firm would dedicate resources for its execution. The search, the identification, selection and implementation require that the firms would dedicate resources for them. These resources may be some of the most expensive and rare resources the firm owns. That would include the time, mind-share and expertise that the high management of the firm, its support and consulting organs may have. If the management dedicates one of its members to deal with the Implementation Process, that dedication comes at a high price. This is true for all firms, as high management resources are always rare, but it is very problematic for SME's as in an SME by nature the redundancy is small and in some cases non-existent.

Some of the process can be outsourced, but not all of it. As shown in the model there are external bodies that can help by outsourcing the search, identification and even the selection, but the internal implementation in the firm will demand management resources. Moreover, even the outsourced segments of the process, can be very difficult for the single firm. Here the model and concept that the author advocates is the shared outsourcing. By sharing the outsourcing the firms do not lose anything and gain much. In an email following an interview in June 2010, Dr. Ilan Peled the General Manager of the Magnet Authority, mentioned that the technologies dealt with by the UA are enabling technologies and not core technologies, and that often there is Technology Transfer among the UA members. These two points strengthen the view of this researcher that there is nothing for the members to loose from sharing the outsourcing, and much to gain, as they do not risk anything that relates to their core activities.

If the technologies are enabling technologies the single firm does not care if others have it, as long as it does. So if it can be assumed that the operation software allows the firm the following advantages: to cut defects found at the customers site by half; thereby reducing the need to ship, retrieve, repair and ship again. That would help the firms' competitiveness, by improving the firm's bottom line by reducing expenses, and transferring some of that benefit to the clients. It can reduce the expenses whether other firms have it or not. This is made even clearer when realizing that these are existing technologies / methodologies, that someone else is already using them, so it is not possible for the firm to have a monopoly on them. In that case it is more important for the firm to own it, than that the others do not. When it is out in the world, it is important to not to be the one without it.

So, enabling, non-proprietary technologies / methodologies, are easily shared by firms, without any reduction in their strategic value. That makes the cooperation between firms in the same sector and market easier. Therefore, the firms perceive the advantage of sharing (actually not sharing, but duplicating) the results while sharing the expenses.

The author argues that, it is more than sharing or reducing resources consumption. For small to medium firms some capabilities for the Implementation Process, or the ability to identify the external party having them, are so expansive, that for them if they were alone the Implementation

Project would never happen. Gaining access to production knowledge in large firms via the special connections of a specific manager in the UA, would not have been possible at all for some of the firms in the UA. For them, the Implementation Process would have been impossible, not just expensive, without the UA.

The Model

The model is a simple one. It is comprised of firms coming together, under governmental (local / regional / national) guidance and with its financial support, in order to access Innovation from sources external to the firms. The organized group optionally formed as a legal entity, acts through subcontractors to perform search, definition, identification and selection jointly, and implementation individually (see Figure 3 - Operation Environment and players). The scope of action is defined by the firms in workgroups, allowing flexibility on the one hand and focus on the other, dictating the focus of the search, identification and selection to the subcontractors (see Figure 4) The results of which, the subcontractors' supply to the UA members. The implementation is up to the individual member and can be assisted by the UA or the subcontractors.

It is important to note that the workgroups are critical, as they allow large numbers of firms to work jointly, while focusing on topics that each member finds important. A more rigid structure, demanding all to be involved in all topics, would make it difficult for SMEs to participate, especially, as participation is costly.

The external subcontractors, can be research organizations, companies or consultants, and may be a combination of some or all of them. The important characteristics of the model is that the entity is expert and external to the members, and that while it may in character be a research entity, this is not its function in the model. The external subcontractor performs the important tasks that the individual firms lack the expertise

Figure 3. Operation environment and players

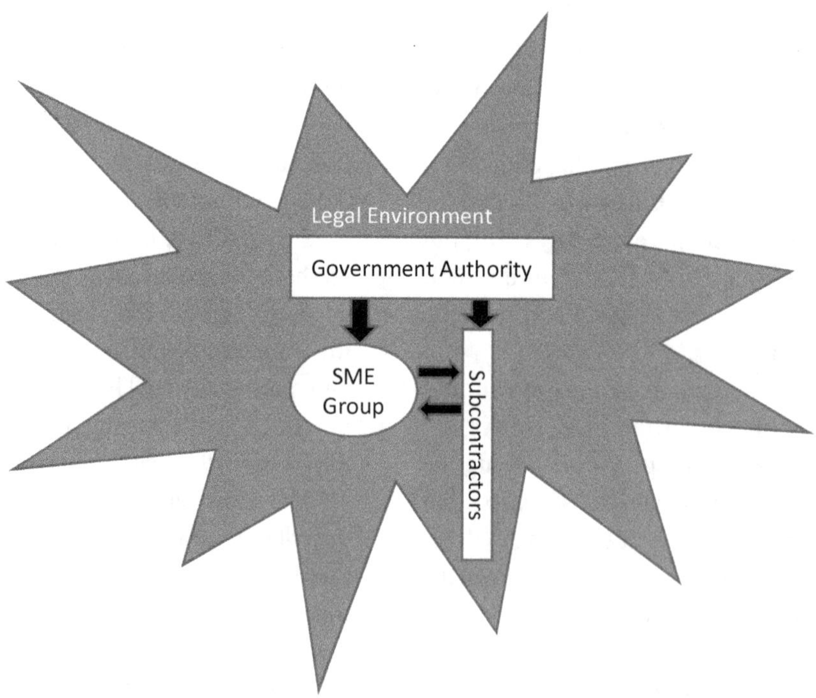

Figure 4. Input output and funding flows

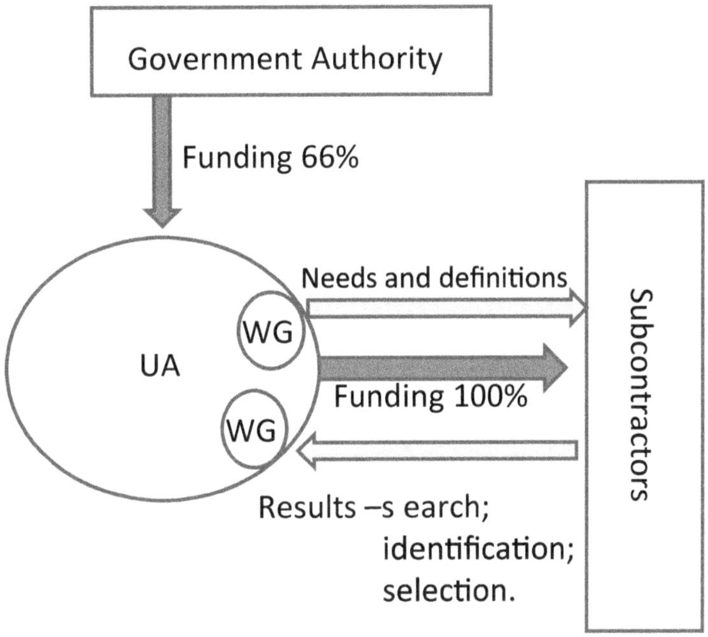

to do or manage on their own. It should deliver by selecting identified solutions to the problems presented by the workgroups, and not create any new knowledge. Being a research entity may help in the identification and selection, but is not mandatory.

The legal aspects of the model need to be discussed here. The easiest and shortest one is the legal form or the legal bonding of the partners involved. There are two crucial points that need to be considered for this aspect in the model. Since the model involves government intervention including supply of public finances, the legal form has to be able to withstand the scrutiny and approval of formal public check-points. Issues such as anti-trust laws, competition, public funding limitations and other will influence the format selected for the coming together of the firms. Whether it is a firm, an association, or just a legal agreement for cooperation – will be determined according to national / local legal regulations and demands. The model has little to add to that. The UA in Israel from the reasons mentioned here selects to establish the entity in the form of a nonprofit association.

The second important point is the IPR. As mentioned earlier the UA does not deal with critical knowledge, but rather enabling technologies and methodologies. However, it is important for the firms coming in to know what they are getting in regarding knowledge. For the firms it is sometimes more important to know that they can protect critical knowledge that they already own, than to get new knowledge that is enabling knowledge. It is important to the firms to know that they are protected, that they can get the knowledge they need, and last but not least that the knowledge they get is legally available to them. Regarding the protection of the knowledge it is important to recall that as in the case study above, in government sponsored programs there may be a requirement to allow all local firms wishing to participate to do so, without reservations due to the existence in the group of competitors. The possibility of competitors in the market being partners in the consortium increases the IPR importance and especially the need of the firms to protect their proprietary knowledge.

The model requires that the firms interact under clear IPR, which discourages opportunism, and provides security for the core knowledge of the firm, as well as for the usability of the knowledge the firms hope to gain.

The ability to gain innovation from other firms in the sector and from third parties is dependent upon the creation of relational assets (Dyer & Nobeoka, 2000). The more traditional and conservative the sector is, the more important that ability. The involvement of a government entity helps create these relational assets (Etzkowitz & Leydersdorrf, 1997). The government body assures the firms that any opportunism exercised by other firms, may have long term and lasting negative impact. It thus reduces the perceived danger of opportunism (existent or not), and allows the firms to create the relational assets, meet, arrange information days and visits etc. at a more relaxed atmosphere.

The funding flow of the support and involvement of the firms in the UA is important for the understanding of the model. Figure 4 describes the support given to the SMEs and the contribution demanded from them. Without the firms contribution it is hard for the UA management to assess which topics are indeed important.

Improving the Attractiveness of Israeli Subcontractors: Case Description

General

In order to better understand the establishment and function of the UA presented in the case study below, some general aspects common to all UAs under the program need to be presented.

The rational of the program in line with the view of Chesbrough, H. (2003) regarding Open Innovation, is that innovation can be achieved through the adoption of existing technologies and methodologies, but new to the firm, from the outside. There is no need to re-develop existing technologies. However, firms need to search for, define, select and implement the required technologies. Therefore, in addition to programs supporting the developments of new technologies and as a complementary measure, the Magnet Authority added the Users Association for Advanced Technologies program. The approach of learning from existing methodologies and technologies is based on search, identification and implementation than research and implementation. This is expressed also in the cases of Technology Transfer among UA members, and as mentioned by Dr. Ilan Peled[2], in the fact that normally the technologies dealt with in the UA are enabling technologies and not core technologies. That fact also assists in reducing the tension related to sharing the technology among the UA members.

The UA operates through workgroups. The membership in the UA is by involvement in at least one workgroup. The fees of participation are also part of the budget of the workgroup. The Magnet Authority supports the UA by supplying 66% of the approved budget in the form of a grant – not a loan (see Figure- 4). The rest of the budget comes from the firms in the form of participation fee in the specific task groups budgets, as explained below. The link of the workgroups to the operation and budget is also evident in the approval process. First a general work-plan and relevant general budget are approved for three years as a framework for the activity. Following that, a yearly detailed work-plan and detailed budget are approved. When the time comes a similar process is followed regarding the extension of the UA for another period – up to additional three years. The work plans do not include funds for the creation of new knowledge but rather for search identification selection and implementation of existing knowledge. The external knowledge can come from various sources ranging from research organization, to consultants. The justification for the form of grant is that the knowledge supplied is not directly related to the development of a specific product or process, but is of a generic type. In most cases, the funds are used for sub-contractors to perform the tasks of the UA workgroups on behalf of the firms, and therefore, are directed to the sub-contractors and not the firms themselves.

The UA rational assumes (Iltam, 2009) that there are barriers for the adoption of new technologies and the implementation of new methodologies. These barriers are problematic especially for SMEs, who lack the resources to sustain the process. The lack of resources is not a total lack, but as mentioned above in most cases the entities are dealing with enabling and not core technologies. The enabling technologies are always relegated to the end of the receiving line regarding resource allocation as they are not deemed crucial. In the case of lack of resources they can be neglected for so long as to go beyond a threshold allowing an easy return and innovation adoption. Therefore, SMEs require more assistance in this specific aspect.

Detailed

This project was managed under ILTAM as the UA. Therefore, before explaining further about the project it is important to explain about the structure of ILTAM.

According to the Web site (Iltam, 2010) the motto of the UA is:" Learning together from the experience of others". The above mentioned goals and regulations show a clear fit with Open Innovation as defined in the literature section. It is up to the firms to embrace as innovation knowledge that comes from the outside. The association is required to disseminate the knowledge among its members, but not to create it, so by definition this is external knowledge and it is definitely innovative to the firms, if not to the sector.

As can be learned from the purposes defined for the UA (ILTAM, 2010) the organization, established as a non-profit association by knowledge intense firms for promoting the competitiveness of its members, by the sharing of existing knowledge, accelerating the implementation of advanced technologies and reduction of development and production costs while increasing the quality of the products.

The membership in ILTAM is institutional; the workers of any member firm are eligible to participate in the events ILTAM organizes and its other activities. ILTAM is following its Motto (by Peter Drucker (Iltam, 2010):

The only competitive advantage of the developed countries is the supply of knowledge workers.

The organization practically intends to reduce the barriers for the introduction of new technologies and to accelerate their rate of implementation.

The methodology used is based on three distinct stages:

1. Creation of a knowledge center constructed from the delegates of interested firms, searching for technological innovations around the world. The association's main role at this stage is to help in bringing in the technologies from abroad, create the brainstorming forums and the mechanisms for cross fertilization among the group members to lower the entry barrier and reduce the risk of implementing new technologies;
2. Distribution of the knowledge in wider forums, including the insights generated at the earlier stage. The main purpose of the association at this stage is to overcome the barriers for widespread usage of the new methodologies and technologies, while adapting them for different firms; and
3. Widespread dissemination to the industry and assistance in implementation to the entire industry.

The method of operation of the association focuses on the assimilation of the knowledge using discussion forums including firms and experts interacting indirectly and informally. There are also wider forums beyond the workgroups. The association also organizes professional seminars bringing experts from abroad to present new technologies to the members, using the connections with international associations (e.g. IEEE). The Association participates in international conferences to import knowledge, and combines academic researchers in the workgroups.

As mentioned above, the association works via workgroups. The workgroups are groups made of 7-15 representatives from interested member firms. They are defined as experts and lead the work within the group. The purpose of the workgroups is to increase the cooperation and sharing of knowledge between the members. The groups also include members of the Academia whose research is relevant to the focus of the group. Groups are formed whenever a new technology or methodology that may have long term implications is identified. The group has to get the approval of the presidency of the association in order to be formed. The work groups are formed for a period of three years.

Once approved the group starts raising awareness among the members by arranging for international seminars and inviting foreign experts. Once the members are aware and join the group, a work-plan is prepared and topics for handling are selected. The work follows the activities mentioned above.

Administratively each group has a list of goals and deliverables. There are the annual meetings (4-6 per year) discussing topics that are relevant to the goals of the group. Additional meetings include 1-2 yearly international conference bringing foreign experts to present relevant topics to the members; and 2 technical meetings with experts presenting relevant topics – group members participating. In order to help facilitate the interaction among the members and with the external world, each group also has a discussion forum where topics can be posted and discussed with people from the outside.

Project Inception

Using the workgroups to identify needs the UA discovered that a major hurdle to the usage of Israeli subcontractors for OEMs was not cost. Cost has an important role, but it seems that there were other reasons, such as timely delivery, management and understanding of customers' needs that caused OEMs in Israel to seek subcontractors abroad. Based on that understanding the program was developed within ILTAM and later proposed to Magnet as a project, approved and executed. The project was run according to magnet regulations (Iltam, 2010) while ILTAM provided the administrative support (Mr. Moshe Salem[3]) and the technical management was done by Dr. Mark Shternheim assisted by Mr. Moshe Nezer both with extensive experience in manufacturing issues. The management structure was very lean, Mr. Salem supplied also via ILTAM accounting and legal services, and the technical management team activity was estimated in total as 2/3 of full time single position.

The plan started as a pilot and was later enlarged based on the results to include direct mentoring for the participating firms, regarding theme selection, translation of themes to an executable work plan and following up progress. Monthly full membership meetings served to share the experience of the members and brought an added value by allowing cross learning of the members from each other. Another factor contributing to the success of the UA was the regular meetings with the Manager of the Magnet Authority Dr Ilan Peled, which included visits to the member firms, joint meetings with the management, which increased transparency and assured the members felt they were watched and that the specific firm success was important to the authority. That encouraged the firms to stick to the goals and timetables and to dedicate the management attention required for attaining the goals.

In 2004, a three years pilot plan called "Improving the attractiveness of production in-land" was launched, including subcontractors and client firms. In February 2007 an extension was approved for two more years and 7 additional subcontractors joined. No new clients joined, as by this point the clients' needs, were well defined, and only additional contractors that could benefit from the work were added. Over the entire plan, 13 subcontractors and 7 client firms participated.

The vision was to bring Israeli high-tech system local manufacturing to a competitive level compared with Asia, EU and the Americas.

UA Topics

The UA activity started by forming the workgroups. The rational was that the basis for future success of the Israeli High-Tech sector as a sector leading the Israeli economy would be based on keeping the manufacture base local. That is to say that the small to medium local subcontractors for manufacture are critical to the sectors success. In order for the local manufacturers to compete with production in the Far East and other places they had to improve their efficiency and effectiveness of the operation to the highest level possible. In order to do that the operation management had to be brought up to the level of the leading producers locally. It was felt that a manufacturing infrastructure based on reliable subcontractors is important for the OEM from the early stages of start-up and onwards. It was realized that the complexity of the products the competition in the world and the worsening economic situation in the world compel the subcontractors to excel in operations in order to survive. It was also realized (Iltam, 2009) that the improvement in operation included methodologies from different disciplines and that made the process more complex. It was also felt that while the subcontractors may have had such difficulties there were international firms present in Israel, with the required expertise, and that there were people with the capabilities to identify and help implement the methodologies – one such person was elected as the technical manager of the UA. Therefore, it was assumed that bringing together the experts and the sub-contractors, as well as the OEM (clients) and subcontractors would help demonstrate the necessity of improvement, bring an incentive to the subcontractors to improve, and help identify the key problem areas.

The topics selected included adopting of operation management techniques, software and internal control, supply management, quality control etc. The technologies, software and methodologies were not developed anew but rather identified and adopted. The UA focused on the implementation within the firm of the results. The involvement of high level management in each of the firms, the commitment of the CEO and the high management were critical to the success and were monitored and encouraged during the projects.

UA Management Structure and Formation

The population of subcontractors for manufacturing, supporting the Israeli High-tech industry, include firms in system assembly, wiring, and metal for electronics. The recruitment of the firms was via manufacturers associations, national networks supported by the Ministry, and personal connections. The recruitment was done in parallel to large high-tech companies and they were asked to bring their subcontractors with them. The subcontractors were presented with the idea and those who showed interest as well as CEO commitment were recruited.

The firms were recruited into workgroups around topics that were identified with the clients as major problem areas of employing Israeli subcontractors. The subcontractors could choose which groups they wanted to join but had to join at least one and show the willingness to invest the additional funding required by the program for participation in the workgroup. That included at least a person year (240,000 New Israeli Shekels (NIS) about 60,000 USD)) and payment for information system application design (80,000 NIS).

Of that budget 66% was supplied as a grant by the authority and the rest had to be funded by the firm. Many of the companies invested more than that specific minimum.

Within the original approved framework, the workgroups prepared the work-plan and derived budget per year.. The plans were grouped into a joined plan with a consolidated budget and were presented to the Magnet Authority for approval. The UA had a steering committee comprised of representatives of the different firms, the technical manager of the UA and the general manager of Iltam. They performed visits to the firms to check up on progress, and if changes in the work-plan or budget were required for recommendation to the Magnet Authority. Additionally there were full assemblies every month on top of the visits. The visits and the full assembly meetings were considered important as they demanded the firms report on progress and so encouraged the firms to work in order to have progress to report on.

UA Results

The UA lasted for 5 years in two parts (3+2), with participation of 13 subcontractors and 7 OEMs. The firms performed 35 improvement projects held 58 full assembly meetings and 147 visits to the firm, as stated in the final 5 years report by the (Iltam, 2009). There were several cases where firms resigned after year one of their project and the main reason identified was that the project was not within the focus of the CEO, and was therefore, not progressing. There were some fears that the firms would not tend to cooperate among themselves, but that did not happen. In the second phase of the project only additional subcontractors were recruited as there was no more need to recruit additional OEM manufacturers. As shown in the final report of the UA (Iltam, 2009) the results were impressive (see Table 1).

Summary of the UA for Improving the Attractiveness of Local Manufacturing

the activity which was run as a UA within the frame work ILTAM, in order to improve the attractiveness of Israeli subcontractors to Israeli OEM manufacturers was deemed very successful. Managing to build up on exiting expertise, knowledge, technologies and methodologies, the firms participating (from the electronics sector) managed to improve on several critical criteria and to increase their attractiveness to the OEM.

The UA worked via workgroups that focused on the identification of the required technologies / methodologies and their implementation within the firms, supported technically by a small and efficient management team at the UA, and the internal commitment of the firms themselves. The activity was further supported financially by the Magnet Authority. The innovation and improved competitiveness were based on applying existing knowledge and not on the manufacture of new

Table 1. Quantified results

Criterion	Improvement rate (min)	Improvement rate (max)
Overall production improvement	80-92%	91-98%
Improvement of initial yield	58-82%	80-92%
Improvement in station yield	65-85%	
Increase in inventory cycle	3 to 6	2.8 to 7
Reduction in errors discovered at client	45%	50%
Reduction in dead inventory	20%	45%

Printed by kind permission from Iltam.

knowledge. It pooled the resources of the firms and allowed them to benefit from their accumulated larger size, in cost reduction and efficiency while not deducting from the specific benefit each one received.

Open Innovation via Collaboration: The Benefits and Drawbacks for SMEs

As explained in the theoretical background and in the section above, Open Innovation via collaboration has its advantages. Same as Collaborative Research has advantages over single firm research (Porath, 2010; Porath, 2008; Kamien, Muller & Zang, 1992; Kamien & Zang, 2000). However, the benefits are even clearer when dealing with non-research models as discussed above. The advantages of participating in a joined effort and reducing resource utilization outweigh the potential dangers from collaboration, as the model presented above focuses on the implementation of existing knowledge, knowledge that is available for other players in the market (therefore in most cases non-proprietary). However, the research finds that the following considerations should be brought to attention:

- The cooperation may allow activities, or access to knowledge / expertise that the firms on their own would never be able to achieve;
- The costs are shared, but the benefits for the individual firm stay the same;
- In the case of existing in the market, non-proprietary technologies / methodologies the biggest risk is being left out and missing on the technologies;
- The risk is reduced by avoiding core technologies;
- Government involvement reduces opportunism and helps create relational assets, making the collaboration more effective and efficient;
- If constructed correctly, the collaboration can drive firms into action (not be the one without progress to report); and
- Budget participation and management commitment are important to assure that the firms will follow the Implementation Process fully and achieve the results and not stop in the middle, wasting resources and creating dissatisfaction.

The case study above presented some tangible measurable benefits that were achieved through a UA embodying the collaboration in innovation discussed here. On a more general way, the project helped maintain economic growth, and the function of a sector and an economic engine driving the economy towards growth. This is important when considering the rational for government involvement in such activities.

FUTURE RESEARCH DIRECTIONS

The model presented here is based on collaboration in Open Innovation, and discussions are based on theoretical background and some substantiation from a single case study. Of course as this case study is a specific example, and further research should be quantified research. Future research should be focused on searching to explain and increase the understanding of the different elements in the collaboration. Statistical analysis and the understanding of each component would contribute to the theoretical explanation of the model, and also to the understanding regarding other theories, such as Organizational theory.

Further qualitative and quantitative research would focus on internal routines and collaboration with external sources, redefining the boarders of the firm (Penrose, 1968); Business Strategy theory dealing with competition / cooperation aspects (Brown & Eisenhardt, 1998); and Network theory regarding cooperation in networks and learning in networks and more. All these should be researched with the practical view of implementing Open Innovation in a more effective way, for the benefit of the SMEs. It is the practicality of implementation that should in direct future research in the field.

CONCLUSION

This chapter in a book dealing with the theoretical side of Open Innovation presented a theoretical model based on an on-going national program. While the program characteristics are influenced by the local environment in which it operates such as: legal, economic, and culture, the model focuses on the more universal aspects, removing the local influences.

The chapter summed some of the theoretical background regarding Open Innovation and related topics and theories. It used a short case study to illustrate some of the relevant mechanisms and some of the relevant points as the basis for further developing the model. The main conclusions are related to the relevance of enabling technologies to sharing the costs, relational assets creation and duplicated benefits.

The characteristics of the model and the relevance to other theories were discussed along with recommended future research venue, especially through statistical validation.

Among the characteristics discussed were the role of the collaboration for Open Innovation as an accompanying program to R&D support; the role of the government body; the advantage for SMEs in coming together for such purposes and the local aspects of the cooperation.

ACKNOWLEDGMENT

This chapter could not have been written without the help, and advice, and guidance of Dr Ilan Peled General Manager of the Magnet Authority in the CSO of the Ministry of Industry Trade and Employment, who helped direct the author towards the interesting aspects of the topic and the right people to talk to. One such person was Mr.. Moshe Salem the General Manager of Iltam who patiently explained to finer aspects of Iltam and graciously allowed the author to benefit from the use of Iltam reports and material. Author's warm thanks to both of them.

REFERENCES

Argawal, A., & Henderson, R. (2001). *Putting patents in context: Exploring knowledge transfer from MIT*. Cambridge, MA: National Bureau of Economic Research (Production series).

Audretsch, D. B., & Feldman, M. P. (1996). R&D spillovers and the geography of innovation and production. *The American Economic Review*, *86*(3), 630–640.

Brown, S. L., & Eisenhardt, K. M. (1998). *Competing on the edge*. Boston, MA: Harvard Business School Press.

Cabral, L. M. B. (2000). R&D cooperation and product market competition. *International Journal of Industrial Organization*, *18*(7), 1033–1047. doi:10.1016/S0167-7187(99)00011-9

D'Aspremont, C., & Jacquemin, A. (1988). Cooperative and non-cooperative R&D in duopoly with spillover. *The American Economic Review*, *78*(5), 1133–1137.

Doz, Y. L., Olk, P. M., & Ring, P. S. (2000). Formation processes of R&D consortia: Which path to take? Where does it lead? *Strategic Management Journal*, *21*(3), 239–266. doi:10.1002/(SICI)1097-0266(200003)21:3<239::AID-SMJ97>3.0.CO;2-K

Dyer, J. H., & Nobeoka, K. (2000). Creating and managing a high performance knowledge-sharing network: The Toyota case. *Strategic Management Journal*, *21*(3), 345–367. doi:10.1002/(SICI)1097-0266(200003)21:3<345::AID-SMJ96>3.0.CO;2-N

Etzkowitz, H. (1999). Bridging the gap: The evolution of industry-university links in the United States. In Branscomb, L. M., Kodama, F., & Florida, R. (Eds.), *Industrializing technology: University-industry linkages in Japan and the United States* (pp. 203–233). Cambridge, MA: The MIT Press.

Etzkowitz, H., & Leydersdorrf, Z. (1997). *Universities in the global knowledge economy: The triple helix university-industry-government relations*. London, UK: Cassell Academic.

EU. (2003). *Communication from the Commission: Investing in research: An action plan for Europe*. COM (2003) 226 final/2.

EU. (2006). Report of the CREST expert group on: Encourage the reform of public research centres and universities, in particular to promote transfer of knowledge to society and industry. *Final report, March.*

EU. (n.d.). *FP7-Description*. Retrieved December 7, 2009, from http://cordis.europa.eu/fp7/

EU-IPR. (n.d.). *Helpdesk*. Retrieved August 10, 2010, from http://www.ipr-helpdesk.org/home.html

EU-SME. (n.d.). *Website*. Retrieved August 10, 2010, from http://ec.europa.eu/enterprise/policies/sme/index_en.htm on August 10th 2010

FP7. (n.d.). *Budget*. Retrieved August 10, 2010, from http://cordis.europa.eu/fp7/budget_en.html

Fontana, R., Geuna, A., & Matt, M. (2006). Factors affecting university – industry R&D projects: The importance of searching, screening and signalling. *Research Policy*, *35*(2), 309–323. doi:10.1016/j.respol.2005.12.001

Harison, E., & Koski, H. (2010). Applying open innovation in business strategies: Evidence from Finnish software firms. *Research Policy*, *39*(3), 351–359. doi:10.1016/j.respol.2010.01.008

ILTAM. (2009Submitted to). *Five years report and recommendations for continued operation.* Magnet Authority.

ILTAM. (n.d.). *Website*. Retrieved June 19, 2010, from http://www.iltam.org/info.php?id=sd_about

Kamien, M. I., Muller, E., & Zang, I. (1992). Research joint ventures and R&D cartels. *The American Economic Review*, *82*(5), 1293–1306.

Kamien, M. I., & Zang, I. (2000). Meet me halfway: Research joint ventures and absorptive capacity. *International Journal of Industrial Organization*, *18*(7), 995–1012. doi:10.1016/S0167-7187(00)00054-0

Katsoulacos, Y., & Ulph, D. (1998). Endogenous spillovers and the performance of research joint ventures. *The Journal of Industrial Economics*, *46*(3), 333–357. doi:10.1111/1467-6451.00075

Katz, M. (1986). An analysis of cooperative research and development. *The Rand Journal of Economics*, *17*(4), 527–543. doi:10.2307/2555479

Kirk, R. E. (1995). *Experimental design: Procedures for the behavioral sciences* (3rd ed.). New York, NY: Brooks/Cole.

Kline, J. J. (2000). Research joint ventures and the cost paradox. *International Journal of Industrial Organization*, *18*, 1049–1065. doi:10.1016/S0167-7187(99)00002-8

Laredo, P. (1998). The networks promoted by the framework program and the questions they raise about its formulation and implementation. *Research Policy*, *27*(6), 589–598. doi:10.1016/S0048-7333(98)00055-9

Laursen, K., & Salter, A. (2004). Searching high and low: What types of firms use universities as a source of innovation? *Research Policy*, *33*(8), 1201–1215. doi:10.1016/j.respol.2004.07.004

Luukkonen, T. (1998). The difficulties in assessing the impact of EU framework programmes. *Research Policy, 27*(6), 599–610. doi:10.1016/S0048-7333(98)00058-4

Luukkonen, T. (2000). Additionality of EU framework programmes. *Research Policy, 29*(6), 711–724. doi:10.1016/S0048-7333(99)00041-4

Ministry of Industry Trade and Employment. (2001). *General manager ordinance 8.5*. Retrieved June 7th, 2010, from www.magnet.org.il

Miyagiwa, K., & Ohno, Y. (2002). Uncertainty, spillovers, and cooperative R&D. *International Journal of Industrial Organization, 20*(6), 855–876. doi:10.1016/S0167-7187(01)00079-0

Olk, P. (1991). *The formation process of research and development consortia*. Unpublished doctoral dissertation, University of Pennsylvania, Pittsburg.

Parkhe, A. (1993). Strategic alliance structuring: A game theoretic and transactions cost examination of interfirm cooperation. *Academy of Management Journal, 36*(4), 794–829. doi:10.2307/256759

Pastor, M., & Sandons, J. (2002). Research joint ventures vs. cross licensing agreements: An agency approach. *International Journal of Industrial Organization, 20*(2), 215–249. doi:10.1016/S0167-7187(00)00070-9

Penrose, E. T. (1968). *The theory of the growth of the firm*. Oxford, UK: Basil Blackwell.

Porath, A. (2006). Legal incentives to economic growth – The case of the Israeli R&D law. *Managerial Law, 48*(3), 322–333. doi:10.1108/03090550610674671

Porath, A. (2008). Preparing country and firms for slow down in the economy—The structured research joint venture (SRJV) option. *Review of International Comparative Management, 9*(5), 68–79.

Porath, A. (2010). *Collaborative Research in Management, inside-out*. London, UK: Sage Publications Ltd.

Ring, P. S., & Van de Ven, A. (1994). Structuring cooperative relationships between organizations. *Strategic Management Journal, 13*(7), 483–498. doi:10.1002/smj.4250130702

Rosenberg, N. (1990). Why do firms do basic research (with their own money)? *Research Policy, 19*(2), 165–174. doi:10.1016/0048-7333(90)90046-9

Spender, J. C. (1996). Making knowledge the basis of a dynamic theory of the firm. *Strategic Management Journal, 17*(Winter Special Issue), 45–62.

Stern, I., & Pozner, J. E. (2007). *Organizational size, performance and frequency-base imitation: A test of competing hypotheses*. ISC Conference, Jerusalem, Israel, 2007.

Thursby, M., Thursby, J., & Mukherjee, S. (2005). *Are there real effects of licensing on academic research? A life cycle view*. NBER Working Paper No. 11497.

Volanski, A. (2005). *Academy in a changing environment: Higher education policy of Israel 1952-2004*. Tel Aviv, Israel: Hakibutz Hameuchad Publishing House Ltd.

Walter, J., Lechner, C., & Kellermanns, F. W. (2007). Knowledge transfer between and within alliance partners: Private versus collective benefits of social capital. *Journal of Business Research, 60*(7), 698–710. doi:10.1016/j.jbusres.2007.01.026

Yun, K. L., Park, Y. S., & Ahn, B. H. (2000). Spillover, competition and better R&D organization. *The Japanese Economic Review, 51*(3), 448–461. doi:10.1111/1468-5876.00162

Zhou, K. Z., & Wu, F. (2010). Technological capability, strategic flexibility, and product innovation. *Strategic Management Journal*, *31*(5), 547–561.

ADDITIONAL READING

A special issue of Strategic Management Journal (2000) dealing with networks. *Strategic Management Journal*, 21(3).

Benfratello, L., & Sembenelli, A. (2002). Research joint venture and firm level performance. *Research Policy*, *31*(4), 493–507. doi:10.1016/S0048-7333(01)00122-6

Brown, S. L., & Eisenhardt, K. M. (1998). *Competing on the edge*. Boston, Massachusetts, USA: Harvard Business School Press.

Chesbrough, H. (2003). *Open Innovation: The new imperative for creating and profiting from technology*. Boston: Harvard Business School Press.

Commission, E. U. (2007). *FP7 in Brief. How to get involved in the EU 7th Framework Programme for research. European Communities*. Luxemburg.

D'Aspremont, C., & Jacquemin, A. (1988). Cooperative and non-cooperative R&D in duopoly with spillover. *The American Economic Review*, *78*(5), 1133–1137.

Das, T. K., & Teng, B. S. (2002). Sustaining Strategic Alliances: Options and Guideline. *Journal of General Management*, *22*(4), 49–64.

Doz, Y. L., Olk, P. M., & Ring, P. S. (2000). Formation processes of R&D consortia: Which path to take? Where does it lead? *Strategic Management Journal*, *21*(3), 239–266. doi:10.1002/(SICI)1097-0266(200003)21:3<239::AID-SMJ97>3.0.CO;2-K

Etzkowitz, H. (1999). Bridging the gap: The evolution of Industry-University Links in the United States. In Branscomb, L. M., Kodama, F., & Florida, R. (Eds.), *Industrializing technology: University-Industry linkages in Japan and the United States* (pp. 203–233). Cambridge, MA: The MIT Press.

Etzkowitz, H., & Leydersdorrf, Z. (1997). *Universities in the global knowledge economy: The triple helix University-Industry-Government relations*. London, UK: Cassell Academic.

EU. (2003). Communication from the Commission; Investing in research: an action plan for Europe. *COM (2003) 226 final/2*.

EU. (2006). Report of the CREST expert group on: Encourage the reform of public research centres and universities, in particular to promote transfer of knowledge to society and industry. *Final report, March.*

EU FP7- description downloaded from http://cordis.europa.eu/fp7/ on December 7th 2009.

Fontana, R., Geuna, A., & Matt, M. (2006). Factors affecting university – industry R&D projects: The importance of searching, screening and signalling. *Research Policy*, *35*(2), 309–323. doi:10.1016/j.respol.2005.12.001

Hegdoorn, J., Link, A. N., & Vonortas, N. S. (2000). Research partnerships. *Research Policy*, *29*(4), 567–586. doi:10.1016/S0048-7333(99)00090-6

Iltam, (2009). Five Years report and recommendations for continued operation. Submitted to Magnet Authority.

ILTAM website Retrieved June 19, 2010 from http://www.iltam.org/info.php?id=sd_about

Laredo, P. (1998). The networks promoted by the framework program and the questions they raise about its formulation and implementation. *Research Policy*, *27*(6), 589–598. doi:10.1016/S0048-7333(98)00055-9

Laursen, K., & Salter, A. (2004). Searching high and low: what types of firms use universities as a source of innovation? *Research Policy*, *33*(8), 1201–1215. doi:10.1016/j.respol.2004.07.004

Miyagiwa, K., & Ohno, Y. (2002). Uncertainty, spillovers, and cooperative R&D. *International Journal of Industrial Organization*, *20*(6), 855–876. doi:10.1016/S0167-7187(01)00079-0

Olk, P. (1991). *The Formation Process of Research and Development Consortia*. Unpublished doctoral dissertation, University of Pennsylvania, Pittsburg.

Pastor, M., & Sandons, J. (2002). Research joint ventures vs. cross licensing agreements: An agency approach. *International Journal of Industrial Organization*, *20*(2), 215–249. doi:10.1016/S0167-7187(00)00070-9

Penrose, E. T. (1968). *The theory of the growth of the firm*. Oxford, UK: Basil Blackwell.

Porath, A. (2006). Legal incentives to economic growth – the case of the Israeli R&D Law. *Managerial Law*, *48*(3), 322–333. doi:10.1108/03090550610674671

Porath, A. (2008). Preparing country and firms for slow down in the economy—The structured research joint venture (SRJV) option. *Review of International Comparative Management*, *9*(5), 68–79.

Rosenberg, N. (1990). Why do Firms do Basic Research (With Their Own Money)? *Research Policy*, *XIX*(2), 165–174. doi:10.1016/0048-7333(90)90046-9

Steinle, C., & Schiele, H. (2002). When do industries cluster? A proposal on how to assess an industry's propensity to concentrate at a single region or nation. *Research Policy*, *31*(6), 849–858. doi:10.1016/S0048-7333(01)00151-2

Walter, J., Lechner, C., & Kellermanns, F. W. (2007). Knowledge transfer between and within alliance partners: Private versus collective benefits of social capital. *Journal of Business Research*, *60*(7), 698–710. doi:10.1016/j.jbusres.2007.01.026

Yun, K. L., Park, Y. S., & Ahn, B. H. (2000). Spillover, competition and better R&D organization. *The Japanese Economic Review*, *51*(3), 448–461. doi:10.1111/1468-5876.00162

KEY TERMS AND DEFINITIONS

Intellectual Property Rights (IPR): The rights, such as ownership, right to use, license sell or rent of intellectual property. Within a consortium (either legal entity or by agreement) the IPR set also the decision process regarding ownership and partnering in the knowledge as well as the above.

Implementation Barrier: The economic, social, managerial or legal barriers that prevent or make it hard to implement innovation, new technologies, and new methodologies in existing organizations. In some sectors the barriers are more difficult to overcome due to specific circumstances.

Implementation Process/Project: The activities leading to a lasting integration of a technology or methodology into the capabilities of an organization. In this chapter the reference is to the process of incorporating an existing technology or methodology into the routines of a firm, in order to improve its competitiveness. The process is performed via a project, and should be completed by the end of the project. The project may be internal or as in the cases referred to in this chapter, assisted by external entities.

Innovation Ingredient: A specific knowledge can be a technology, methodology or any other form, which represents a progress regarding the existing situation in the organization. The knowledge may be new, but it can be just new to the organization seeking to adopt it. The knowledge maybe protected, licensed or free, but it has to be available for the organization to implement.

Inertia: The tendency of organizations to oppose and slow down changes. This tendency makes adaptation to external changes by internal changes difficult, and tends to preserve the internal routines. While this is not the physics inertia but an organizational characteristic, it does require energy and time to overcome.

Chief Scientist Office: The CSO is a unit in the Israeli ministry of Industry Trade and Labor charged with the encouragement of the civil industrial R&D in Israel. The CSO budget has two sources – the share of the ministry budget, and income from return on revenues from the firms it supported. The CSO operates several programs (Magnet, Technology Incubators and more) and is also responsible for the Israeli industrial participation in bi-national and international programs (FP7, Eureka and more). The CSO funding for generic type of research is in the form of a grant (small portion of support) and for product oriented R&D in the form of conditional loan, which in time generates for it the second source of income.

Magnet Authority: A subunit within the CSO in charge of generic research programs. It was officially formed in 1994 with a single program – the R&D consortia (very similar in character to the EU FP Integrated Projects of FP6), but soon additional programs were added to its responsibility. The UA was added in 1995 and others later on. The unit is very small and in order to reduce overhead it retains services from the CSO – Evaluators and other support services. For more info please use http://www.magnet.org.il/default.asp?id=26.

FP7: The 7th Framework Programme for Research operated by the EU Commission and managed by DG Research in order to advance the policies of the EU, among the most important one is the Lisbon Agenda aimed at making the EU a "Knowledge Based Society". FP7 has a total budget of over 50 billion Euros for the years 2007-2013, and it supports research based or research linked activities. For more information please see the websites in the bibliography and in the recommended reading above.

Triple Helix Concept: A theoretical concept published by Etzkowitz and Leydersdorrf, (1997), describing the inter-relationships between Industry-Academia-Government based on the Double Helix of the DNA, as three strands entwined and influencing each other.

ENDNOTES

1. Free translation from Hebrew by the author.
2. In email correspondence following an interview in June 2010.
3. Based also on personal interview with Mr. Moshe Salem held on June 29th 2010 at his office and supplementary Q&A by mail and phone.

Chapter 8
Innovating Elite Undergraduate Education through Quality Continuous Improvement: A Learning Enterprise's eTransformation Perspective

Kam Hou Vat
University of Macau, Macau

ABSTRACT

This chapter investigates a mechanism of organizational innovation serving to make sense of a maturing university community involving educational executives, academic staff, and students in the occasion of a new campus development, starting in the year 2009 and realizing in the year 2013, under the auspices of a national policy benefiting the long-term development of higher education in the Macau Special Administrative Region (Macau SAR) of China. It is understood that the university as a public institution should not be operated like a business enterprise, running on profit-making initiatives; yet, without the enterprising context, the transformation of the existing infrastructure could hardly be innovated effectively, especially regarding the productivity of its staff, both academic and administrative. As a university with a staff count of less than 1000 and a history of close to thirty years, the University of Macau (UM) is ready to steward an elite undergraduate education marked by a quality learning experience that could become her branding value in the immediate future. The question is how innovatively UM could scale up in this opportunistic growth to excel for the local community. This case study is aimed to investigate from the perspective of a learning enterprise, a reflective way of forward thinking to record the author's observation and interpretation of what is entailed in this process of upbringing a relatively young university in this age-old city, Macau, famous for its rich heritage of East (Chinese) meeting West (Europeans – Portuguese). Of specific interest is the proper context of open innovation in university governance for organizational transformation. The chapter examines the accountability framework for undergraduate curriculum reform and by treating the electronic transformation (e-transformation) as one of the open innovation strategies, the chapter explores the e-transformation of the university environment, based on holistic concerns of the campus community.

DOI: 10.4018/978-1-61350-519-9.ch008

The challenge is to identify the organizational context of innovation, which lies in the realm of electronic governance (e-governance), referring mainly to the decisions that define expectations, enable empowerment, and verify performance of the systems in support of community engagement and shared responsibilities in campus development, providing a sense-making perspective to interpret what is entailed in the organizational innovation of the university in this precious occasion of campus relocation. In practice, the lessons learned behind the e-transformation of the learning enterprise should benefit all walks of our local community, including the community of the small and medium sized enterprises (SMEs).

INTRODUCTION

In their book, *Engines of Innovation: The Entrepreneurial University in the Twenty-First Century*, Thorp and Goldstein (2010) makes an interesting point: "What is most exciting about innovation is that it begins with a problem; the bigger the problem, the more significant the innovation needed. From the very beginning of our work, we have been convinced that for research universities to realize their full potential, they must attack the world's biggest problems, and this notion is increasingly being embraced throughout academia (p.2)." In this light, the University of Macau (UM), in its 10-year development plan from 2008 to 2017, aims to become a leading university in the country (China), and beyond. It aims to produce all-round graduates with the potential to become pillars of the Macau Special Administrative Region (SAR) serving beyond Macau in the mainland of China as well as in the global community.

This chapter is going to examine what quality concerns UM has put into her radar screen of continuous improvement to support the elite undergraduate education ideal, from the perspective of a learning enterprise, receiving financial support from both the SAR Government, and from donors in the local and global community. In particular, the study is looking into an innovation opportunity granted by the Central Government of China to the SAR Government and thus to the University in early 2009, to develop, to govern and to manage a new green-field campus (to be completed by 2013) twenty times the size of UM's current campus in an island (Hengqin), just next to Macau, located in the mainland of China, to be connected to the SAR through a dedicated channel for convenient transportation (see more from http://www.umac.mo/new_campus_project/).

In the future campus, the buildings there will house those who will create well-paying, sustainable jobs, and provide Macau (and China) with a sustainable competitive advantage in the global economy. And such buildings are most likely our research labs, classrooms of the future, and innovation centers, where big ideas are hatched and subsequently translated into reality. Indeed, as Thorp and Goldstein (2010, p.2) characterizes: "The superhighways that will change the world are not asphalt but electronic, and rather than connecting the world's great cities, the new superhighways provide access to most of the world's knowledge." Thereby, UM sets off globally for a higher purpose in organizational innovation to connect the world's knowledge in her new Hengqin campus.

Understandably, as this study involves the investigation of some innovative learning culture through e-transformation of organizational development in the learning enterprise, such as teaching, learning, research and assessment, in the context of SMEs, the same learning experience could actually be rendered as transferable means to accomplish such business initiatives as project management in research and development, as well as product innovation. Many an SME in advanced economies have been learning to use information and communication technology (ICT) to innovate and transform their processes, products, services and business models, significantly improving productivity and competitiveness (Hanna, 2010). Indeed, such an initiative echoes clearly the need of an open innovation strategy in today's knowledge society, being more a norm than an exception in this Internet age.

BACKGROUND

In January 2009, following a visit to Macau paid by Mr. Xi Jinping, Vice President of the People's Republic of China, UM was informed of the possibility of building her Hengqin campus and that this campus would serve as a pilot project being part of a broader collaboration between Guangdong, a province in southern China, and Macau, in the joint effort to develop Hengqin, currently an island with green fields just next to Macau, and to be substantially developed in the next 10 to 15 years through a nationally coordinated scheme of city planning. On June 27, 2009, the Standing Committee of China's National People's Congress (NPC) officially approved the University of Macau's Hengqin Campus Proposal, submitted by the Macau SAR Government in March 2009.

This approval delegates authority to the SAR Government to apply the SAR's legal and administrative systems to UM's Hengqin Campus for a duration of four decades, starting from right after the new campus has been completed and transferred to the SAR and renewable afterwards. Indirectly, this approval has increased the territory of Macau by one square kilometer, the size of UM's Hengqin campus. The new campus will offer an exciting academic environment and advanced teaching, learning, and research facilities to enable UM to pursue her objectives of offering quality undergraduate education and promoting innovation in scientific research, with the goal to develop UM into a world-class university.

Although there is no precedent to follow with such a joyful surprise, UM has expressed the commitment to preserve her traditions and values, and to strive toward excellence in the scholarship of teaching, in the scholarship of discovery (research), in the scholarship of application (services), in the scholarship of integration (interdisciplinary bridging), and in the scholarship of leadership. To uphold such an institutional commitment, UM has also initiated efforts in fund-raising from the local and global community to joint hands in accomplishing the mission of elite undergraduate education. Thereby, besides financial support from the Macau SAR, UM as a learning enterprise is running an organization transformation scheme similar to that of a social enterprise, absorbing resources from the society to provide educational services to the community.

Meanwhile, with the advent of Internet technologies, UM is also exploring emerging technologies and examining operational examples of what an electronic institution for education might look like. With the latest round of electronic transformation (e-transformation) efforts, UM is looking for ways to improve administrative efficiency, implement faster access and response to information, enhance skills and knowledge development, and improve student learning experience. Most importantly, UM is also tackling issues in redesigning existing organizational frameworks, techniques for overcoming barriers to, and successfully implementing change.

A Clear and Vital Vision

The undergraduate experience at UM is led by a clear and vital vision, elite undergraduate education, which is embraced as UM's identification of the late Ernest Boyer's (1990) exposition of the four domains of scholarship plus the scholarship of leadership (Kouzes & Posner, 2007) as perceived by the author.

- **Scholarship of teaching:** The scholarship of teaching is the development and improvement of pedagogical practices (Braxton, Luckey, & Helland, 2002). It is believed that effective teachers engage in scholarly teaching if they undertake assessment and evaluation to promote improvement in their own teaching practice. Indeed, scholarly teaching activity becomes scholarship of teaching when faculty members make their teaching public by opening it to review and critique by peers in their disciplines through publications and presentations.

- **Scholarship of discovery:** This is traditional research which UM considers as the scholarship of the creation of new knowledge. It requires creative and critical thought, research skills, publication in peer-reviewed journals and books, and presentations at disciplinary conferences.
- **Scholarship of application:** This form of scholarship involves the use of a scholar's disciplinary knowledge to address important individual, institutional, and societal problems. Scholars who engage in this type of work, also called the scholarship of engagement and outreach, must be able to solve problems of importance to diverse stakeholders such as policymakers, and community members, and to communicate effectively with their audiences in language understandable to persons without disciplinary expertise.
- **Scholarship of integration:** This form of scholarship makes connections within and among disciplines. When disciplinary and interdisciplinary knowledge is synthesized, interpreted, and connected, the work brings new insight to original research (Braxton, Luckey, & Helland, 2002). The products include policy papers, reflective essays, research translations, popular press publications, synthesis of the literature on a topic, and textbooks.
- **Scholarship of leadership:** This form of scholarship is meant to develop leaders for elite undergraduate education. The goals are to enable faculty to provide curriculum direction, improved teaching, educational research and development, as well as institutional leadership at all levels of undergraduate education. It is a program in the making (Gruppen, Frohna, Anderson, & Lowe, 2003), using a variety of educational methods and provides a broad curriculum in educational theory, assessment and evaluation, research design and methods, teaching-skills development, and educational leadership.

A Framework of Accountability for Elite Education

For more than a quarter of a century, the Association of American Colleges and Universities (AACU: www.aacu.org) has repeatedly called for the academy to take responsibility for assessing the quality of student learning in college. Indeed, thoughtful and forceful leadership from within the academy itself on both assessment and accountability is more essential today than ever in many a university campus around the globe. It is convinced that educational leaders within UM are already doing good work to define what academic excellence really means in today's world. Yet, it is vitally important that UM should build on this foundation to find improved ways to demonstrate achievement of academic excellence by students and institutions. Namely, UM must respond to calls for greater accountability in ways that strengthen as well as document the quality of student learning in college.

It is believed that such accountability efforts should focus on the knowledge, skills, and capacities that are most important for today's students. They also should take into account the best campus practices already developed to advance and assess these outcomes. Borrowing from AACU's vision developed in its various reports, *Integrity in the College Curriculum* (AAC, 1985, 1990), *Greater Expectations: A New Vision for Learning as a Nation Goes to College* (AACU, 2002), and *College*

Learning for the New Global Century (AACU, 2007), the University of Macau has incorporated a quality improvement context consistent with the three elements of concerns as follows (AACU, 2008, p.1):

- A clearly articulated, collective conception of the qualities of a college-educated person
- Intentionality and coherence in educational programs to cultivate those qualities
- Assessment to determine the extent to which the desired learning has been achieved

It is important for any university to enable her campus community to answer legitimate questions about what their students are learning in college, including clear aims and outcomes of higher learning. The lack of evidence on student learning outcomes is always damaging. Yet, assessing what students have learned in colleges and universities requires a sophisticated understanding both of context and of how knowledge and skills are to be used. This means that we must hold ourselves accountable for assessing students' best work, not generic skills and not introductory levels of learning, with full respect for the diversity of institutional missions and students' educational goals in the contemporary academy. In the context of UM, the elite undergraduate education model must be accountable for a set of key learning outcomes that all college graduates should achieve regardless of their field of study or choice of residential college. The important step is to establish clarity about the kinds of learning that make a difference for all college graduates over time: as thoughtful people, as participants in the community, and as responsible all-round citizens in society.

To ensure that every student experiences the intellectual and social benefits of a well-designed and stimulating undergraduate education, UM's version of liberal education (AACU, 2007), called General Education (GE) is charged with the mission to reflect the many social, economic, cultural, and technological changes that occur in the world over the years. GE at UM aims to be inclusive and to provide an empowering education to widely diverse students. In the 20th century, many came to contrast liberal education with professional education and to regard it as, by definition, not practical. But in today's knowledge economy, a good liberal education embraces science and new technologies, hands-on research, global knowledge, teamwork, cross-cultural learning, active engagement with the world beyond the academy, and a commitment to lifelong learning.

These forms of learning should provide a strong foundation for success in a dynamic world, and the outcomes are also essential as a foundation for civic participation and for a meaningful life. In contrast to liberal education, professional education, often narrow and situation-specific, is found to be insufficient for a world characterized by complexity, conflicting judgments, and accelerating change. In fact, students in technical fields need and deserve the complementary benefits of a liberal education to help them make sense of the social and environmental contexts in which they will use their skills, and to get themselves prepared for lifelong work rather than just an initial job. In this regard, new frameworks for educational accountability should focus on students' high level of achievement in the college outcomes that characterize a liberal education.

Indeed, creating a framework of accountability in support of a specific learning culture is not confined to higher education, but extensible to the SME business communities. Such a learning culture implies an environment where learning takes place each day, all day, fundamentally changing the way we think about how, what, and when we learn, and how we can apply learning to practice. Such a culture should facilitate key aspects of organizational learning: the modern business context and the importance of learning at every juncture; the organic and adaptive approaches organizational leaders can take to design enduring success; and the expanding role of individuals within organizations and the implications for business leaders, educators, technologists, and learners. Literally, the same framework of accountability should help identify the steps many SMEs must take to remain competitive for years to come, and it should explain how learning strategies applied to all aspects of every job can provide solid returns and substantial results.

Focusing on Key Educational Objectives

Unquestionably, the public has questions about the quality of education that colleges and universities are providing, and it deserves to know how well students are doing. According to AACU's (2008) recommendations, it is important for leaders of higher education to embrace a set of highly valued and widely affirmed educational goals, establish high standards for each such goal, and assess their achievement across the curriculum. In their 2007 report, *College Learning for the New Global Century*, AACU proposes as a framework for assessment and accountability, the following essential learning outcomes, quite consistent with UM's ideal of elite undergraduate education. There is emerging consensus across many professions, the business community, civic leadership, and the academy that these educational capabilities are valuable for work, citizenship, and a satisfying life (AACU, 2008, p.6 and pp. 15-19).

- *Knowledge of human cultures and the physical and natural world*

Such capabilities are acquired through study in the sciences and mathematics, social sciences, humanities, histories, languages, and the arts. They are focused by engagement with big questions, both contemporary and enduring.

- *Intellectual and practical skills, including*:
    1. Inquiry and analysis
    2. Critical and creative thinking
    3. Written and oral communication
    4. Quantitative literacy
    5. Information literacy
    6. Teamwork and problem solving

Such capabilities are practiced extensively, across the curriculum, in the context of progressively more challenging problems, projects, and standards for performance.

- *Personal and social responsibility, including*:
    1. Civic knowledge and engagement – local and global
    2. Intercultural knowledge and competence
    3. Ethical reasoning and action
    4. Foundations and skills for lifelong learning

Such capabilities are anchored through active involvement with diverse communities and real-world challenges.

- *Integrative and applied learning, including*:

Synthesis and advanced accomplishment across general and specialized studies.

Such capabilities are demonstrated through the application of knowledge, skills, and responsibilities to new settings and complex problems.

It should be noted that this set of essential learning outcomes was developed by AACU through a multi-year dialogue with hundreds of colleges and universities about needed goals for student learning; analysis of a long series of recommendations and reports from the business community; and analysis of the accreditation requirements for engineering, business, nursing, and teacher education. The findings are documented in various publications of the Association of American Colleges and Universities (AACU): *Greater Expectations: A New Vision for Learning as a Nation Goes to College* (2002), *Taking Responsibility for Quality of the Baccalaureate Degree* (2004), and *Liberal Education Outcomes: A Preliminary Report on Achievement in College* (2005).

Affirmatively, these outcomes for student learning are not arbitrarily chosen. They are valuable outcomes because they prepare students to bring knowledge, experience, and reflective judgment to the complexity of the contemporary world. They are important for a globally engaged democracy, for a dynamic, innovation-fueled economy, and for the development of individual capability (AACU, 2007, p.6). They give graduates a strong foundation to deal with issues that are challenging, unscripted, and often vigorously contested. They teach students to find and evaluate evidence and to take into account competing perspectives as they form judgments about significant questions. They help develop both a respect for the value of human diversity and a set of internal values that serve as a compass in an era of accelerating change. The outcomes recommended, which can be achieved in different ways, across highly diverse institutional contexts and fields of study, are practically compatible with UM's ideal of elite undergraduate education, especially for her General Education program. It is indeed important that such forms of learning, essential for all students, must be fostered across the entire educational experience.

Cultivating and Assessing Essential Learning Outcomes

According to the AACU report, *Our Students' Best work: A Framework for Accountability Worthy of Our Mission* (2008, 2nd edition, p.8), a comprehensive accountability and assessment framework should include the following elements in order to cultivate and assess students' essential learning outcomes as mentioned in the previous section:

- **Orientation:** This should be provided for the student during the first year about the institution's expectations for important learning outcomes, benchmark assessments of each student's demonstrated accomplishment, and expected further progress in relation to these outcomes.
- **A plan of study:** In coordination with the student's academic adviser, this plan should transparently connect the expected outcomes to the student's choice of activities, courses, and major field(s) of study.
- **Milestone assessments:** As students progress in their studies, such assessments should be tied to key outcomes with timely feedback to the student and his or her adviser. Oftentimes, these assessments should be aligned between two-year (first two years) and four-year (first and second two years) study program so that successful transfers are possible, and they can be compiled in an electronic portfolio that demonstrates each student's progress.
- **Capstone or culminating experiences:** The requirements of capstone experiences allow students to actively demonstrate their cumulative accomplishments in college education. The capstone – which can be the centerpiece of an electronic portfolio – is a critical element of this framework because it provides a designated place in the regular curriculum where students do their best work. It should be conceived as both a culminating integrative experience and as the centerpiece of the effort to assess sophisticated learning.

Indeed, even though the outcomes characteristic of college education (in the context of UM, elite undergraduate education), can be described generically, they must be cultivated and assessed in context. Namely, whatever the field of study, a student's progress in achieving related education outcomes ought to be assessed periodically from the initial to the final year, in both general education and the chosen major field(s) of study. The 2007 National Survey of Student Engagement in the US (NSSE, 2007) reports that 61 percent of college seniors are expected to complete a capstone or culminating experience of some kind. Typically, capstones are completed in the student's major field; however, many institutions also require capstone experiences in general education. Meanwhile, many other institutions and programs already require students to compile portfolios of their work as a requirement for graduation. Increasingly, many universities across the United States and Canada are putting such portfolios online to both encompass the multiple modes through which achievement can be demonstrated and to address the mobility of students across programs and study and institutions.

In fact, capstone projects and electronic portfolios (e-portfolios) provide promising anchors for a meaningful approach to educational accountability. They provide contexts in which student work can be assessed for the crosscutting outcomes of general education as described earlier, and for conceptual knowledge and skills appropriate to the students' selected major(s). The primary focus will be a fuller reading of the available evidence on student's cumulative achievement of the key learning outcomes. The best evidence should come from assessment of students' authentic and complex performances in the context of their most advanced studies: research projects, community service projects, portfolios of student work, supervised internships, and many others.

Higher education institutions should use a common framework of college education to report externally on students' level of achievement. Yet, they should also help the public understand that the standards for advanced accomplishment take different forms in different fields. The key accountability question to ask of universities is whether they currently expect all their students to undertake complex projects and capstone assignments that are assessed for advanced learning outcomes, be they for general education or for professional education.

According to the suggestions from AACU (2008, p.9), the set of accountability questions that should be asked of every higher education institution, include the following:

1. Are all students expected to produce advanced, culminating work?
2. Is this culminating work assessed for broad general education outcomes as well as knowledge relevant to the specific field?
3. Have standards or rubrics been established and made public for what is expected at this advanced level in each program?
4. Are examples of this advanced work and the related standards regularly peer reviewed in the context of accreditation?
5. Have milestone assessments been established that prepare students to meet advanced standards and, where relevant, to plan for successful transfer from one institution to another?
6. Does the curriculum effectively prepare students to meet the standards that will be expected in milestone and culminating assessments?

Summarizing and Reporting Results to the Public

To respond with integrity to calls for greater accountability, and to demonstrate the level of student achievement of key learning outcomes, many a higher education institution is willing to adopt the recommendations offered by AACU as stated in the 2008 report, *Our Students' Best Work: A Framework for Accountability Worthy of our Mission* (pp.12-14). The rationale is simple:

namely, it is not enough for an institution to assess its students in ways that are grounded in the curriculum; colleges and universities also must provide useful knowledge to the public about goals, standards, accountability practices, and the quality of student learning. Common rubrics will be needed to summarize levels of student achievement across different academic fields and institutions and for particular groups of students. In this regard, the VALUE rubric released by AACU in 2007 upon launching its initiative called *Valid Assessment of Learning in Undergraduate Education* (VALUE: http://www.aacu.org/value/index.cfm) is designed to explore the development of assessment rubrics for the broad range of essential learning outcomes stated earlier, and to articulate expectations and criteria for student achievement from admission through advanced levels of performance. The VALUE project helps test the application of rubrics to actual student work completed in their courses and co-curricular activities.

By evaluating work that students produce in their regular courses and co-curricular activities, the VALUE approach to assessment is found more directly useful to faculty as they seek to improve courses and programs, and assess student growth and development over time. Meanwhile, by gathering and disseminating work through electronic portfolios (e-portfolios), the same student work can be assessed at course, program, and institutional levels. More conveniently, with e-portfolios, student work from on-and-off campus and from all the institutions a student may have attended can also be included, thus presenting an overview of student accomplishment over time and space. A summary report, including results from many students majoring in different disciplines, to an accreditation body, to a government official, or to the general public can be prepared that summarize the data across the institution. When such data are made available, each campus can take steps to engage faculty and students in interpreting the meaning and implications of assessment outcomes. Faculty members should use the findings as a basis for discussion and as a catalyst for needed changes in educational programs. Elaborated below are AACU's (2008, pp.12-14) ten recommendations for creating a new accountability framework for student learning. It is found to be very instrumental to an institutional effort in assessing student learning:

1. *Make liberal education the new standard of excellence for all students.* Liberal education (as a broader embodiment of general education; http://www.aacu.org/leap/what_is_liberal_education.cfm) should become the new standard of excellence for all students, whatever their major or anticipated career. The set of essential learning outcomes as described earlier in the chapter, are important indicators of what students need to accomplish as citizens, as workers in a particular profession or field, and as thoughtful, creative, responsible human beings.
2. *Articulate locally owned goals for student learning outcomes.* For higher education to be accountable for intended learning outcomes individual institutions (and systems) must translate these outcomes into goals and language that are meaningful in local contexts. Goals for student accomplishment should be developed and articulated in dialogues that include both faculty members and members of the wider community. Clarity about essential learning outcomes is the foundation of both a robust educational program and of a responsible framework. To meet the highest possible standard of excellence, campus goals for student learning should also be challenging and constantly evaluated.
3. *Set standards in each goal area for basic, proficient, and advanced performance.* One of the important benefits of having clearly described goals and levels of achievement

is that students themselves will begin to understand the standards for quality in different fields and become more capable of assessing their own learning. Likewise, levels of performance should be specified in concrete and detailed ways so that student work can be reviewed and judged similarly by different individuals.

4. *Develop clear and complementary responsibilities between general education and departmental and other programs for essential learning outcomes.* It is important for students to experience their general education and major as integrated and coherent. Although specific responsibility may be assigned respectively to general education and professional education to support essential learning goals, it is best if there is discussion and understanding among faculty and other professionals about what is expected in all parts of the curriculum and co-curriculum. This is especially important if students are encouraged to make connections among courses and programs, and if advanced courses intentionally build on prior work. Needless to say, it does little good to agree on valued goals for student learning if responsibility for cultivating them is not well planned.
5. *Charge departments and programs with responsibility for the level and quality of students' most advanced work.* Once the goals for student learning have been articulated at the campus level, they should be translated into program-specific goals for student accomplishment. In particular, such goals should be set for general education so that they respect the aims, design and character of specific study programs on campus. In addition, student learning goals should be articulated within the context of academic majors. Put it more simply, each program should translate campus-wide goals for liberal/general education into goals appropriate to the field. Programs also need to articulate field-specific goals for their students, communicating how and why these standards contribute to effective accomplishment in that field.
6. *Create milestone assessments across the curriculum.* Assessment of student progress over time requires that campuses distinguish among advanced, proficient, basic, and below basic levels in relation to specific goals. Students should be taught to gauge their progress against high expectations for their most advanced work. It is intended that no student should learn for the first time about shortfalls in meeting proficiency standards at the point of graduation. Thereby, assessments of student progress in achieving goals should be built into the ongoing curriculum and embedded in designated courses or assignments in general education, departmental majors, and other programs such as the residential college systems.
7. *Set clear expectations for culminating work performed at a high level of accomplishment.* Each department should identify expected proficiency standards and culminating work – encompassing liberal/general education and major-specific outcomes – that will both cultivate advanced knowledge and skill and demonstrate students' cumulative learning. Culminating work may include research projects, supervised internships, capstone courses, public performances, licensure or other validated tests in a field, and/or cumulative e-portfolios providing examples of student achievement in relation to specific goals.
8. *Provide periodic external review and validation of assessment practices and standards.* There should be periodic evaluation by external reviewers of the goals, the proficiency standards, and work samples submitted by students to meet standards. Such external reviews provide validation of both the goals

and standards. A representative sample of student performances in different fields will provide sufficient evidence for external feedback.

9. *Make assessment findings part of a campus-wide commitment to inquiry and educational improvement.* Accountability efforts should be part of a continuous improvement scheme, engaging students with the best quality in achieving the related educational goals. Each campus and program should review the quality and level of students' best work, and seek ways to ensure that the educational program provides repeated opportunities for students to practice and reach expected levels of learning. Campus reward systems should incorporate the importance of faculty members' intellectual and professional leadership in both assessment and educational improvement.
10. *Increase public visibility and transparency of learning goals and student achievement levels.* According to AACU (2008, p.14), each college and university should make public on its Web site the following: a) general and program goals for student learning; b) proficiency expectations for rating levels of student achievement in relation to these goals; c) a description of the kinds and range of performances that are used in assessing student progress (with links to different programs); and d) a report on student achievement levels (e.g., advanced, proficient, basic, and below basic) in relation to each goal.

Assessing for Student Learning as UM's Core Mission

One of the important missions to support elite undergraduate education at the University of Macau (UM) is to support assessment for student learning, especially, to evangelize the practice of outcomes-based assessment (OBA) (Driscoll & Wood, 2007; Allen, 2004; Huba & Freed, 2000) in preparation for assessing various academic programs, as part of UM's institutional research efforts. Assessment can answer important questions, questions about the learning of individual students, the effectiveness of a single course or program, or even the entire institution. The basic assumption is that evidence of student learning should be used for multiple levels of assessment, in which precision in formulating the questions of interest should help pinpoint the level of analysis, determine the appropriate methods, and guide data sampling, aggregation, interpretation, and use. In the context of OBA, the best evidence of learning comes from direct observation of student work rather than from an inventory (say, list of courses completed) or summary of self reports.

According to the AACU report (Miller & Leskes, 2005), *Levels of Assessment: From the Student to the Institution*, course-embedded assignments provide the most valid evidence for all levels of analysis because they are closely aligned with faculty expectations and with the teaching-learning process. The ways of sampling, aggregating, and grouping the evidence for analysis will depend on the original questions posed. The questions will also determine how the data are interpreted to produce action. Internally, faculty members and support staff should accomplish aggregation by describing standards, translating them into consistent scoring scales, and anonymously applying the resulting rubrics to the evidence at hand. Such a process does not assign a grade to an individual student but rather attempts to understand better the learning process and how to improve its effectiveness (Leskes & Wright, 2005; Leskes & Miller, 2005). Elaborated below are representative investigation of the different levels of assessment which have been found conducive to the development of various assessment programs for elite undergraduate education initiatives.

Assessing Student Learning Within Courses

The basic questions of concerns would probe what individual students are learning and how well they are meeting the goals of a course, be they related to disciplinary content or to using transferable intellectual and practical skills. Table 1 shows the setup of our investigation.

Assessing Student Learning Across Courses

The questions of interest should probe what and how well individual students are learning during the progression of a particular program (say, the major or general education) or over their years at college. Table 2 shows the setup of our investigation.

Assessing Courses

The questions of interest, both formative and summative, should address the achievements of an entire class or the effectiveness of individual or multiple-section courses (Table 3).

Assessing Programs

The questions of interest mostly summative in nature (with only a small portion of formative elements) should guide assessment of our academic programs in place, such as general education or disciplinary majors (Table 4).

Table 1. Setup of the investigation within courses

Items of Interest	Details Expressed
Typical assessment questions	• Is the student learning as expected? • Has the student's work improved over the semester? • How well has the student achieved the learning outcomes set for the course? • What are the student's strengths and weaknesses? • How well is the instructor communicating with and engaging the student?
Sources of evidence	All student work embedded in the course (say, quizzes and exams, papers, projects, presentations, and portfolios) can provide evidence. This is the level of assessment at which instructor-assigned grades typically provide feedback to students about their progress and success.
Aggregation of data	Aggregation is often sequential as evidence is collected for each student during the course to track individual learning and improvement. Typically a final course grade holistically sums up a semester of learning.
Data uses	• As formative and/or summative feedback to students so they can understand their progress in the course and ways to improve learning • For feedback to the course instructor on how well he or she is communicating with and motivating each student, which can shape subsequent lessons and assignments within the course.
Responsibilities	• Individual students are responsible for the effort they exert, the quality of their work, and meeting the instructor's expectations. They are more likely to fulfill these responsibilities when consistently informed of learning goals and academic norms. By teaching students how to conduct self- and peer-assessments, the professor can improve student understanding of the learning process. • Individual instructors are responsible for setting expectations and making them transparent to students. As educators, our professional responsibility extends to the quality of our own teaching and to monitoring how well the pedagogical methods we employ assist students in learning. While the holistic assignment of grades (an A, B, or F) is one way to evaluate student work, such grades represent averaged estimates of overall quality and communicate little to students about their strengths, weaknesses, or ways to improve. A better way to aid learning is through analytical assessments, which can be as simple as written comments on student papers or as structured as the use of a detailed rubric for an assignment; such analysis can reveal precisely which concepts a student finds challenging.

Table 2. Setup of the investigation across courses

Items of Interest	Details Expressed
Typical assessment questions	• Has the student's work improved and/or met standards during the program or since admission to college? • How well has the student achieved the disciplinary outcomes of the major program? • How well has the student achieved the general learning outcomes of the institution across four years?
Sources of evidence	• Embedded work in individual courses, for example quizzes and exams, papers, projects, presentations • Portfolios that assemble samples of the student's work in a number of courses • Capstone experiences or projects • Student self-reflection on the learning process • Relevant externally developed exams (say, licensure) NB: Typical grades can provide some holistic feedback to the student but are difficult to interpret across courses except at very broad levels (like GPA) or to disaggregate into learning outcomes, such as how the student has learned to communicate orally.
Aggregation of data	Given appropriate formats and data, students can aggregate evidence of their own learning (say, of a particular skill such as writing) across courses, programs, or their entire time in college to track improvement. Conventionally, departments aggregate an individual's grades across courses when they require, for example, that their majors must maintain a minimum GPA of 2.5 in disciplinary courses.
Data uses	• As formative and/or summative feedback to students so they can understand their progress over time and ways to improve learning • For feedback to program faculty on how well individual students are achieving the goals and outcomes.
Responsibilities	• Individual students are responsible for the quality of their work and for gathering evidence of their learning. They are also responsible for integrating their learning over time and across courses. • Collectively faculty members share the responsibility for clarifying goals and outcomes and providing rubrics for student self assessment. • Individually faculty members are responsible for objectively assessing the assembled work samples or the test results and providing both holistic and analytic feedback to the student.

Assessing the Institution

The questions of interest for institution-level assessment are often designed for internal improvement purpose, but they are also intended to meet external accountability demands. It is not uncommon that results from internal improvement questions can also serve the external accountability purpose (Table 5).

THE LEARNING ENTERPRISE INNOVATION CHALLENGE

The modern university comprises many activities. It teaches students, conducts research for various clients, provides opportunities for self-actualization, engages in economic development, stimulates social change, and in Macau, even renders mass charity opportunity, such as participating in the annual charity walk, raising fund for the needy. It is not uncommon that the organization of the contemporary university compares in scale and complexity with many major corporations; hence, it could also be called a learning enterprise.

In fact, the very complexity of the university has made substantive involvement in the broader governance of the learning enterprise an innovation challenge for all participants. The increased complexity, financial pressures, and accountability of universities demanded by the government, the media, and the public at large require stronger management than in the past (Balderston, 1995). Yet, as Duderstadt (2004) has pointed out, as

Table 3. Assessing courses

Items of Interest	Details Expressed
Typical assessment questions	• How well is the class collectively achieving the course's intended learning outcomes and objectives (at any point, at the end)? How well is the class collectively achieving general or transferable learning outcomes and objectives? • Are the assignments helping students achieve the expected level of knowledge or skills? • How well are students prepared for the following courses in the sequence? • Is the course level appropriately targeted for the ability or abilities of the students when they begin? • With what degree of consistency do different sections of a course achieve similar outcomes? • How well is the course fulfilling its purpose in a larger curriculum?
Sources of evidence	• Embedded assignments of students in the course: papers, exams, projects, journals, portfolios • Externally or commercially developed tests, as long as they are well aligned with the teaching and learning of the course • Course portfolios constructed by the instructor that include syllabi, expectations, and examples of student work • For multi-section courses, common assignments that provide evidence across sections NB At the course level, traditional student grades are unlikely to provide sufficiently detailed insights to answer the questions unless tightly tied to explicit analytical standards and scoring rubrics.
Aggregation of data	• To assess individual courses: Sampling the work of all students in a course can reveal how well the course content and assignments are helping students achieve the expected outcomes. • To assess multiple-section courses: Common assignments across sections (or common requirements such as a student or course portfolio) can be sampled, averaged, compared, discussed, or otherwise reviewed by the faculty involved and/or by departments or committees to ensure consistency across sections. • To assess both individual courses and multi-section courses: Student portfolios and end-of-course reflections can provide evidence of both cognitive and affective learning outcomes aggregated at the level of the individual students.
Data uses	• For formative feedback so instructors can improve learning • For summative feedback to inform planning for the future by an instructor or a course committee • to support cross-sectional analysis of how consistently multi-section courses are achieving important learning outcomes or the purposes of the course in a sequence
Responsibilities	Instructors and committees are responsible for setting expectations for the course, establishing common standards for multi-section courses, understanding how the course fits into a coherent pathway of learning, and using analysis of the evidence to improve teaching and course design.

universities develop the administrative staffs, policies, and procedures to handle such issues, they also have created a thicket of paperwork, regulations, and bureaucracy that could possibly weaken the authority and attractiveness of academic leadership.

Broad participation in university governance is often hampered by bureaucratic policies, procedures, and practices, as well as by the anarchy of committee and consensus decision making. Put it simply, the academic tradition of extensive consultation, debate, and consensus building before any substantive decision is made or action taken poses a particular challenge in this regard because such a process is frequently incapable of keeping pace with the profound changes occurring in higher education, let alone the creative anarchy arising from a faculty culture that prizes individual freedom.

In a rapidly changing society, many university presidents believe, although they are understandably discreet in stating, that one of their greatest challenges is protecting their institutions from the deteriorating quality of their student learning. This is an accountability issue not to be ignored, but also squarely exhausting to defend, as clarified

Table 4. Assessing programs

Items of Interest	Details Expressed
Typical assessment questions	• Do the program-related courses, individually and collectively, contribute to its outcomes as planned? • How well does the program fulfill its purposes in the entire curriculum? • How well do the program sub-requirements (distributive requirements in general education) contribute to the overall purposes? • Does the program design resonate with its expected outcomes? • Are the courses organized in a coherent manner to allow for cumulative learning? • Does the program advance institution-wide goals as planned?
Sources of evidence	Direct evidence of student learning from many sources can contribute to program-level assessment: • assignments from individual courses • student portfolios built over the program duration • entering student tests or assignments • capstone projects • results of common assignments • commercial tests Selected assignments from other programs can be re-scored (given a second reading) by program faculty (say, to assess the general education program's success in developing such institution-wide goals as communication, quantitative literacy, critical thinking, or ethical responsibility). Given the number of potential data sources and the amount of evidence that could be amassed, careful planning is needed to identify the important points for sampling and analysis. Such program assessment may likely involve several sources of evidence gathered at the point of entry, a midpoint, and at the end of the program. Each point data is particularly valuable as a summative indicator of how well the program, taken as a whole, is achieving its goals. Individual student grades are not informative at this level.
Aggregation of data	Course-level assessments of the courses in a program can be analyzed individually or collectively to reveal whether program goals are being achieved; sampling might be prudent in a large program. Information about the sub-requirements in a program (say, distribution areas) can be aggregated to the program level (say, general education). Sampling of student portfolios considered excellent, average, and sub-par can vividly portray growth in student performance from beginning to the end of a program. Disaggregated data can reveal how sub-groups of students are succeeding in the program. Some external, commercially available assessments can be compared to norms (say, the Major Field Tests from ETS).
Data uses	• to confirm the purpose of the program (say, its place in the entire curriculum or connection to mission) • to check alignment of program design with program outcomes • to discern how well the program, from its beginning to end, fosters cumulative learning of the desired outcomes • to discover how well the program as a whole enables students to achieve end-point levels of competence for all program outcomes • to identify superfluous and/or missing curricular and co-curricular elements in the program
Responsibilities	Responsibility largely rests on the program faculty, collectively and individually. Collectively, the faculty assumes responsibility for the entire program achieving its and relevant institution-wide goals and outcomes. Individual instructors are responsible for advancing the program and institutional goals embedded in their courses. Faculty members cooperate in establishing program standards and scoring rubrics for the quality of work expected.

by the extensive amount of work demonstrated in (Miller & Leskes, 2005). In Macau, there has been a pronounced shift in recent years toward a greater emphasis on oversight and public accountability for student learning. This is particularly the case with the governing board (university council) of the University of Macau, as the first public university in the SAR.

Table 5. Assessing the institution

Items of Interest	Details Expressed
Typical assessment questions	• What do the institution's educational programs add up to in terms of student learning? • How well are the institution's goals and outcomes for student learning being achieved? • How much have our students learned over their college years? • How well does the institution educate students for the complexities of the twenty-first century? • What evidence is there that the institution is fulfilling its educational mission? • How can institutional effectiveness be demonstrated authentically to external stakeholders?
Sources of evidence	A significant body of evidence from multiple sources will be required to answer institution-level questions. Documentation of how well students are meeting institution-wide goals and outcomes requires a clear statement of these learning expectations. The picture of student learning will be based primarily on summarized data from program assessments, supplemented by results from appropriate exams (say, those taken for graduate or professional school admissions, licensure, or certification). Sampling student work, both at the entry and graduation levels, can serve to answer value-added assessment questions. Some selected course-level assessments – particularly those from common experience courses such as a required core – could contribute to the institution-wide picture.
Aggregation of data	Much of the data will already have been aggregated when analyzed for institution-level assessment: aggregated by courses, by programs, or by student cohort. For example, sampled, aggregated, and summarized student achievement of the desired learning outcomes in a freshman general education course could be compared to sampled, aggregated, and summarized achievement in a senior course. Or an analysis of the cohort completing the Collegiate Learning Assessment instrument could reveal the level of critical thinking in the graduating class. Constructing both narrative and quantitative summaries of the stories from programs will shape the broad picture of teaching and learning at the institution. Disaggregated data can reveal how well sub-groups of students are succeeding.
Data uses	• to reveal what students know and can do when they graduate in order to guide the design of the institution's undergraduate program • to understand the value added by an institution's undergraduate program • to discover the interactions among various programs (say, general education and the majors), especially in how they help students achieve institution-wide learning goals • to guide and support decisions about resource allocation, faculty hiring, and professional development • to demonstrate to external stakeholders the institution's effectiveness in educating students
Responsibilities	The responsibility for institution-level assessment rests with administrators working in close collaboration with the faculty, student affairs professionals, and other campus staff members. Collaborative groups should design an ongoing comprehensive program of institutional assessment, use data to improve learning, keep student success a top priority, ensure linkages to strategic planning and resource allocation, and communicate with external groups.

Nonetheless, though faculty involvement in academic matters is essential for program quality and integrity, it is not surprising that faculty values are not necessarily well aligned with those required to manage a complex institution. For example, faculty members tend to be individualistic (independence), highly entrepreneurial lone rangers (freedom) rather than the team players required for institutional management (responsibility and accountability). They also tend to resist strong, visionary leadership and firmly defend their personal status quo. In short, it is frequently difficult to get faculty commitment to (or even interest in) broad institutional goals that are not congruent with personal goals, even though those areas of faculty governance are clearly within their unique competence such as curriculum development, student learning, academic values, and ethics. One of the key challenges to effective university governance, according to Duderstadt (2004), is to make certain that all of the constituencies of shared governance – governing boards, administration, and faculty – understand clearly their roles and responsibilities.

Meeting the Governance Challenge

The academic tradition of extensive consultation, debate, and consensus building before any substantive decision can be made or action taken is perceived to present a governance challenge in many a university today (Tierney, 2004; Marginson & Considine, 2000; Ramsden, 1998). This challenge somehow derives from a voluntary culture of the university which is compatible with the process of consultation, communication, and collaboration as visible from the organic campus community, rather than that of command, control, and communication so familiar in many a mechanistic corporation in the history of industrial development.

Nonetheless, this organic process of campus harmonization is hardly capable of keeping pace with the profound changes facing effective governance of the public university. Literally put, not everything could be improved by making it more democratic, as commented by Duderstadt (2004, p.149). Interestingly, this tension between the rationality of collectively organizing to achieve declared organization goals and the ultimate recalcitrance of human beings as members of organizations (Checkland & Holwell, 1998), was long referred to, by the writings of a pioneering sociologist, Ferdinand Tonnies in his major work *Gemeinschaft and Gesellschaft* (1887) (translated as *Community and Association* by Loomis (1955)), in which two types of society or organizations were depicted.

Making Sense of Gemeinschaft and Gesellschaft in University Governance

Gemeinschaft, referred to as the natural living community (in this light, the university), is the campus group with which the faculty are affiliated. It is a big family in which we are considered as brethren and sisters. On the other hand, *Gesellschaft*, referred to as the workspace in which men and women join in some complicated contracted sense – as when we choose to become employees of a company or a public sector organization. Although both *Gemeinschaft* and *Gesellschaft* serves as the basis of a typology for analysis rather than a description of actual organizations, the interesting point is that actual organizations in the world – so artificially designed and created – still continue to possess some of the natural characteristics of the family, thanks to the ultimate autonomy and unpredictability of human beings: no *Gesellschaft* is ever simply a rational machine whose members willingly combine together totally to pursue organizational goals.

In trying to make sense of a university capable of better governance of her institutional goals, we have to understand that our university, as a public institution, is a social unit, or collectivity, concerned with some collective action, and there are associated social practices which relate to this. To be a member of an organization is to have a contractual relationship with it, whether a legal contract of employment or a more complex psychological contract, or both. Members of an organization in a role structure are assumed to share an image of their organization in terms of its context, goals and objectives, its structures, processes and resources, and the measures of performance, which indicate whether or not the goals are being achieved. Within this framework, the members make their contributions to decision making in pursuit of objectives.

Nonetheless, it is the readiness of people, members and non-members (wider community in the case of a university) of the organization alike, which causes the organization to exist. People talk and act as if there were a collective entity, which could behave like a conscious being, with the ability to decide to do things and then make them happen. Oftentimes, there are conceptualizations based on the interests and agendas of individuals, or sub-groups within the organizations as well as the overall declared, public, official account of the organization. The existence of these different interests and agendas means that the organization as a whole, the collectivity, has constantly to seek accommodations among conflicting interests upon which action can be based. It cannot simply assume consensus. And, following Vickers (1965), the action is here expressed more richly as managing a changing set of relationships, rather than taking rational decisions to achieve goals.

Learning to Lead in Higher Education

A quick look at the remarkable pace of change required in the private sector (measured usually in months, not years) often suggests that universities must develop more capacity to move rapidly. This will require university leaders to occasionally make difficult decisions and take strong action without the traditional consensus-building process. Universities need to better define those areas in which the special competence of the faculty requires their consent (say, academic programs and policies), those areas in which faculty advice will be sought and considered but not considered authoritative (say, funding priorities), and those areas in which faculty does not need to be consulted at all (say, parking policy).

Indeed, the leadership of the university must be provided with the authority commensurate with its responsibilities (Keller, 2004). Academic leaders, whether at the level of department chairs (or heads), deans, vice-presidents, or event the president, should have the same degree of authority to take actions, to select leadership, and to take risks and move with deliberate speed, that their counterparts in the corporate world and government enjoy. As James Duderstadlt (2004, p.150) has commented, "The challenges and pace of change faced by the modern university no longer allow the luxury of consensus leadership, at least to the degree that building consensus means seeking the approval of all concerned communities before action is taken. Nor do our times allow the reactive nature of special interest politics to rigidly moor the university to an obsolete status quo, thwarting efforts to provide strategic leadership and direction." For example, as shared by Duderstadt, deans often expect the provost to offer additional resources to gain their cooperation on various institution-wide efforts. Obviously, this bribery culture is quite incompatible with the trend toward increasing decentralization of resources.

Hence, it is logical to expect that the leadership and management of universities increasingly will need to rely on lines of true authority just as their corporate counterparts do. Namely, presidents, executive officers, and deans will have to become comfortable with issuing clear orders or directives from time to time. Thereby, throughout the organization, subordinates will need to recognize that failure to execute these directives will likely have significant consequences, including possible removal from their positions, given all the consistent respect for faculty responsibility in academic programs and academic freedom. Unquestionably, collegiality will continue to be valued and honored; however, the modern university simply must accept a more realistic balance between responsibility and authority.

Recognizing and Developing Faculty Performance

To better balance authority and responsibility, as Ramsden (2000) puts it, leadership strength should be rebuilt at middle levels within the university. This may be done by involving some degree of restructuring in the organization of the university to better respond to its responsibilities, challenges, and opportunities (Menand & Gates, 2010). In this regard, there should be more effort made to identify the administration as a broader body than simply as the executive officers of the university, including deans, chairs, directors, and coordinators. It is important to get this broader group to be perceived as spokespersons for university objectives for such simple reason that the contemporary university is a public corporation that must be governed, led, and managed with competence and accountability to benefit its various stakeholders (Tuchman, 2009), including students, alumni, government body, and community supporters, such as fund-raisers. Literally, the complexity and importance of today's universities requires capable management and administration supported by trained professionals, with firm backgrounds in academic involvement. And it is convinced that such experience can be provided only by those who have toiled in the vineyards of teaching, research, and assessment as faculty members. The key to achieving adequate competence and accountability in university governance is to infuse in all of its components the perspectives of practicing faculty members, just as at the level of individual academic units through the use of various faculty committees to address crucial issues such as faculty hiring and promotion, student admission and performance, and curriculum and degree program development. In the context of university management, this can be achieved by appointing capable faculty members to lead administrative positions, provided that they are given the training necessary to manage complex organizations and functions in a competent and accountable fashion.

Safeguarding the Fundamental Values of Higher Education

In his book, *The Last Professors: The Corporate University and the Fate of the Humanities,* Frank Donoghue (2008), an Ohio State English professor, sees a troubling new conception of higher education emerging among administrators whose thinking reflects the bottom-line calculations of business executives, not the intellectual ideals of liberal arts scholars. Indeed, the history of the university, not just in the United States, is one of a social institution created and shaped by public needs, public policy, and public investment to serve a growing community, such as in Macau. Yet, according to Zemsky and Wegner (1998), in few places within the academy, either at the level of governing boards or in government higher education policy, does there appear to be a serious and sustained discussion of the fundamental values so essential to the nature and role of the university at a time when it is so desperately needed.

Duderstadt (2000) points out that as scholarship became increasingly professionalized and specialized, any coherent sense of the purposes and principles of a university became fragmented. Values such as tolerance, civility, and personal and social responsibility have been largely absent from the academic curriculum. Most of our undergraduates experience little discussion of values in their studies. Our graduate schools focus almost entirely on research training, with little attention given to professional ethics or even preparation for teaching careers, for that matter. Still, in any serious consideration of how our universities are governed and led, it is important always to begin with the basics, to reconsider carefully the key roles and values of the university that should be protected and preserved during a period of change. For example, how would an institution prioritize among roles such as educating the young (undergraduate education), preserving and transmitting our culture (libraries, visual and performing arts), basic research and scholarship, and serving as a responsible critic of society? In a similar vein, what are the most important values to safeguard? Clearly, academic freedom, openness to new ideas, a commitment to rigorous study, and an aspiration to achieving excellence would be on the list for most institutions. We all need to do our shares to preserve the core missions, characteristics, and the values. It is only through a concerted effort to understand the important traditions of the past, the challenges of the present, and the possibilities for the future can we enable our institutions to thrive during a time of such change.

Engineering the Entrepreneurial University

In the emerging digital economy of the 21st century (Tapscott, 1997), organizations co-evolve with their environments. With the rapid advances in networking technologies and the commercialization of the Internet, many a university around the globe, are actively reflecting on their organizational design and operating philosophy to devise new strategies, develop new capabilities, design new organizational structures, and deploy new governance models (Eifert & Puschel, 2004; Rivard, Aubert, Patry, Pare & Smith, 2004). Oftentimes, the governance model in a public sector organization could be considered as a prototype being creatively constructed and deconstructed over time. The disruptive effect of fast technological innovations and adoptions has literally redefined the essence of governing effectiveness, from sense and respond, to anticipate and lead. If we are planning to sense the changes and then respond, it might be already too late according to the enterprise experience (Lumby, 2001; Duke, 2002); rather, anticipating the coming changes and executing with speed to lead, is a fact of life today. Like an enterprise, it is a governing necessity to be equipped with the ability to respond quickly and effectively to change. This necessity has rendered the concept of electronic governance (e-governance) (Riley, 2003, 2001) indispensable, relating to the practical rendering over an electronic environment, such practices as consistent management, cohesive policies, responsive processes and decision-rights for different areas of responsibilities. Of particular interest is the electronic medium (say, an e-campus) to support e-governance activities, referring mainly to the decisions that define expectations, enable empowerment, or verify performance of the people or organizational units involved. Our framework of analysis here examines the university's electronic transformation (e-transformation) efforts to turn its bricks-and-mortar entity into its clicks-and-mortar counterpart, amidst the technological advances of the Internet and the World Wide Web.

Understanding E-Governance in University Setting

Today, the connotation behind the "e" transformation of governance is an important policy issue and surely one that will influence how the governing body and the governed community will interact in the coming decade (Oliver & Sanders, 2004). Clearly, the Internet and the information and communication technologies (ICTs) have an important part to play in the process. Still, as the "e" prefix becomes affixed to more and more aspects of governance, it is helpful to clarify the implications between governance and e-governance. Governance is often used to refer to the outcome of the interaction among the governing body, the public service, and the governed community, throughout the political process of policy development, program design, and service delivery. As for their "e" counterparts, Riley (2001) provides a useful description for their connotation: In its simplest sense, e-governance is about the use of emerging ICTs to facilitate the process of governing and public administration. In the context of a university environment, it is about providing the campus community with the ability to choose the manner in which they wish to interact with their governing body. And it is about the choices the governing body makes about how ICTs will be deployed to support the choices preferred by the campus community.

In practice, the transformation of governance structures and processes, enabled by technology, will be embedded in new modes of information exchange, perhaps, more integrated and distributive approaches to operations and service delivery, and a more open and participatory system of policy making. By capitalizing on the Internet revolution, the university authorities can create new channels of communication and new methods for participation via suitable e-governance measures.

It is foreseeable that the changing environment, coupled with faculty, student, and local community demands, would continue to encourage the university governing body's involvement in e-governance initiatives and related uses of ICTs.

Accepting E-Transformation Challenge in E-Governance Initiatives

As a result of the e-governance initiatives, many a campus is actively transforming their bricks-and-mortar entity into its clicks-and-mortar counterpart. We call such a transition effort the electronic transformation of the organization, or simply the e-transformation effort (Henderson & Venkatraman, 1993; Hoque, 2000). Obviously, such an effort requires an objective methodology (Vat, 2000, 2002), which must be instrumental to creating an efficient electronic organization (e-organization) model that could enable the organization to launch and to learn based on some innovative initiative, and then incorporate the lessons learned and launch again. Consequently, the organization transformation could be considered as the essence of a learning organization (Senge, 1990; Garvin, 1993; Duke, 2001; Shattock, 2003), implying its constant efforts to better itself for any coming challenges. An example in the context of our university e-governance is to consider the challenge of managing a learning university (Duke, 2002) and putting the university online (Cornford & Pollock, 2003).

A university comprises valuable human capital coming from its teams of knowledge workers, who have a strong formal education, have learned how to learn, and have a habit of continuing to learn throughout their lifetime. Nevertheless, such human capitals as the university's intellectual assets could be made more visible only through their application and reuse (Conklin, 1996; Stewart, 1997). These then are good reasons to stewarding people's intellectual knowledge, however implicit it may be, and making it available within and outside the organization whose competitive edge comes from having and effectively using such knowledge. The idea of electronic portfolio (e-portfolio) systems (Aalderink & Veugelers, 2006; Dalziel, Challan, & Sutherland, 2006), as part of a university initiative to improve teaching and learning, fits timely to advance this goal in higher education. However, this vision requires e-transformation efforts on the part of the conventional university to take advantage of not only the new technological, but also the renewed governance opportunities. The result could eventually be an essential constituent of an e-governance initiative, which comprises an electronic form of the original university renewed based on the working model of a virtual corporation (Davidow & Malone, 1992; Cheng, 1996; Hedberg, Dahlgren, Hansson, & Olve, 1997), to enable a re-engineered vision of the university's education, administration and governing process.

Innovating Quality E-Portfolios for Continuous Improvement

As online technologies and information resources rise in salience with the advent of the Internet, we are witnessing the emergence of a multi-faceted techno-pedagogic reality in the development of online support for student learning. The e-portfolio model of education (Vat, 2010, 2009, 2008) could be considered as one example phenomenon as a result of several important converging forces. Such forces are causing the education community to re-examine where learning takes place and how it could be assessed, how work and knowledge should be managed, who we, education practitioners, really are as we present ourselves to the world, and how we use technology for teaching and learning. This idea of the e-portfolio is said to be a flashpoint "at the converging of imperatives and opportunities in the management of learning for human and social capital development" (Jafari & Kaufman, 2006, p.xxvi).

Technically, its context ranges from the simple conceptualization of e-portfolio as a means of capturing student progress through a program of study, involving student work, student reflection, and faculty comments related to activities of teaching and learning (Henry, 2006), to the technological potential (Plater, 2006) which allows faculty and institutions to actually enable each student to have a personally managed, meaningful, coherent, integrated lifelong record of learning that demonstrates competence, transcends educational levels, and is portable across institutions of learning. In fact, e-portfolios are more than storage devices of the learner's best work (O'Brien, 2006). They provide the means for students to set learning goals, monitor and regulate their progress toward these goals (a form of self-directed learning), as well as develop their self-assessment skills. Practically, e-portfolios should serve as the student's pathway from classroom to career.

The Personalized Aspect of E-Portfolios

As life-long learners, we are always looking for tools to transform our learning experience, to enable learners to become autonomous and enjoy a truly personalized development path. It is believed that the e-portfolio is one of the most significant tools for achieving this goal. It should support the realization of a portfolio-based career, and act as an instrument for social inclusion, allowing us to "tell our story" and celebrate our achievements (Flanigan & Amirian, 2006). In fact, the e-portfolio could facilitate a continuum in the learning space where someone starting an e-portfolio at school, college, university, or work would not have to throw away the investment of years when moving from one episode of life to another. The e-portfolio should be our faithful digital companion, reflecting our digital identity and supporting our learning, and enabling transactions with others in a variety of social networks. For instance, in the professional circles, e-portfolios could become the indispensable tools for reflective practitioners extracting learning from the workplace, and sharing their reflections with their peers to contribute to the development of different professional learning communities.

The Learning Aspect of E-Portfolios

In a typical learning environment, there are many roles the e-portfolio can play, examples of which include the means of assessing student learning, the means of showcasing outstanding student achievements, and the means of ensuring learner accountability (Acosta & Liu, 2006; Sherman, 2006). Yet, whichever role the e-portfolio might play, there is one aspect that all e-portfolios have in common: namely, the learners must create portfolio elements or artifacts to be presented within the portfolio itself. As instructor or facilitator of e-portfolio learning, the design of sample e-portfolio requirements to document and communicate the learning of skills reflected in the learning process becomes critical. Examples include a learning contract with specific lesson plan detailing what the expected learning should be and the way to demonstrate the acquired learning. Such e-portfolio requirements should delineate the specific artifacts to be created by the learners to complete the process of learning. Indeed, this act of "creation" would necessitate the learning and/or application of a variety of skills related to the learning episode. Importantly, using the e-portfolio requirements as an aid of setting personal learning goals becomes a form of instructional scaffolding. Oftentimes, learners need to articulate clearly the goals of every piece of new learning experience by demonstrating the series of created artifacts to be included in the e-portfolio as evidence of the lessons learned. In this regard, examples of similar works from different learners could be collected into the e-portfolio repository for comparison and evaluation.

The Design Aspect of E-Portfolios

The advent of Web technology has brought about the currency of e-portfolio, which can not only be considered as an effective way to assess student learning, but also as a vehicle for knowledge development and for career building (Napper & Barrett, 2004). The key behind the e-portfolio movement lies in the empowerment of the learner to take charge of his or her own learning (Ramsdon, 2003; Barrows, 1988). Specifically, the e-portfolio scheme of learning shifts the locus of control from what we faculty teach to what students learn (Acosta & Liu, 2006); namely:

- **Enable students to determine what they need to learn through questioning and goal setting:** It is believed that students should work to identify their knowledge and skill deficits, and to develop strategies in the form of personal learning goals for meeting those deficits. The emphasis is to foster a sense of students' ownership in the learning process. In particular, e-portfolios emphasize analysis and reflection, and the development process, but not merely the product of learning. This process perspective not only raises the cognitive bar, but also shifts the locus of control from not so much what the instructor is doing, to what the student is doing to meet learning objectives. Moreover, the student can reflect on his or her learning and can demonstrate learning to persons outside of the immediate learning environment with the production of relevant electronic artifacts. For example, interested employers could review a student's resume, group project contributions, and other items of interest the student wants to make accessible. Likewise, if teachers, through the e-portfolio support environment, can guide the students in identifying what they already know and what they need to learn, then knowledge gaps and mistakes can be viewed in a positive way such as another opportunity to learn. And students can assume more responsibility in addressing their own learning needs during any instructional episode.
- **Enable students to manage their own learning activities:** It is believed that students must be enabled to develop their learning plans, which should describe priorities, instructional tactics, resources, deadlines, roles in collaborative learning situations, and proposed learning outcomes, including presentation and dissemination of new knowledge and skills, if applicable. Traditionally, these instructional events are arranged by teachers to be obeyed by students, in order to accomplish a specified set of pre-determined objectives. Yet, it is not advantageous for students to learn to be self-directed. To manage their own learning activities, students must be guided and supported by the teacher, through the e-portfolio environment, slowly taking on more and more responsibility of their own learning. For example, collaborative learning, inside and outside of the academy, is another feature of the new portfolio model, which should document such efforts as peer-to-peer projects promoting teamwork and communication skills, student-mentor projects (say, internships in the industry) giving students the opportunity to experience the world of work for better understanding of their future profession and workplace culture, student service-learning projects offering students first-hand understanding of societal issues and problems. Whichever type of projects the student is involved, he or she should maintain housekeeping of his or her e-portfolio and allow peers, mentors, and the community to give input, while the instructor at school provides the opportunity for the interactions, and assesses the intended learning outcomes.

The Curriculum Aspect of E-Portfolios

It is anticipated that the e-portfolio, as a tool to transform teaching and learning, should become a catalyst for curriculum change and a new model of assessment at the University of Macau (UM), which should connect the educational mission and objectives with the needs of society. It should also bring students closer to their future profession, and carry learning into students' future careers and possibly into their lifelong devotions. Thereby, the e-portfolio review process should serve as the feedback mechanism to update the academy on the skills required by students as they enter society. Put it simply, if students are immersed in projects that extend into the dynamic workplace and community (rather than the limitations of the campus) then they must demonstrate not only applicability of knowledge, but also flexibility and adaptability. The pedagogical challenge then is to set up connections between academic objectives and societal needs that will update the curriculum by incorporating current global perspectives. It is also expected that faculty members will then be in discussions with interested parties in the community to determine student outcomes. Therefore, the assessment of a course, program of study, and the related discipline will be somehow corroborated with persons outside of the academy. In this regard, the deliberation of an e-portfolio scheme of student learning, including its elements of flexibility should always be an important area of concerns.

FUTURE TRENDS FOR CONTINUOUS INNOVATION

Today, the predominant university model is still the combination of traditional teaching and academic research as mapped out by Wilhelm von Humboldt, the founder of the University of Berlin in 1809 (Denning, 1997). But, change is unavoidable in the 21st Century. This change is mainly driven by the new technological possibilities, and the new learning environments which result (Tsichritzis, 1999). We have witnessed the convergence of computing, communication, and document management technologies, as well as certainly the computer networks, pervading in the Internet. Such technologies have an enormous potential for transforming education to meet the growing need for customized, on-demand learning.

Yet, a new model of knowledge production, delivery, and presentation, is needed to combine an individualized approach, flexibility, and ease of dissemination without sacrificing the effectiveness of learning. This model must also offer the learners, the technological and pedagogical possibilities to collaborate with participants and experts all over the world via the World Wide Web, and to access online resources integrated into the study materials. However, it is well understood that if our universities were to absorb this model and to retain the longstanding position as our intellectual watering holes, the university today must be redefined with new concepts. And our discussion so far attempts to expound the re-engineering required through looking at university education as an enterprise for learning, the prime candidate for being the application of today's electronic organization transformation. The key to the success of any learning enterprise is innovation. Our university needs to make innovation happen now and ever.

Yet, what does innovation look like? Or, how do we know innovation when we see it? A quick look around the world reveals that innovation has many forms, including a better or new product, such as an automobile; a combination of products, such as an Internet phone; a fashion, such as a designer dress; a business model, such as the virtual university; a service, such as a school that offers customized (or personalized) distance learning or tutoring; an attribute, such as oil that helps clean your car engines; and a package, such as a paint that can never spills. Indeed, there are hundreds of competing definitions of innovation, most of which are narrowly focused and somewhat restrictive because they associate innovativeness with a particular element or attribute. Words such as "technological" or "breakthrough" are common examples.

Yet, innovation is increasingly not restricted to any one type; it is often a combination of forms, which, when put together produce hybrid solutions. Thereby, in an organization setting (say, a university), innovation should have a much wider role and application, and needs a more operational definition to identify how it works in practice. For example, all types of innovation employ some form of useful novelty aimed at making things better or new. Namely, useful innovation is intended to create some form of tangible or intangible value. More importantly, an innovation is always specific to the situation and time in which it was given rise. Innovation has a transformative quality because it will both replace existing products and services, as well as replace ways of doing things, and in turn, it will be replaced by a subsequent innovation. In this light, innovation is always involved in an endless cycle of emerging and dissolving (Degraff & Quinn, 2007), and it comes with two partners: creativity and change. To put these concepts in context, we can visualize how an engine works: creativity is the spark the ignites the fuel; change is the heat that the combustion produces; and innovation is the engine turning the heat into power and moving the vehicle up the road towards a specific destination. Thereby, creating a learning culture in support of innovation for continuous improvement in actualizing organizational development, is a very legitimate goal that is so much an integral part of modern-day organization e-transformation efforts, be they applied to the university setting or to the SME context. Yet, we need to ask: What organizational engines of innovation are supposed to be in place?

Engines of Innovation

Any engines of innovation must be designed with some blueprint, which is the ultimate map in that it contains the totality of an engine's forms and functions. Borrowing from the organic metaphor of Burns and Stalks (1977), the blueprint of the human engine for innovation is often referred to as a genome (or a genome map), which, at its inception, contains all the information it needs to grow into a mature organism. This map exists in every cell of the organism and provides specific instructions for each cell to integrate with other cells. It is both whole and complete in all its forms, and yet part of a larger system.

On researching the innovation context for organizations, it is much appreciated to discover the work done behind the term *Innovation Genome* (DeGraff & Quinn, 2007, pp. 8-32), which is used to describe how the entire system of organizational innovation functions at all levels: the individual, the organization, and the larger strategic environment where value is recognized by markets and consumers. One of the key pieces of the *Innovation Genome* is a map, which shows four different approaches to innovation, known often as the four-quadrant model. Each quadrant represents characteristics and practices that produce different forms of value. The quadrants operate essentially the same way for individuals, organizations, and markets. The strengths, weaknesses, and interactions of these four quadrants determine an organization's ability to produce specific forms of innovation in specific situations. The four quadrants, recognizable by the key measures, workplace environments, organizational practices, and leadership behaviors associated with each of the four types considered as best practices, include: the Collaborate, the Create, the Compete, and the Control quadrants.

The Collaborate Quadrant

According to DeGraff and Quinn (2007), the Collaborate quadrant is based on a social approach in which the leaders believe that an enterprise has as much purpose to build relationships, nurture community, and empower individuals as it does to produce goods and services. To accomplish this goal, the leaders must ensure that new ideas benefit the community and that any communication and partnerships must be sensitive to the needs, abilities, and ambitions of everyone in the organization. Put it simply, the Collaborate approach to innovation is based on a human-relations view of organizations that emphasizes the need for individuals to unite in a positive way to build cooperative communities of practice. This profile connects individuals to a greater good or a high-principled mission. Namely, empowering people to do what they believe to be right is essential for the Collaborate leader. Mutual trust and integrity are cornerstones of this culture, which creates highly committed individuals. Leaders develop their teams and team members through organizational learning practices, education, and coaching. This network also includes customers, who are treated as part of the family. Subsequently the organizations are considered as good to work for because of the emphasis on balance between job duties and personal life.

The Create Quadrant

The Create quadrant is a generative environment; i.e., it keeps regenerating itself such that one idea always leads to another and to another and so on. In a sense, the Create approach to innovation is about producing many radically different ideas instead of one big one. It is highly responsive to turbulent and fast-changing conditions. When the future is unclear or rapidly changing, the ability to launch a wide array of experiments and speculate on new markets provides a strategic advantage. Organizations in the Create quadrant excel at brainstorming, creating elaborate strategic plans, starting spinout ventures under the radar, and assembling a diverse array of characters that are willing to let go of perfection in order to create something new. Breakthrough products, services, ideas, and people are found in this fluid and ever-shifting environment. Essentially, the Create form of leadership judges their success on the innovativeness and future readiness of their products, services, and ideas. They keep an eye on future trends, judging which way they think the wind will blow and then applying their imaginations to the difficult task of making the wind blow in the desired direction. Individuals who work in a Create organization are usually involved in all aspects of the enterprise. Power and responsibility flow from individual to individual or from team to team according to their capability and the project at hand. The glue that holds Create organizations together is a shared pursuit of a grand vision, seemingly impossible goals, radical innovation, and a culture of risk taking.

The Compete Quadrant

The Compete quadrant is based on a survival-of-the-fittest approach to innovation. This is a quadrant where high achievers thrive. In most cases, Compete companies are publicly traded business and therefore must demonstrate short-term profitability for shareholders. They believe that business is a zero-sum game – everyone is either a winner or a loser. High-achieving competitors, who thrive in this quadrant, are most comfortable in a results-oriented environment. They enjoy challenges and the hard work needed to win. The Compete form of leadership motivates employees by articulating clear objectives, often in the form of strategic moves to beat the competition. Indeed, the Compete environment maintains an outward focus on meeting customer demand; Compete companies judge their success by their market share, revenue, brand equity, and profitability. There are a number of practices such organizations gravitate toward, and many are aggressive in nature, including mergers and acquisitions, performance management scorecards, and pay-for-performance plans, which provide rewards to those who achieve or produce the most.

The Control Quadrant

The Control quadrant is based on a technological or engineering approach to innovation. It takes a systematic view such as continuous improvement "Getting better every day in every way" through the discipline of applying multi-step processes, which are known to work with little risk. This approach has a safety-net feature to it, and it is particularly useful for large complex organizations that need to create products and services that must have a hit at the first launch. The Control form of leadership tends to be methodical, pragmatic, and precise. Control leaders excel in a workplace that has clearly delineated roles and responsibilities, systems and processes, and policies and procedures that ensure things are done correctly. They are clear-thinking realists. Yet, the Control quadrant is often overlooked as a form of creativity because implementing incremental changes in systems, structures, and standards may not look creative enough, but in the real world, many industry giants, like Toyota, often use a safe and predictable method to win the race. Such companies have often perfected many of the industry's foremost system and process tools for design and development. Like Toyota, they virtually break processes down to the most elemental level in order to understand how something works and how it can be made better. Thereby, Control leadership focuses inward and requires discipline. It is concerned with improving quality while at the same time cutting costs. It is especially valuable in industries that require standardized procedures, rule reinforcement, and consistent products, such as transportation vehicles. Control measures serve to eliminate errors and increase the likelihood of expected outcomes.

A Path to Innovation

Undoubtedly, recognizing what results an organization wants and the practices, competencies, and the leadership types associated with these outcomes is the first step toward crafting an approach to making innovation work where expected. It is believed that the *Innovation Genome* from DeGraff and Quinn (2007) should serve as a guide to respond to challenges in multiple situations by providing a way to gauge an organization's progress and to ensure that it gets the outcomes expected. Ironically, the major impediment to making innovation happen in many an organization is the inability to get through its own systems, particularly those designed to aid innovation.

It is convinced that innovation must be perceived as a democratic endeavor, something that every one is expected to do. We need leaders to demonstrate ownership for all aspects of innovation, as travel guides to those yet uninitiated in the ways of innovation. Such leaders are often self-authoring people who add creativity to ordinary organizational activities by capturing the essence of a transformation toward an objective. Indeed, creativity is the fuel, or the raw material of innovation. It is a capability that individuals and organizations must recognize and capture in all its forms. Once an organization has realized this capability, it will be set on a path toward systemizing a culture of innovation. The organization then would develop the expertise necessary to put all these pieces together to lead itself on a path toward innovation. In the end, it is believed that leading innovation is really about leading people. It is about finding a way to engage people to draw innovation from themselves and bring it to life within their work settings.

In the context of the University of Macau, it is the author's observation that the innovation model makes explicit the need for dedicated management activities that explore and track the abstract requirements needed to realize some objective while simultaneously, but independently, investigating and specifying the concrete means for satisfying the abstract requirements. It can be expressed as a framework for accommodating dynamic changes in both requirements and available means (or services) in an organizational and technological setting, in which the means for reaching a goal are continually and routinely evaluated in relation to explicit criteria. That way, management could ensure that requirements are satisfied as appropriately as possible. It is convinced that this innovation practice could be applied to the setting of e-transformation efforts in the new campus when new technologies, processes, and methods, could be adopted in the case of developing the e-portfolio systems of creating and delivering educational services. This is also the innovative context of any technology change management, which could be summarized using the IDEAL model (Gremba & Myers, 1997), quite compatible with the *Innovation Genome* from DeGraff and Quinn (2007), even though it was originally conceived as a life cycle model for software process improvement based on the capability maturity model (CMM) for software at the CMU-SEI (Paulk, Weber, Curtis, & Chrissis, 1994), and has since been adapted for broader applications.

IDEAL provides a usable and understandable approach to continuous innovation by outlining the steps necessary to establish a successful improvement program. The model consists of five stages. Initiating (I) is to lay the groundwork for a successful improvement effort. Diagnosing (D) is to determine where we are relative to where we want to be. Establishing (E) is to plan the specifics of how we will reach our destination. Acting (A) is to do the work according to the plan. Learning (L) is to learn from the experience and improve our ability to adopt new measures (technologies) in the immediate future. The strength of the IDEAL stages is their ability to help an organization develop a sustainable and resilient culture for innovation and the corresponding capabilities to afford the same. They could be put to the best use in getting larger groups of individuals to execute the innovation playbook of an organization, to unleash their creativity, for some timely changes, in order to meet the challenges of a complex array of dynamic forces, operating in a predictable innovative way. Once an organization gets familiar with the IDEAL stages of innovation, it should soon realize how this model could be extremely versatile and applicable in a number of different settings, such as identifying high-potential leadership areas, launching a winning project, integrating best practices for innovation into existing organizational processes, and changing organizational culture and competency.

REMARKS FOR CONTINUING CHALLENGE

The turbulence generated by the integration of information technology into higher education has provoked much wishful thinking, among educational planners, faculty members, and university administrators, regarding many promising e-transformation projects, such as Web-based course management, groupware for faculty or students, and cooperative environments for teaching, research, administration and system support. Unavoidably, it will be necessary for academics to adjust their teaching, and for administrators to re-examine the business of running a university with the advent of the Internet technologies. Yet, collaboration matters. It is intended and understood that many a design of learning experience should not be an erratic response to technical advances, but rather it must be a genuine attempt to achieve specific educational goals (knowledge development and transfer). According to Johnson and Johnson (1989), if universities were to succeed in the emerging marketplace of the educational industry, they would need to be mindful of nurturing some basic conditions throughout the process of providing educational services to students.

The Renewed Mindset for an e-Portfolio-Based Education

In contrast to the traditional 'direct transfer' model in which the instructor is assumed to be the sole source of knowledge and skills, the renewed learning model could be an interactive, collaborative knowledge building process (Harasim, 1990, 1999). In the traditional linear model of education, learning design proceeded from defining objectives to lesson planning to course delivery. Educators first engaged in a comprehensive learning needs analysis process, often based on assessments done by others about competencies and learning objectives. Comprehensive course syllabi were developed. Finally, the course was delivered as planned. Associated with this linear approach were a set of teaching strategies characterized by being predominantly one-way, centralized, and broadcast-oriented. When students appeared bored and unengaged in this type of program, the solution was to find ways to use new media to make the one-way broadcast more entertaining. Much early online learning was nothing more than a way to generate a broadcast of an expert and his or her multi-media slides with good production values. However, an expert lecturing to a group of passive students is engaging in didactic one-way teaching no matter how excellently that lecture is delivered, say from a stage in an auditorium, or via broadcast television to students sitting in their living rooms. Today, we need a renewed mindset for education. Teaching and learning is currently seen as an ongoing process rather than a program with a fixed starting and ending point and the importance of widespread participation by learners in the design of their own learning has also been widely recognized (Kimball, 1995). ICT (information and communications technologies) are particularly well suited to a more dynamic approach to managing education. Good teachers have always been open to changing their lesson plans based on student input. New media makes it easier. And online environments can provide spaces for continuing conversation among students and teachers about what is working and what is not working in the course setting. The idea of participatory course design is not to be neglected. The new campus environment should provide an opportunity to support collaborative learning in ways we have not been able to do before.

The New Roles for Teachers and Students

Instead of performing as the sage on the stage transmitting knowledge to a class of innocent students, in the innovative, collaborative learning environment, teachers' roles are to be defined in terms of mediating learning through dialogue and collaboration where knowledge is created in the community of students rather than being transferred from the individual.

More specifically, the idea of mediating could include such aspects of facilitating, modeling, and coaching (Chung, 1991; Whipple, 1987). Facilitating involves creating rich activities for linking new information to prior knowledge, providing opportunities for cooperative work and collective problem solving, and offering students a multiplicity of authentic learning tasks. Modeling serves to share with students not only the perceived content to be learned, but also the important meta-cognitive skills of higher-order thinking, in the process of communication and collaboration. Coaching involves giving hints or cues, providing feedback, redirecting students' efforts, and helping them use a strategy. A major principle of coaching is to provide help only when students need it so that students retain as much responsibility as possible for their own learning.

In fact, we need to teach students to rely less on teachers as the source of knowledge. We need to help them learn to learn as self-directed groups of active, autonomous, and responsible individuals. In the renewed learning settings, students are expected to assume their new roles as collaborators and active participants. It may be useful to think how these new roles influence processes and activities before, during, and after each episode of learning. For example, before learning, students set goals and plan learning tasks. During learning, they work to accomplish tasks and monitor their progress. After learning, they assess their performance and plan for future learning.

In practice, students constantly need help from the teachers to help them fulfill such new roles. Students must learn to become teachers of their own. Indeed, a frequent formula (Dilworth, 1998) that action learning proposes has been quite useful in constantly reminding students of their new role in the new learning scenario. Namely, L = P + Q + R, where L (learning) equals P (programmed instruction) plus Q (questioning) plus R (reflection). Here P represents the knowledge coming through textbooks, lectures, case studies, computer-based instructions, and many others. This is an important source of learning but carries with it an embedded caution flag. That is, P is all based in the past. Q means continuously seeking fresh insight into what is not yet known. This Q helps avoid the pitfall of imperfectly constructed past knowledge. By going through the Q step first, we are able to determine whether the information available is relevant and adequate to our needs. It will point to areas that will require the creation of new P. R simply means rethinking, taking apart, putting together, making sense of facts, and attempting to understand the problem. Following the use of this formula, action steps are planned and carried out with constant feedback and reflection as the learning takes place. It can provide for the mature students elevated levels of discernment and understanding through the interweaving of action and reflection.

CONCLUSION

It is convinced that the organizations that will truly excel in the future will be the organizations that discover how to tap people's commitment and capacity to learn at all levels in an organization. To harvest the knowledge and experience of people and make it available to the organization as a whole, ICT technologies need to be managed differently to support dialogue rather than mere databases. In particular, communication technologies are needed to support a learning environment, which could stimulate and nurture the complex network of interpersonal relationships and interactions. Also, people must be allowed to make choices about whom they need to communicate and learn with without regard to traditional organizational boundaries, distance and time.

In other words, they need to manage their own learning to form new groups and teams as requirements develop and change. The new framework for managing the university should be about managing the learning process as well as managing course contents. The kinds of questions we need be asking ourselves are not only about how to plug one type of technology into another, but also about how to use technology to leverage resources and group dynamics in new ways to make fundamental changes in every part of the learning process.

It is not surprising to recognize that the easy part of this process might be the technology, and the tough part is to invent and to innovate with the organizational context to create new models of experiences for knowledge sharing with the technology. Humboldt (Denning, 1997) did the most to spread the notion that universities are places of research, whose ultimate value, according to Tsichritzis (1999), lie in innovations. And an innovation occurs when the standard practices of a community of people shift, so that they are more effective at what they do. Consequently, we are optimistic that any new campus development, just like any new business enterprise in the making, should offer ample opportunities for the whole community of the organization to grow and to become leaders in innovation for the common good starting from right now and beyond.

REFERENCES

AAC. (1985, 1990). *Integrity in the college curriculum: A report to the academic community*. Washington, DC: Association of American Colleges.

AACU. (2002). *Greater expectations: A new vision for learning as a nation goes to college*. Washington, DC: Association of American Colleges and Universities.

AACU. (2004). *Taking responsibility for quality of the baccalaureate degree*. Washington, DC: Association of American Colleges and Universities.

AACU. (2005). *Liberal education outcomes: A preliminary report on student achievement in college*. Washington, DC: Association of American Colleges and Universities.

AACU. (2007). *College learning for the new global century*. Washington, DC: Association of American Colleges and Universities.

AACU. (2008). *Our students' best work: A framework for accountability worthy of our mission*. Washington, DC: Association of American Colleges and Universities.

Aalderink, M., & Veugelers, M. (2006). E-portfolio and educational change in higher education in the The Netherlands. In Jafari, A., & Kaufman, C. (Eds.), *Handbook of research on e-portfolios* (pp. 358–369). Hershey, PA: Idea Group Reference. doi:10.4018/978-1-59140-890-1.ch032

Acosta, T., & Liu, Y. (2006). ePortfolios: Beyond assessment. In A. Jafari & C. Kaufman (Eds.), *Handbook of research on ePortfolios* (pp. 15-23). Hershey, PA: Idea Group Reference.

Allen, M. J. (2004). *Assessing academic programs in higher education*. Bolton, MA: Anker Publishing Company, Inc.

Balderston, F. E. (1995). *Managing today's university: Strategies for viability, change, and excellence*. San Francisco, CA: Jossey-Bass.

Barrows, H. S. (1988). *The tutorial process*. Springfield, IL: Southern Illinois University Press.

Boyer, E. L. (1987). *College: The undergraduate experience in America*. San Francisco, CA: Jossey-Bass.

Boyer, E. L. (1990). *Scholarship reconsidered: Priorities of the professoriate*. San Francisco, CA: Jossey-Bass.

Burns, T., & Stalker, G. M. (1977). *The management of innovation*. London, UK: Tavistock.

Checkland, P., & Holwell, S. (1998). *Information, systems, and information systems: Making sense of the field*. Chichester, UK: John Wiley and Sons.

Cheng, W. (1996). The virtual enterprise: Beyond time, place and form. *Economic Bulletin*. Singapore International Chamber of Commerce, 5-7 February.

Chung, J. (1991). Collaborative learning strategies: The design of instructional environments for the emerging new school. *Educational Technology*, *31*(12), 15–22.

Conklin, E. J. (1996). *Designing organizational memory: Preserving intellectual assets in a knowledge economy*. White Paper. Washington, DC: Touchstone Consulting Group, Inc. Retrieved from http://www.touchstone.com/tr/whitepapers.html

Cornford, J., & Pollock, N. (2003). *Putting the university online: Information, technology, and organizational change*. Buckingham, England: SRHE and Open University Press.

Dalziel, C., Challen, R., & Sutherland, S. (2006). E-Portfolio in the UK: Emerging practice. In Jafari, A., & Kaufman, C. (Eds.), *Handbook of research on e-portfolios* (pp. 370–377). Hershey, PA: Idea Group Reference. doi:10.4018/978-1-59140-890-1.ch033

Davidow, W. H., & Malone, M. S. (1992). *The virtual corporation: Structuring and revitalizing the corporation for the 21st century*. Harper Business.

DeGraff, J., & Quinn, S. E. (2007). *Leading innovation: How to jump start your organization's growth engine*. New York, NY: McGraw-Hill.

Denning, P. J. (1997). A new social contract for research. *Communications of the ACM, 40*(2), 132–134. doi:10.1145/253671.253755

Dilworth, R. L. (1998). Action learning in a nutshell. *PIQ, 11*(1), 28–43.

Donoghue, F. (2008). *The last professors: The corporate university and the fate of the humanities*. Fordham University Press.

Driscoll, A., & Wood, S. (2007). *Developing outcomes-based assessment for learner-centered education: A faculty introduction*. Sterling, VA: Stylus.

Duderstadt, J. J. (2000). *A university for the 21st century*. University of Michigan Press.

Duderstadt, J. J. (2004). Governing the twenty-first-century university: A view from the bridge. In Tierney, W. G. (Ed.), *Competing conceptions of academic governance: Negotiating the perfect storm* (pp. 137–157). Baltimore, MD: The Johns Hopkins University Press.

Duke, C. (2002). *Managing the learning university*. Philadelphia, PA: SRHE and Open University Press.

Eifert, M., & Puschel, J. O. (2004). *National electronic government: Comparing governance structures in multi-layer administrations*. New York, NY: Routledge.

Flanigan, E. J., & Amirian, S. (2006). ePortfolios: Pathway from classroom to career. In A. Jafari & C. Kaufman (Eds.), *Handbook of research on ePortfolios* (pp. 102-111). Hershey, PA: Idea Group Reference.

Garvin, D. A. (1993). Building a learning organization. *Harvard Business Review, 71*(4), 78–91.

Gremba, J., & Myers, C. (1997). The IDEAL model: A practical guide for improvement. Pittsburgh, PA: Software Engineering Institute. Retrieved from http://www.sei.cmu.edu/ideal/ideal.bridge.html

Gruppen, L. D., Frohna, A. Z., Anderson, R. M., & Lowe, K. D. (2003). Faculty development for educational leadership and scholarship. *Academic Medicine, 78*(2), 137–141. doi:10.1097/00001888-200302000-00007

Hana, N. (2010). *Enabling enterprise transformation: Business and grassroots innovation for the knowledge economy (innovation, technology, and knowledge management)*. New York: Springer.

Harasim, L. (1990). Online education: An environment for collaboration and intellectual amplification. In Harasim, L. (Ed.), *Online education: Perspectives on a new environment* (pp. 39–66). New York, NY: Praeger.

Harasim, L. (1999). A framework for online learning: The Virtual-U. *IEEE Computer, 32*(9), 44-49.

Hedberg, B., Dahlgren, G., Hansson, J., & Olve, N. (1997). *Virtual organizations and beyond: Discover imaginary systems*. John Wiley & Sons Ltd. Henderson, J., & Venkatraman, N. (1993). Strategic alignment: Leveraging information technology for transforming organizations. *IBM Systems Journal, 32*(1), 4–16.

Henry, R. J. (2006). ePortfolio thinking: A provost perspective. In A. Jafari & C. Kaufman (Eds.), *Handbook of research on ePortfolios* (pp. 54-61). Hershey, PA: Idea Group Reference.

Hoque, F. (2000). *e-Enterprise: Business models, architecture, and components*. Cambridge, UK: Cambridge University Press.

Huba, M. E., & Freed, J. E. (2000). *Learner-centered assessment on college campuses: Shifting the focus from teaching to learning*. Boston, MA: Allyn and Bacon.

Jafari, A., & Kaufman, C. (Eds.). (2006). *Handbook of research on ePortfolios*. Hershey, PA: Idea Group Reference. doi:10.4018/978-1-59140-890-1

Johnson, D. W., & Johnson, R. T. (1989). *Cooperation and competition: Theory and research*. Edina, MN: Interaction Book Company.

Keller, G. (2004). A growing quaintness: Traditional governance in the markedly new realm of US higher education. In Tierney, W. G. (Ed.), *Competing conceptions of academic governance: Negotiating the perfect storm* (pp. 158–176). Baltimore, MD: The Johns Hopkins University Press.

Kimball, L. (1995). Ten ways to make online learning groups work. *Educational Leadership*, *53*(2), 54–56.

Kouzes, J. M., & Posner, B. Z. (2007). *The leadership challenge*. San Francisco, CA: Jossey-Bass.

Leskes, A., & Wright, B. (2005). *The art and science of assessing general education outcomes*. Washington, DC: Association of American Colleges and Universities.

Loomis, C. P. (1955). *Translation of Tonnies' Gemeinschaft and Gesellschaft (1887)*. London, UK: Routledge and Kegan Paul.

Lumby, J. (2001). *Managing further education: Learning enterprise*. Thousand Oaks, CA: SAGE Publications, Inc.

Marginson, S., & Considine, M. (2000). *The enterprise university: Power, governance and reinvention in Australia*. Cambridge, UK: Cambridge University Press.

Menand, L., & Gates, H. L. Jr. (2010). *The marketplace of ideas: Reform and resistance in the American university (issues of our times)*. New York, NY: W.W. Norton & Company.

Miller, R., & Leskes, A. (2005). *Levels of assessment: From the student to the institution*. Washington, DC: Association of American Colleges and Universities.

Napper, V. S., & Barrett, H. C. (2004). Information and technology: Assessment and ePortfolio. *Proceedings of the Society for Information Technology & Teacher Education Conference* (pp. 39-40).

NSSE. (2007). *Experiments that matter: Enhancing student learning and success*. Bloomington, IN: National Survey of Student Engagement.

O'Brien, K. (2006). ePortfolios as learning construction zones: Provost's perspective. In A. Jafari & C. Kaufman (Eds.), *Handbook of research on ePortfolios* (pp. 74-82). Hershey, PA: Idea Group Reference.

Oliver, E. L., & Sanders, L. (2004). *E-government reconsidered: Renewal of governance for the knowledge age*. Saskatoon, Canada: Saskatchewan Institute of Public Policy.

Paulk, M. C., Weber, C. V., Curtis, B., & Chrissis, M. B. (1994). *The capability maturity model: Guidelines for improving the software process*. Reading, MA: Addison Wesley.

Plater, W. M. (2006). The promise of the student electronic portfolio: A provost's perspective. In Jafari, A., & Kaufman, C. (Eds.), *Handbook of research on ePortfolios* (pp. 62–73). Hershey, PA: Idea Group Reference. doi:10.4018/978-1-59140-890-1.ch007

Ramsden, P. (1998). *Learning to lead in higher education*. London, UK: Routledge. doi:10.4324/9780203278116

Ramsden, P. (2003). *Learning to teach in higher education* (2nd ed.). London, UK: Routledge Falmer.

Riley, T. B. (2001). *Electronic governance: Living and working in the wired world*. Commonwealth Centre for Electronic Governance, August. Retrieved from http://www.electronicgov.net/pubs/research_papers/index.shtml

Riley, T. B. (2003). E-government vs E-governance: Examining the differences in a changing public sector climate. *International Tracking Survey Report* (No. 4). Commonwealth Centre for Electronic Governance. Retrieved from http://www.electronicgov.net

Rivard, S., Aubert, B. A., Patry, M., Pare, G., & Smith, H. A. (2004). *Information technology and organizational transformation: Solving the management puzzle*. Amsterdam, The Netherlands: Elsevier.

Senge, P. M. (1990). *The fifth discipline: The art and practice of the learning organization*. London, UK: Currency Doubleday.

Shattock, M. (2003). *Managing successful universities*. Berkshire, UK: Open University Press.

Sherman, G. (2006). Instructional roles of electronic portfolios. In Jafari, A., & Kaufman, C. (Eds.), *Handbook of research on ePortfolio* (pp. 1–14). Hershey, PA: Idea Group Reference. doi:10.4018/978-1-59140-890-1.ch001

Stewart, T. A. (1997). *Intellectual capital: The new wealth of organizations*. New York, NY: Doubleday.

Tapscott, D. (1997). *The digital economy: Promise and peril in the age of networked intelligence*. New York, NY: McGraw Hill.

Thorp, H., & Goldstein, B. (2010). *Engines of innovation: The entrepreneurial university in the twenty-first century*. The University of North Carolina Press.

Tierney, W. G. (2004) (Ed.). *Competing conceptions of academic governance: Negotiating the perfect storm*. Baltimore, MD: The Johns Hopkins University Press.

Tsichritzis, D. (1999). Reengineering the university. *Communications of the ACM*, *42*(6), 93–100. doi:10.1145/303849.303867

Tuchman, G. (2009). *Wannabe U: Inside the corporate university*. Chicago, IL: University of Chicago Press.

Vat, K. H. (2000). Designing knowledge infrastructure for virtual enterprises in organizational learning. In *Proceedings of the Tenth Annual Business Information Technology Conference (BIT2000)*, Nov. 1-2, Manchester, England, (CD-ROM Paper No. 45).

Vat, K. H. (2002). Developing learning organization strategy for online education: A knowledge perspective. In *Proceedings of the Fifth Annual Conference of the Southern Association for Information Systems (SAIS2002)*, (pp. 291-298). Savannah, GA: Southern Association for Information Systems, Mar. 1-2.

Vat, K. H. (2008). E-portfolio and pedagogical change for virtual universities. In Putnik, G. D., & Cunha, M. M. (Eds.), *Encyclopedia of networked and virtual organizations* (pp. 508–515). Hershey, PA: Information Science Reference. doi:10.4018/978-1-59904-885-7.ch067

Vat, K. H. (2009). An e-portfolio scheme of flexible online learning. In Rogers, P., Berg, G., Boettcher, J., Howard, C., Justice, L., & Schenk, K. (Eds.), *Encyclopedia of distance learning* (2nd ed., pp. 941–949). Hershey, PA: Information Science Reference. doi:10.4018/978-1-60566-198-8.ch132

Vat, K. H. (2010). Developing student e-portfolios for outcomes-based assessment in personalized instruction. In Kats, Y. (Ed.), *Learning management system technologies and software solutions for online teaching: Tools and applications* (pp. 259–290). Hershey, PA: Information Science Reference. doi:10.4018/978-1-61520-853-1.ch015

Vickers, G. (1965). *The art of judgment*. London, UK: Chapman and Hall.

Whipple, W.R.(1987). Collaborative learning: Recognizing it when we see it. *AAHE Bulletin, 4-6*. (ERIC Documentation Reproduction Service No. ED289396).

Zemsky, R., & Wegner, G. (1998). A very public agenda. *Policy Perspectives, 8*(2).

ADDITIONAL READING

Ackoff, R., & Enery, F. (2008). *On purposeful systems: An interdisciplinary analysis of individual and social behavior as a system of purposeful events*. London, England: Aldine Transaction (A Division of Transaction Publishers).

Aleman, Ana M., & Wartman, K.L. (2009). *Online social networking on campus: Understanding what matters in student culture*. New York: Routledge.

Coleman, D., & Levine, S. (2008). *Collaboration 2.0: Technology and best practices for successful collaboration in a Web 2.0 world*. Cupertino, CA: HappyAbout.info.

Conner, M., & Clawson, J. G. (2004) (Eds.). *Creating a learning culture: Strategy, technology, and practice*. Cambridge: Cambridge University Press.

DiBiase, D. (2002). *Using e-portfolios at Penn State to enhance student learning: Status, prospects, and strategies*. Last retrieved on 2011JAN29 at https://www.e-education.psu.edu/files/e-port_report.pdf

Ehin, C. (2009). *The organizational sweet spot: Engaging the innovative dynamics of your social networks*. New York: Springer.

Fahey, L., & Randall, R. M. (1998)(Eds.). *Learning from the future: Competitive foresight scenarios*. New York: John Wiley & Sons, Ltd.

Gay, G., & Hembrooke, H. (2004). *Activity-centered design: An ecological approach to designing smart tools and usable systems*. Cambridge, Massachusetts: The MIT Press.

Gorelick, C., Milton, N., & April, K. (2004). *Performance through learning: Knowledge management in practice*. New York: Elsevier (Butterworth Heinemann).

Hanna, N. (2009). *e-Transformation: Enabling new development strategies (Innovation, technology, and knowledge management)*. New York: Springer.

Hemlin, S., Allwood, C. M., & Martin, B. R. (2004) (Eds.). *Creative knowledge environments: The influences on creativity in research and innovation*. Northampton, MA: Edward Elgar.

Maeda, J. (2006). *The laws of simplicity: Design, technology, business, life*. Cambridge, Massachusetts: The MIT Press.

McCarthy, J., & Wright, P. (2004). *Technology as experience*. Cambridge, Massachusetts: The MIT Press.

Merchant, N. (2010). *The new how: Creating business solutions through collaborative strategy*. Sebastopol, CA: O'Reilly.

Oden, H. W. (1999). *Transforming the organization: A social-technical approach*. London: Quorum Books.

Ritchie, R., & Crick, R. D. (2007). *Distributing leadership for personalized learning*. London, England: Continuum International Publishing Group.

Rouse, W. B. (2007). *People and organizations: Explorations of human-centered design*. Boboken, NJ: John Wiley & Sons.

Rovai, A. P. (2009). *The Internet and higher education: Achieving global reach*. Oxford, England: Chandos Publishing.

Rowley, D. J., & Sherman, H. (2001). *From strategy to change: Implementing the plan in higher education*. San Francisco, CA: Jossey-Bass.

Sallis, E., & Jones, G. (2002). *Knowledge management in education: Enhancing learning and education*. London, England: Kogan Page Limited.

Savin-Baden, M. (2008). *Learning spaces: Creating opportunities for knowledge creation in academic life*. Bershire, England: Open University Press.

Scharle, A., & Szabo, A. (2000). *Learner autonomy: A guide to developing learner responsibility*. Cambridge, England: Cambridge University Press.

Scharmer, C. O. (2007). *Theory U: Leading from the future as it emerges – The social technology of presencing*. Cambridge, Massachusetts: The Society for Organizational Learning, Inc.

Sharpe, B., & van der Heijden, K. (2007). *Scenarios for success: Turning insights into action*. Chichester, England: John Wiley & Sons, Ltd.

Vaishnavi, V., & Kuechler, W., Jr. (2008). *Design science research methods and patterns: Innovating information and communication technology*. New York: Auerbach Publications (Taylor & Francis Group).

van der Heijden, K., Bradfield, R., Burt, G., Cairns, G., & Wright, G. (2002). *The sixth sense: Accelerating organizational learning with scenarios*. Chichester, England: John Wiley & Sons, Ltd.

Walker, W. R., & Herrmann, D. J. (2005) (Eds.). *Cognitive technology: Essays on the transformation of thought and society*. London: McFarland & Company, Inc., Publishers.

KEY TERMS AND DEFINITIONS

Continuous Improvement: A process of organizational development based on an unwavering commitment to progress, examples of which could include such activities as identifying core beliefs, creating shared vision, using data to determine gaps between the current reality and the shared visions, identifying the innovations that will most likely close the gaps, developing and implementing an action plan, as well as endorsing collective accountability.

Electronic Transformation (E-Transformation): A process of organizational renewal based on transforming the organization from its bricks-and-mortar entity into its clicks-and-mortar counterpart. This is done through the application of suitable information and communication technologies (ICTs). In the Internet era, one example use of the ICTs is to Web-enable the operations and services of the organization, such that members of the organizations could experience the organizational operations and services through virtual encounter.

Electronic Portfolios (E-Portfolios): Electronic records to keep track of some developmental process or processes, such as student learning throughout a period of time, say, throughout the undergraduate years of study, including such artifacts as learning records (homework, projects, presentations), as well as showcases of major achievements such as final year (capstone) works before graduations, together with institutional authentication of the student work, plus reflective comments from students, and feedback from teachers course by course.

Engines of Innovation: An organization mechanism to support individuals' initiative to innovate: namely, the use of new ideas, new methods, and new practices to achieve various organizational goals. It may be some platform where organization members could use to enhance interpersonal exchanges through the application of some ICTs, to support virtual encounter of distributed communities of practice in the organization.

Framework of Accountability: An organizational structure to support such initiatives whose success or failures must be accounted for, based on clearly defined measures of interest, and based on relevant data collected on a longitudinal basis, in support of the some formative assessment of the situation of concerns, such as whether student learning has been improved in the setting of higher educational institutions.

Innovation: The use of a new idea or method or practice, say, in the application of a new computer technology in organizational development.

Learning Enterprise: An organization that helps transfer learning from individuals to a group, provide for organizational renewal, keep an open attitude to the outside world, and support a commitment to knowledge. It can also viewed as an enterprise that focuses on developing and using its information and knowledge capabilities in order to create higher-value information and knowledge, to modify behaviors to reflect new knowledge and insights, and to improve bottom-line results.

Chapter 9
Sources of Innovation and Success in the Food Service Industry

A.T. Juntunen
University of Helsinki, Finland

ABSTRACT

The purpose of the study in this chapter is to investigate and analyze the sources of innovation and success in food service industry. It also discusses of what kind of network management capabilities were needed in these SME cases. This chapter will demonstrate that the SMEs have a good chance to succeed if they can harness the external knowledge in their start up process. This chapter contributes to the prior strategic management and business networks research by analyzing the selected SME cases in the food service industry.

INTRODUCTION

For years, the idea of internal, closed R&D was considered to be the only source for companies to success, but today the companies are more and more harnessing external knowledge while leveraging their in-house ideas and R&D (Chesborough, 2006). Due to the Internet and global competition, SMEs are able to gain access to wide range of information related to the start up business and specific industry challenges and competition situation. Currently, the vast amount of information available online includes information and websites for grants, loans, industry and area specific information like, local unemployment, nearby schools and government agencies, cost of living, income and crime rates. SMEs' R&D strategy can benefit from external research capabilities to create knowledge diffusion.

DOI: 10.4018/978-1-61350-519-9.ch009

According to Henry Chesbrough (2006, pp.45-52) companies need to leverage the abundance of knowledge outside their own companies to be successful.

The food service business is one of the fastest growing industries in the world and the new technologies and social media bring new actors in the global markets. Food service industry generates approximately 360 billion USD turnover per year in the United States (Simpson & Carevic, 2004), and globally the turnover is 1.3 trillion USD (Webber, 2004). There are increasingly complex network structures emerging in this industry. Food service companies have been characterized as customer-centric, knowledge-based, and service-intensive. They also use new technologies more and more in order to be competitive. (See e.g. Morland, Wing, Roux, & Poole, 2002 of food service locations and demography; Johns & Pine, 2002 of consumer behavior; Taylor, 2002 of standardization and differentiation; Rodgers, 2008 of technological sophistication)

However, many SMEs in food service business sector fail because they do not have the financing or research laboratories in comparison to bigger manufacturers (Rodgers, 2007, p.900). The success (or failure) of the SMEs depends on the differentiation (see e.g. Hudson, 1994 of restaurants and cafes' differentiation), lack of innovation production technologies (Hudson, 1997), lack of originality like e.g. superior cooking (Van Stamm, 2003), finding their own niche (Gardner & Wood, 1991: van Stamm, 2003; Poulston & Yiu, 2011) or lack of funding to do the necessary research in their industry area (Rodgers, 2007; Hudson, 1994), and thus, be competitive. What are then the sources of innovation and success? This study analyzes and discusses about the innovation sources that are used in SMEs in a network context. It also discusses about the managerial capabilities are required in order to success. This chapter contributes to the prior strategic management and business networks research by analyzing the existing successful SMEs in the food service industry. The key questions in this research is to analyze the variety of SMEs' business cases that have succeeded in their business goals and what have been their sources of innovation and success and managerial capabilities required. These topics will be discussed in the following sections.

This chapter will first discuss on the previous research in the area of business networks and open innovation. Secondly, the research strategy and methods are presented. The case descriptions are described next. Finally, the chapter ends with the conclusion and future research views.

BACKGROUND AND KEY CONCEPTS

The paper builds on the prior strategic management literature, Resource-Based View of the firm (RBV), economic sociology and social networks research, and industrial network approach by examining the previous research of open innovation and business networks. This section also presents the concept definitions used in this chapter.

There is a vast amount of earlier research on business networks, and therefore, this chapter presents only the most relevant ones. The amount of literature also results in multiple concept definitions, like the researchers in strategic management and Resource-Based View (RBV) (See e.g. Brandenburger & Nalebuff, 1996; Jarillo, 1993; Parolini, 1999: Gregoire, Wade, Julland & Antia (2001); Kajalo, Rajala & Westerlund, 2007; Makkonen, 2008; Möller & Rajala, 2007, Möller, 2010) consider networks as strategic networks and value nets, whereas economic sociology and social networks research (Powell et al., 1996) and industrial network approach (e.g. Håkansson & Ford, 2002; Harrison, 2004: Håkansson & Snehota, 1995; Westerlund & Svahn, 2008) argue of the character, relationships and evolution of business networks. In earlier researches, business networks are defined as a set of relationships between companies where they engage in multiple two-way relationships to bring products and services to the market (e.g. Aldrich, 1998; Parolini, 1999).

Achrol (1997) stated that one of the fundamental shifts in the 21st century is from dyadic business relationships towards business networks with multiple actors involved. This entailed value creation with multiple partners participating in the network. Moreover, the last decade has shown a vast growth in organizational collaboration and creation of business networks with multiple goals like, R&D, competitive alliances, distribution channel networks and buyer-supplier configurations (see for example, Achrol & Kotler, 1999; Amit & Zott, 2001; Cartwright & Oliver, 2000; de Man, 2004; Frels, Shervani, & Srivastava, 2003; Gulati, 1998; Powell, Koput, & Smith-Doerr, 1996; Spekman, Isabella, & MacAvoy, 2000; Srinivasan, Lilien, & Rangaswamy, 2006). The network perspective suggests that the actors in business networks are interconnected and interacting with each others to create value (Ford, Gadde, Håkansson, & Shenota, 2003; Brass, Galaskiewicz, Greve, & Tsai, 2004).

The value creation is an important concept. Porter (1985) used the value chain concept primarily to refer to the firm-level activities through which a firm produces value for its customers. The value system describes the combination of the value activities of multiple actors forming "value creating" end products (see also Anderson & Narus, 1999; Doz & Hamel, 1998; Norman & Ramirez, 1993). These value activities are performed using sets of resources (see Parolini 1999, pp.59-63). The value activities are defined from the customer's perspective in the Value Network approach (Parolini 1999, pp.70-71). The division of value activities can be different. One or two companies can take care of the whole value system or it can contain several actors and industries. A value-system perspective argues that value creation spans organizational and industry boundaries (Amit & Zott, 2001).

Figure 1. The various sources of innovations and new knowledge

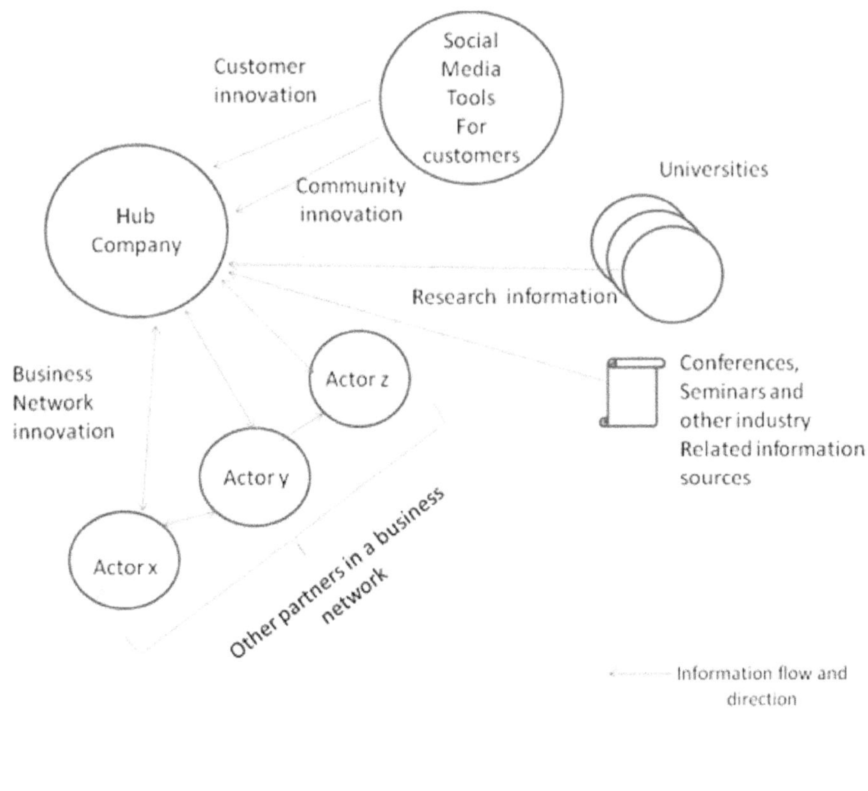

A hub company in this study (illustrated in the Figure 1) is the company that orchestrates the other companies (e.g. suppliers, providers and partners). The hub is an organization that gives direction and goals for the business network, negotiates roles in the network, and decides of the structure of the network. Ritter, Wilkinson and Johnston (2004) argued that no single hub can provide direction or control to any network. However, many studies consider that it is possible for a single firm to manage networks (see e.g. Dyer, 1996; Dyer & Nobeoka, 2000; Dyer & Singh, 1998, 2000; Lorenzoni & Lipparini, 1999).

Open innovation concept includes the transition from traditional closed, internal innovation process towards open innovation using external partners and exploiting external knowledge in creating new business or products and services. These external partners and knowledge sources can be constant partners in a jointly agreed business network or they can be temporary sources of information and knowledge for the SMEs. In addition, to be competitive and survive in a network economy, companies need diversified knowledge and resources to develop new and innovative products and processes. Specifically, they need to connect their organization through vertical and horizontal networking strategies to other market players in order to gain new knowledge beyond their immediate organizational boundaries (Håkansson & Snehota, 1995; Kogut & Zander, 1997; Powell et al., 1996; Teece et al., 1997). In Teece's (1986) view, the firm must bear the responsibility for how best to harness innovation opportunities and be profitable.

Understanding the Business Models

A business model has two functions: value creation and value capture. Also, the importance of knowledge dissemination requires that a company opens its business model to its partners in order to let more ideas and information flow from the outside and let more information flow from the inside (Chesbrough, 2006c). Therefore, it is essential to understand the business models and the value creation in them. The business model determines the value and the success of the company.

SMEs seek benefits by creating business models in which networks play a crucial role. In the inter-organizational context, business managers require new managerial capabilities in order to cope with the increasing competition and the external sources of innovation. In addition, SMEs need to find ways to link the value-creating and -capturing elements of the business model with network management of different network actor regardless of their role in the service innovation, production and delivery.

In addition, Axelsson and Angdal (2000) and Ellis (2000) emphasise the role of the social relationships as the initiator of the SMEs' expansion and internationalization. Large, diversified firms enjoy more opportunities to exploit different network ties than smaller, less-connected companies (Komulainen, Mainela, Tahtinen, 2004), but larger firms also have more complex and entwined business relationships.

The previous research on business models rests in many aspects on strategy discussion and draws on strategic concepts and issues (see e.g. Osterwalder, 2004; Morris et al, 2005; Tikkanen et al, 2005). The business model defines the portion of the value that the business creates internally (Chesbrough, 2006b, p.177), and what part of the business is either outsourced, or created and delivered by the other network partners. According to Chesborough (2006a, 2006b), the firm's R&D projects can be launched internally or externally, and new knowledge and information can enter the project in its various stages. He labeled this model as an open innovation because there are many ways how the new ideas and knowledge can flow into the process and many ways of how it can flow out into the market.

INNOVATION IN A BUSINESS NETWORK CONTEXT

The following section will discuss about innovation sources in a business network context. This study argues that essential to any successful business is the underlying system through which it produces value. This value–system construct affects on the ways the SME is able to gather external knowledge for innovations and new ideas in order to renew its business or to make a startup business a successful one.

Sources of Innovation

The Figure 1 illustrates the several sources of innovation, new knowledge and ideas that SMEs can use and utilize to create new products, services or a new business. The existing businesses can utilize social media, like Facebook and blogs, to advertise, inform and gather new ideas and information from their fans. Fans here refer to the fans of a certain Facebook site that the company has started for its business. The fans that have joined there can get information of new services and products, games, competitions, events etc related to the business. They can also contribute to the company by participating and innovating new products and services. Companies, like Dunkin Donuts "create Dunkin's next donut" (http://www.dunkindonuts.com/donut/) and Bush beans (http://www.bushbeans.com/beaninfo/entertowin.php) in food industry, have used their website and the social media like Facebook and Twitter to create a fan base, new products and collect new recipes from their fans for their own business use.

Furthermore, communities can create new innovations for businesses. Social media sites offer community based information and consumer groups that can be exploited to create new ideas and to get customer feedback of existing services and products.

An example of making customers work for you and market your business is Groupon. It has created a website (Groupon.com) in which a customer can get a certain discount or a deal for a certain product or service if they gather a certain amount of other customers interested in the same deal. This is a new marketing strategy that many companies from different industries can use. It relies on the customer's own interest and contacts to gather commonly interested other customers, or friends, to buy the particular service or product.

In addition, a firm can collect new knowledge and information from his partners in a business network: suppliers, co-creators of services and products, delivery companies, etc. A firm can also become a partner with a larger company, and utilize its vast resources to stimulate greater information exchange and knowledge dissemination. The benefit for a larger company is to gain exclusive access to a patent or to specific product idea that would fit into their own product and service portfolio

An organization can also use research information done by universities and research centers to gain external knowledge for his business. Firms can also gather information from industry-specific events, like seminars and conferences.

In summary, there are multiple sources of innovation and external knowledge that any existing or startup business can use to create a success in its chosen business area

Research Strategy and Methodology

The methodological choices of this study are guided by the basic aim of expanding existing knowledge evolving around SMEs and their business network and knowledge management development in the context of food service. Due to the novel nature of this phenomenon and the complexity of business networks in particular, the researcher has chosen a case study as the method of choice. Since we are dealing with a relatively new area of study in which the researcher has only little or no control over the events that are occurring in a real-life context, a case study is a very appropriate method (Stake, 2000, pp. 435-454).

The constructionist and relativist philosophy of science is widely utilized in the research on business networks (Tikkanen, 1996). The significant point in such a research is the emphasis on subjectivity and situation-specificity. Thus, the interpretations are strongly influenced by the researcher's analytical and theoretical frameworks and the context of the research.

This chapter presents a qualitative case study (Yin, 1994), and it is based on document analysis of several SMEs. The documents include the company websites and other public information. Case studies are argued to be valid tools for the accomplishment of various aims including modifying and advancing of an existing theory (Leblebici et al., 1991) or creating a new one. According to Eisenhardt (1989), four to ten cases usually are adequate for multiple case studies, when aiming at theory generalization, like in this particular multiple SME-case study setting. The motive for choosing case companies with different types of business models and different niches in the food service industry is to provide a solid basis for a cross-case analysis that would reveal whether, and what kinds of, differences existed in the start up and managerial capabilities of different businesses and what had made them successful in their own business area.

All the interviews were taped with the interviewee's permission and then transcribed and analyzed accordingly. The researcher employed qualitative data analysis in order to thematize the material (see Miles & Huberman, 1984) and used internal and public documents, food service industry reports and company visits to triangulate the respondents' answers, as suggested in literature (Patton, 1987; Yin, 1994). Data triangulation was used between the information sources mentioned above. In practice, data triangulation was first used to compare the different perspectives presented by each of the interviewees and, subsequently, to compare the interviews with other sources, such as industry reports, in order to validate the observations and interpretations. The results are presented in the next section.

The identities of the companies or the respondents are not revealed for reasons of confidentiality.

Descriptions of the Case Companies

All the cases presented here are SMEs. They have started by one or two people and grown since the initial startup. These companies have succeeded well since the startup. The following sub-sections present and analyze the specific cases in the food service industry.

Company A is an online baking company. Its business model includes gathering new product ideas and charity events from customers. An online bakery requires much less overhead than a storefront bakery. The owner partnered with a larger bakery that could handle the volume anticipated. Company A is using both the family recipes and taking ideas from customers. The startup business is also doing specific orders, like allergen- and gluten-free dough. The owner did the original research and information gathering by travelling around the country and visiting bakeries. Working as partly non-profit, the company has managed to get a successful startup and charity business. Fifteen percent of every purchase (exclusive of shipping) goes to one of the nonprofits the buyer selects. This business model opened up to gather information from customers, existing industry leaders and partnered with a larger, existing company to create a successful business. The niche of this B2C business is the charity and taking an old-fashioned, traditional business online. The new technologies, like Internet and social media, has enabled this business model to be successful. The business strategy is outside-in meaning that the company takes ideas from the customers and outside world instead of concentrating in in-house knowledge.

Table 1. The company/ interview / documentation data summary table

Company	Number of start-up personnel And year	Number of personnel currently	Number of actors (companies) collaborating in the business network	Business area	Niche, competitive advantage	Interview duration	Document, information source
1. Online bakery	1/2009	5	59	online bakery, charity	Old-fashioned business transformed to online business	1 hr	Company website, other company documents
2. Goat farm	2/2007	3	68	dairy business	Organic business, own designer products	1 hr	Company website, public documents
3. Gadgets	1/2009	1	2	New gadgets	Innovative design	-	Company website, public documents
4. online bakery	1/2009	10	6	Online bakery	Old-fashioned business transformed to online business, Products just for men	-	Company website, public documents
5. Brewery	1/2007	50	7	Brewery	Local taste, sharing culture and taste of the founding city	-	Company website, public documents
6. Ice cream company	1/1995	80	87	Ice cream	New innovative flavors	1 hr	Company website, public documents

Company 2 is an organically managed goat dairy farm. The owners wanted to develop an artisanal cheese as part of their business idea. The result of the cheese business is a new, gorgeous, edible black crust cheese. By taking this strategically well-planned approach to design own cheese brand, they found a niche that no other artisanal goat cheese maker has in this business. And the result is that their products are on the menus of some of the finest restaurants in New York. They have also invented other niche products like soaps. Their B2B business model included going back to basics, starting a farm business, but finding a new niche product area and making it exclusive. This approach needed business research capabilities, financial resources and knowhow to make new innovative products. Partnering skills were required to market and to deliver the end product to restaurants and shops. The innovativeness included finding different usage of the goat milk, creating new products and delivering the products to their business customers. They also used innovative marketing using television, e.g. Planet Green Network, videos, books and social networking like Facebook and Twitter to introduce their green business idea and not just playing product advertisements on television. The business strategy is inside-out meaning that the company has own innovative ideas and know how to produce new products with the resources they have. They are customer-centric business and listen to the customers as well. Customers are also their sources of new product ideas.

Company 3 is an innovation of a specific product. The innovator needed to find a partner to make and to deliver his innovation. He thought that his idea of a plastic lime bottle opener would be a good fit with the Corona beer products. He succeeded finding the partners and turned his plastic new product innovation's patent idea into a multimillion-dollar licensing deal with an international company. The innovator in this small business did not start with the business model, but with a specific innovation and then started finding partners to patent, to make prototypes and the final products and a company to deliver the product to customers. This small business needed managerial capabilities in finding suitable partners, contracting and gathering information of patents and of possible partners in this business.

Company 4 is an online bakery. It had a business idea for creating products for men only. The owner gathered the information from existing bakeries and makes them as his partners by working and consulting them to create recipes. He also focused on building a strong online presence. The result was a successful business with 90% of the customers is women, not men like he had originally planned. This business required managerial capabilities in contracting, finding partners and good recipes for products. Moreover, it shows that the original idea might be good even if the customer target segment can change from the original idea to the existing business. The new technologies, like Internet and social media, has enabled this business model to be successful. The business strategy is inside-out meaning that the company has own innovative ideas and uses own resources to create the competitive advantage.

Company 5 is a brewer company. The owner of this company had a long experience in the brewer industry and he had worked for both a craft beer brewery and an import. His niche in this business was to brew beer using local flavors like guava. The end products are more expensive, but have caught on far beyond the startup area and spread the interest throughout the country. The new company sold about 1,000 barrels in 2009. By the end of the first quarter of 2010, it had already exceeded that threshold and was on pace to sell more than that per quarter throughout the year, marking more than a 400 percent growth in 12 months. This business model relied on experience gained from working for other brewer companies in different positions and finding a niche. The owner also managed to use the customers as marketers of the new products. The owner knew the business area and the competitive market extremely well, and that was the basis of his innovation and success. The business strategy is inside out meaning that the company relies on its own innovativeness and ideas in their product research and development.

Company 6 is an ice cream business. It has created 40 new ice cream flavors and different mixtures of flavors. The owner of this company had working experience in this business area. After nearly 20 years in the dairy business, the owner started out his own ice cream business in 1995 and wound up creating successful new ice cream flavors. His business model was based on the experience of the business, and knowing what customers like. He was successful in licensing his flavors and gathers now millions from his licensing agreements with multiple manufacturers. His managerial capabilities included the knowledge of the manufacturers in this business and the potential customers. The business strategy is inside-out but also outside in meaning that the company has own innovative ideas but they also listen to the customers and what they like and change accordingly.

Challenges

The barriers of open innovation and success in SMEs seems to be the marketing resources, lack of interest in the administration and bureaucracy, financial resources in startup situation, lack of resources (technical, human), property rights and how to acquire them, and customer acceptance and how to be known to create the financial stability and to guarantee the continuance of the firm.

Moreover, there are also managerial capabilities required in business management and succeeding in it. These include finding the suitable partners, contracting, and hiring or training capable employees.

The study also finds, the age of the entrepreneurs in these cases differ from experienced and seniors to young people. But, the age does not seem to be a barrier to innovate or success in a small business.

CONCLUSION

Among the studied cases, some of the businesses have used the new technologies and social media successfully as part of their business model or turning their business completely online, but some of the businesses are more traditional ones, like the brewery or the gadget innovation business. They all use open innovation model, the external information of different sources to create a diffusion of knowledge. These SMEs let more ideas and information flow from the outside to assist in innovation, production, delivery and marketing of services.

The studied SME-cases seem to have a strong need for self-experimentation and rapid adaptation of new ideas or marketing strategies in order to survive and prosper in the highly competitive environment of the food industry business. They also have network management skills to benefit of their ideas and innovations and harness the external knowledge and information from different sources.

These SMEs have also noticed that the strategy of doing everything by themselves is not working in today's competitive environment, and therefore, they have done research and selected what part of the business they want to keep in-house and what parts can be outsourced to partners. Moreover, they managed to articulate the value created in their business, identified the market segment, their revenue logic, and defined the business structure of the business to create and deliver the product or service to the customers. They also determined their own position in the business network to support their value share and guaranteed that no other company was capable of taking their hub position and their own business from them. They identified their position in the markets, their competitors and complementors, if any.

One factor in the growth of successful companies is the formation of marketing alliances and networks among the industry companies, which means that firms use inter-firm relationships to facilitate their business models. Networking seems to be characteristic when new knowledge is required, and it is based on outsourcing and the use of external experts. Outsourcing can include logistics, product delivery and recipe or other new product concept creation. Franchising is another possibility for established businesses to expand their knowledge base and markets. Also, it seems that horizontal merging and partnerships in the food service business are especially common in attempts to create a successful business.

In this respect, the main theoretical contribution of this study is the identification of the network management capabilities in different types of food service business models and finding out the sources of innovation in these businesses. Thus, the study of capabilities in connection with open innovation contributes to industrial network theory (Achrol & Kotler, 1999; Gulati, 1998; Håkansson, 1982; Powell, 1990) and to the resource-based view of resources and capabilities (Hooley et al., 1998; Teece et al, 1997; Rosenbröijer, 1998). The findings of this study point out that the emphasis on managerial capabilities varies by type of business, and that capability creation and acquisition are linked with the business network context. Furthermore, this study reveals that particular capabilities are created

internally, like the training of new employees and franchising skills, while others are acquired through business network and interaction with external relationships, like the knowledge of the specific business area, and competitors, complementors and collaborators in that area. In addition, the start up-businesses use different business strategies to succeed in their goals.

FUTURE RESEARCH DIRECTIONS

The objectives of this chapter are to provide interesting facts that permits more exploration. Further research, addressing some of the issues summarized in the previous section, such as their use of new technologies, social networks, marketing and business strategies, marketing alliances, network management, and product innovation process could prove to be important for SMEs in other business sectors. These include the management of business networks and also harnessing innovations and new ideas with the help of the emerging new technologies and social media.

REFERENCES

Achrol, R. S. (1997). Changes in the theory of interorganizational relations in marketing: Toward a network paradigm. *Journal of the Academy of Marketing Science, 25*(1), 56–71. doi:10.1007/BF02894509

Achrol, R. S., & Kotler, P. (1999). Marketing in the network economy. *Journal of Marketing, 63*(Special Issue), 146–163. doi:10.2307/1252108

Aldrich, D. (1998). The new value chain. *Information Week, 700*, 278–281.

Amit, R., & Zott, C. (2001). Value creation in e-business. *Strategic Management Journal, 22*, 493–520. doi:10.1002/smj.187

Axelsson, B., & Agndal, H. (2000). Internationalisation of the firm: A note on the crucial role of the individual's contact network. *Proceedings of the 16th IMP Annual Conference*, Bath, UK.

Brandenburger, A. M., & Nalebuff, B. J. (1996). *Co-opetition*. New York, NY: Doubleday.

Brass, D. J., Galaskiewicz, J., Greve, H. R., & Tsai, W. (2004). Taking stock of networks and organizations: A multilevel perspective. *Academy of Management Journal, 47*(6), 795–817. doi:10.2307/20159624

Cartwright, S. D., & Oliver, R. W. (2000). Untangling the value web. *The Journal of Business Strategy, 21*(1), 22–27. doi:10.1108/eb040055

Chesborough, H. (2006a). *Open innovation: A new paradigm for understanding industrial innovation. Center for Open Innovation. IMIO.* Presentation.

Chesborough, H. (2006b). *Open innovation: Researching a new paradigm.* Oxford University Press.

Chesborough, H. (2006c). *Open business models: How to thrive in the new innovation landscape.* Oxford University Press.

de Man, A.-P. (2004). *The network economy. Strategy, structure and management.* Cheltenham, UK: Edward Elgar.

Eisenhardt, K. (1989). Building theories from case study research. *Academy of Management Review, 14*(5), 532–550.

Ellis, P. (2000). Social ties and foreign market entry. *Journal of International Business Studies, 31*(3), 443–470. doi:10.1057/palgrave.jibs.8490916

Ford, D., Gadde, L.-E., Håkansson, H., & Shenota, I. (2003). *Managing business relationships.* Chichester, UK: John Wiley & Sons Ltd.

Frels, J. K., Shervani, T., & Srivastava, R. K. (2003). The integrated networks model: Explaining resource allocations in network markets. *Journal of Marketing, 67*, 29–45. doi:10.1509/jmkg.67.1.29.18586

Gardner, K., & Woods, R. (1991). Theatricality in food service work. *International Journal of Hospitality Management, 10*(3), 267–278. doi:10.1016/0278-4319(91)90059-Q

Gregoire, Y., Wade, M., Hulland, J., & Antia, K. (2001).Resource redeployment in an ecommerce environment: A resource-based view. *American Marketing Association Conference*, Long Beach, CA.

Gulati, R. (1998). Alliances and networks. *Strategic Management Journal, 19*(4), 293–318. doi:10.1002/(SICI)1097-0266(199804)19:4<293::AID-SMJ982>3.0.CO;2-M

Håkansson, H., & Ford, D. (2002). How should companies interact in business environments. *Journal of Business Research, 55*, 133–139. doi:10.1016/S0148-2963(00)00148-X

Håkansson, H., & Snehota, I. (Eds.). (1995). *Developing relationships in business networks*. London, UK: Routledge.

Harrison, R. (2004). *IMP as fashion: Past, present and future*. Retrieved January 27, 2011, from http://www.impgroup.org/paper_view.php?viewPaper=4482

Hooley, G., Broderick, A., & Möller, K. (1998). Competitive positioning and the resource-based view of the firm. *Journal of Strategic Marketing, 6*(2), 97–115. doi:10.1080/09652549800000003

Jarillo, J. C. (1993). *Strategic networks: Creating the borderless organization*. Oxford, UK: Butterworth-Heinemann.

Järvensivu, T. (2007). *Values-driven management in strategic networks: A case study of the influence of organizational values on cooperation*. Helsinki School of Economics (academic dissertation).

Johns, N., & Pine, R. (2002). Consumer behaviour in the food service industry: A review. *International Journal of Hospitality Management, 21*(2), 119–134. doi:10.1016/S0278-4319(02)00008-7

Kajalo, S., Rajala, R., & Westerlund, M. (2007). Approaches to strategic alignment of business and information systems: A study on application service acquisitions. *Journal of Systems and Information Technology, 9*(2), 155–166. doi:10.1108/13287260710839238

Komulainen, H., Mainela, T., & Tähtinen, J. (2004). Social networks in the initiation of the high tech firm's internationalization. *Proceedings of the 20th IMP Conference September, 2- 4, 2004*, Copenhagen, Denmark.

Leblebici, H., Salancik, G. R., Copay, A., & King, T. (1991). Institutional change and the transformation of interorganizational fields: An organizational history of the US radio broadcasting industry. *Administrative Science Quarterly, 36*(3), 333–363. doi:10.2307/2393200

Makkonen, H. (2008). Beyond organizational innovation adoption – A conceptual and empirical analysis. *Journal of Business Market Management, 2*(2), 63–77. doi:10.1007/s12087-008-0023-z

Miles, M. B., & Huberman, A. M. (1984). *Qualitative data analysis: A source book of new methods*. Newbury Park, CA: Sage Publications.

Möller, K. (2010). *Value networks and open innovation*. Topics in Open Innovation Series, Haas Business School, UC Berkeley, September 21, 2010.

Möller, K., & Rajala, A. (2007). Rise of strategic nets - New modes of value creation. *Industrial Marketing Management, 36*(7), 895–908. doi:10.1016/j.indmarman.2007.05.016

Osterwalder, A. (2004). *The business-model ontology – A proposition in design science approach*. Academic dissertation, Ecole des Hautes Etudes Commerciales, Université de Lausanne.

Parolini, C. (1999). *The value net: A tool for competitive strategy*. Chichester, UK: John Wiley & Sons Ltd.

Patton, M. Q. (1987). *Qualitative evaluation and research methods*. Newbury Park, CA: Sage Publications.

Poulston, J., & Yiu, A. Y. K. (2011). Profit or principles: Why do restaurants serve organic food? *International Journal of Hospitality Management, 30*, 184–191. doi:10.1016/j.ijhm.2010.04.004

Powell, W. W., Koput, K. W., & Smith-Doerr, L. (1996). Interorganizational collaboration and the locus of innovation: Networks of learning in biotechnology. *Administrative Science Quarterly*, *41*, 116–145. doi:10.2307/2393988

Rodgers, S. (2009). The state of technological sophistication and the need for new specialised tertiary degrees in food services. *International Journal of Hospitality Management*, *28*(1), 71–77. doi:10.1016/j.ijhm.2008.04.001

Rosenbröijer, C.-J. (1998). *Capability development in business networks*. Doctoral dissertation, Swedish School of Economics and Business Administration, Helsinki.

Simpson, R., & Carevic, E. (2004). Designing a modified atmosphere packaging system for foodservice portions on non-respiring foods: Optimal gas mixture and food/headspace ratio. *Food Research International*, *14*(4), 257–272. doi:10.1111/j.1745-4506.2004.tb00194.x

Spekman, R. E., Isabella, L. A., & MacAvoy, T. C. (2000). *Alliance competence: Maximizing the value of your partnerships*. New York, NY: John Wiley & Sons.

Srinivasan, R., Lilien, G. L., & Rangaswamy, A. (2006). The emergence of dominant designs. *Journal of Marketing*, *70*(2), 1–17. doi:10.1509/jmkg.70.2.1

Stake, R. E. (2000). Case studies. In n. K. Denzin & Y. S. Lincoln (Eds.), *Handbook of qualitative research*, 2nd ed, (pp. 435-54). Newbury Park, CA: Sage Publications.

Taylor, S. (2000). Is McDonaldization inevitable? Standardization and differentiation in food and beverage organizations. In Wood, R. (Ed.), *Strategic questions in food and beverage management* (pp. 48–64). doi:10.1016/B978-0-7506-4480-8.50007-3

Teece, D. J. (1986). Profiting from technological innovation: Implications for integration, collaboration, licensing and public policy. *Research Policy*, *15*, 285–305. doi:10.1016/0048-7333(86)90027-2

Teece, D. J., Pisano, G., & Shuen, A. (1997). Dynamic capabilities and strategic management. *Strategic Management Journal*, *18*(7), 509–533. doi:10.1002/(SICI)1097-0266(199708)18:7<509::AID-SMJ882>3.0.CO;2-Z

Tikkanen, H. (1996). Pohjoismaisen verkostolähestymistavan tieteenfilosofiset perusteet. *The Finnish Journal of Business Economics*, *45*(4), 384–403.

Tikkanen, H., Lamberg, J. A., Parvinen, P., & Kallunki, J. P. (2005). Managerial cognition, action and the business model of the firm. *Management Decision*, *43*(6), 789–809. doi:10.1108/00251740510603565

Webber, M. (2004). Bold review required in the foodservices! *The Leading Edge*, *43*, 28–29.

Westerlund, M., & Svahn, S. (2008). A relationship value perspective of social capital in networks of software SMEs. *Industrial Marketing Management*, *37*(5), 492–501. doi:10.1016/j.indmarman.2008.04.003

Yin, R. K. (1994). *Case study research. Designs and methods* (2nd ed.). Newbury Park, CA: Sage Publications.

ADDITIONAL READING

Adler, P. S. (2001). Market, Hierarchy, and Trust: The Knowledge Economy and the Future of Capitalism. *Organization Science*, *12*(2), 215–234. doi:10.1287/orsc.12.2.215.10117

Adler, P. S., Goldoftas, B., & Levine, D. I. (1999). Flexibility versus efficiency? A case study of model changeovers in the Toyota production system. *Organization Science*, *10*(1), 43–68. doi:10.1287/orsc.10.1.43

Alajoutsijärvi, K., Möller, K., & Rosenbröijer, C.-J. (1999). Relevance of focal nets in understanding the dynamics of business relationships. *Journal of Business-To-Business Marketing*, *6*(3), 3–35. doi:10.1300/J033v06n03_02

Anderson, J. C., & Narus, J. A. (1999). *Business Market Management: Understanding, creating and delivering value*. Prentice Hall.

Anderson, P., & Tushman, M. L. (1990). Technological discontinuities and dominant designs: A cyclical model of technological change. *Administrative Science Quarterly*, *35*(4), 604–633. doi:10.2307/2393511

Araujo, L. (1998). Knowing and learning as networking. *Management Learning*, *29*(3), 317–336. doi:10.1177/1350507698293004

Araujo, L., Dubois, A., & Gadde, L.-E. (2003). The multiple boundaries of the firm. *Journal of Management Studies*, *50*(5), 1255–1277. doi:10.1111/1467-6486.00379

Bakingfood.com. www.bakingfood.com.

Birkinshaw, J., Nobel, R., & Ridderstråle, J. (2002). Knowledge as a contingency variable: do the characteristics of knowledge predict organization structure? *Organization Science*, *13*(3), 274–289. doi:10.1287/orsc.13.3.274.2778

Bradach, J., & Eccles, R. (1989). Price, authority, and trust: From ideal types to plural forms. In Scott, W. R. (Ed.), *Annual Review of Sociology* (*Vol. 15*, pp. 97–118).

Brown, J. S., & Duguid, P. (2001). Knowledge and organization: A social practice perspective. *Organization Science*, *12*(2), 198–213. doi:10.1287/orsc.12.2.198.10116

Cartwright, S. D., & Oliver, R. W. (2000). Untangling the value web. *The Journal of Business Strategy*, *21*(1), 22–27. doi:10.1108/eb040055

Castells, M. (1996). *The rise of the network society*. Cambridge, MA: Blackwell.

Chesborough, H. & David Teece. (1996, Jan.-Feb.). When is Virtual Virtuous: Organizing for Innovation. *Harvard Business Review*, 65–73.

Chesborough, H. (2000, Sept./Oct.). Morten Hansen, Nitin Nohria, & Donald Sull (2000), Networked Incubators: Hothouses of the New Economy. *Harvard Business Review*, *78*(5), 74–84.

Cova, B., & Cova, V. (2002). Tribal marketing: The tribalisation of society and its impact on the conduct of marketing. *European Journal of Marketing*, *36*(5/6), 595–620. doi:10.1108/03090560210423023

Cravens, D. W., Shipp, S. H., & Crawens, K. S. (1994). Reforming the traditional organization: The mandate for developing networks. *Business Horizons*, *37*(4), 19–28. doi:10.1016/0007-6813(94)90043-4

Crowston, K., & Scozzi, B. (2002). Open source software projects as virtual organisations: competency rallying for software development. *IEE Proceedings, 149(1)* (Online No 20020197).

Dougherty, D. (1992). Interpretative barriers to successful product innovation in large firms. *Organization Science*, *3*(2), 179–202. doi:10.1287/orsc.3.2.179

Doz, Y., Santos, J., & Williamson, P. (2001). *From global to metanational*. Boston: Harvard Business School Press.

Doz, Y. L., & Hamel, G. (1998). *Alliance Advantage: The art of creating value through partnering*. Boston, MA: Harvard Business School Press.

Dyer, J. H. (1996). Specialized supplier networks as a source of competitive advantage: Evidence form the auto industry. *Strategic Management Journal, 17*(4), 271–292. doi:10.1002/(SICI)1097-0266(199604)17:4<271::AID-SMJ807>3.0.CO;2-Y

Dyer, J. H., & Nobeoka, K. (2000). Creating and managing a high-performance knowledge-sharing network: The Toyota case. *Strategic Management Journal, 21*, 345–367. doi:10.1002/(SICI)1097-0266(200003)21:3<345::AID-SMJ96>3.0.CO;2-N

Dyer, J. H., & Singh, H. (1998). The relational view: Cooperative strategy and sources of inter-organizational competitive advantage. *Academy of Management Review, 23*(4), 660–679.

Dyer, J. H., & Singh, H. (2000). Using alliance to build competitive advantage in emerging technologies. In G. S. Day P. J. H. Schoemaker & R. E. Gunther (Eds.), *Wharton on managing emerging technologies*. New York: John Wiley & Sons, Inc.

Granovetter, M. (1973). The strength of weak ties. *American Journal of Sociology, 78*, 1360–1380. doi:10.1086/225469

Grant, R. M. (2007). *Contemporary strategy analysis concepts, techniques, applications* (6th ed.). Oxford, UK: Blackwell Publishing.

Håkansson, H., & Persson, G. (2004). Supply chain management: the logic of supply chains and networks. *International Journal of Logistics Management, 15*(1), 11–26. doi:10.1108/09574090410700202

Hinterhuber, A. (2002). Value chain orchestration in action and the case of the global agrochemical industry. *Long Range Planning, 35*(6), 615–635. doi:10.1016/S0024-6301(02)00160-7

Hudson, B. (1994). Innovation through acquisition. *The Cornell Hotel Administration Quarterly, 35*(3), 82–87.

Hudson, B. (1997). Industrial Cuisine Revised. *The Cornell Hotel and Restaurant Administration Quarterly*, (June): 81–86.

Husler, C., & Ronde, P. (2007). The impact of cognitive communities on the diffusion of academic knowledge: Evidence from the networks of inventors of a French university. *Research Policy, 36*, 288–302. doi:10.1016/j.respol.2006.11.006

Jarillo, J. C. (1993). *Strategic networks: Creating the borderless organization*. Oxford, UK: Butterworth-Heinemann.

Lambe, C. J., Spekman, R. E., & Hunt, S. D. (2002). Alliance competence, resources, and alliance success: Conceptualization, measurement, and initial test. *Journal of the Academy of Marketing Science, 30*(2), 141–158. doi:10.1177/03079459994399

Lipparini, A., & Fratocchi, L. (1999). The Capabilities of the Transnational Firm: Accessing Knowledge and Leveraging Inter-firm Relationships. *European Management Journal, 17*(6), 655–667. doi:10.1016/S0263-2373(99)00056-0

Morland, K., Wing, S., Roux, A., & Poole, C. (2002). Neighborhood Characteristics Associated with the Location of Food Stores and Food Service Place. *American Journal of Preventive Medicine, 22*(1), 24–29. doi:10.1016/S0749-3797(01)00403-2

Normann, R., & Ramirez, R. (1993). From Value Chain to Value Constellation. Designing Interactive Strategy. *Harvard Business Review, 71*(4), 65–77.

Porter, M. E. (1985). *Competitive advantage: Creating and sustaining superior performance*. New York: Free Press.

Van Stamm, B. (2003). *Managing Innovation, Design & Creativity* (p. 90). West Sussex, England: Wiley.

Weick, K. (1995). *Sensemaking in organizations*. Thousand Oaks, CA: SAGE Publications, Inc.

Yin, R. K. (1994). *Case Study Research - Design and Methods*. Beverly Hills, CA: Sage Publication.

Zollo, M., Reuer, J. J., & Singh, H. (2002). Interorganizational routines and performance in strategic alliances. *Organization Science*, *13*(6), 701–713. doi:10.1287/orsc.13.6.701.503

KEY TERMS AND DEFINITIONS

Activities: Seem to appear in sequences or in chains (Porter 1985). In this view, company is seen as a set of activities that are performed to design, produce, market, deliver and support its product.

Actor: An organization or an individual that perform activities and control resources and create value via transformation of resources.

B2B: Business to Business.

B2C: Business to Customers.

Business Network: Any intentional collaborative form of a clearly defined group of actors having observable connections, processes and rules of conduct with each other. Participants of a business net are aware of their connections and they have chosen to participate based on perceived value, benefits and/or strategic advantages.

Hub: A hub company is the leading company in a business network. It orchestrates the other actors and guides their actions. It decides the goals of the business network and the strategy how to reach the goals. There can be several other actors in a business network, but only one hub company.

Network Position: Defined by the characteristics of the company's relationships and the benefits and commitments that arise from them.

Networking: This term encompasses all of the interactions of a company or individual in the network.

Relationship: The pattern of interactions and the mutual behavioral conditions over time between the actors in a business network.

Value System: Sets of value chains based on activities. Value system can also be defined as sets of organizations.

Value Network: The term describes a value system as a set of activities in a network form instead as a set of organizations.

Section 3
Entrepreneurships, Strategies and Crisis Management

Chapter 10
Dynamic Learning as Entrepreneurial Action in the Context of Open Innovation:
An Instrumental Case from a Communities-of-Practice Perspective

Nicholas Theodorakopoulos
Aston University, UK

Catarina Figueira
Cranfield University, UK

Nada Kakabadse
University of Northampton, UK

Andrew Kakabadse
Cranfield University, UK

ABSTRACT

Recent scholarly discussion on open innovation put forward the notion that an organisation's ability to internalise external knowledge and learn from various sources in undertaking new product development is crucial to its competitive performance. Nevertheless, little attention has been paid to how growth-oriented small firms identify and exploit entrepreneurial opportunities (i.e. take entrepreneurial action) related to such development, in an open innovation context, from a social learning perspective. This chapter, based on an instrumental case-firm, demonstrates analytically how learning as entrepreneurial action takes place, drawing on situated learning theory. It is argued that such learning is dynamic in nature and is founded on specific organising principles that foster both inter- and intracommunal learning.

DOI: 10.4018/978-1-61350-519-9.ch010

INTRODUCTION

Entrepreneurial action entails creating new resources or combining existing resources in new ways to develop and commercialise new products, move into new markets/service new customers and/or introduce new organising processes (Sexton and Smilor, 1997; Hitt *et al.*, 2001; Gartner *et al.*, 2003). Growth-oriented small firms, the driving engine of economies across the world, are conceived of typifying entrepreneurial action, related to identifying and exploiting successfully entrepreneurial opportunities (see for instance Davidsson *et al.*, 2002). By default, such entrepreneurial action is integrally related to innovatory activity (Schumpeter, 1934).

Gibbons *et al.* (1994) advanced the idea of a mode 2 knowledge production, a new way to generate knowledge within a dynamic, interactive system of multiple and diverse actors who are both users and co-producers of knowledge. In a similar vein, it has been argued that many firms, in order to sustain their ability to introduce new products to the market successfully, have shifted to a model of open innovation that exploits the knowledge of a wide range of actors (Chesbrough, 2003). In contrast to closed innovation, firms innovating in an open innovation model are able to use external ideas and knowledge in conjunction with internal R&D efforts to enhance their absorptive capacity (Cohen & Levinthal, 1990) and sustain innovation. Boundary-spanning linkages and internal mechanisms for effective knowledge sharing such as cross-functional teams are central to the notion of open innovation. The former allow access to external knowledge, whilst the latter enable its effective integration with extant knowledge base at a group/organisational level (Chesbrough, 2003, 2006; Chesbrough & Crowther, 2006). Overall, open innovation can be seen as a holistic approach to innovation management, "which refers to systematically encouraging and exploring a wide range of internal and external sources for innovation opportunities, consciously integrating that exploration with firm capabilities and resources, and broadly exploiting those opportunities through multiple channels" (Bröring & Herzog, 2008, p335).

From a social constructionist/practice-based perspective, Situated Learning Theory (SLT) (often referred to as Communities of Practice (CoP) theory) (Lave and Wenger, 1991; Brown and Duguid, 1991; Brown and Duguid, 1998; Wenger, 1998; Wenger, 2000) offers a potent theoretical lens for enhancing understanding of entrepreneurial action in an open innovation context. Importantly, it enables the examination of the phenomenon on the platform of social, participative practice, in stark contrast with conventional cognitive learning approaches (Geraldi *et al.*, 1998). SLT's centrepiece, the notion of CoP, provides the embedding generative framework for the development of new knowledge, emphasising the need to understand learning and knowing related to innovation as social micro-processes (Lave and Wenger, 1991; Brown and Duguid, 1991; Wenger, 1998; Brown and Duguid, 1998; Brown and Duguid, 2001a, b; Tsoukas, 2002; Huysman, 2004; Tsoukas and Mylonopoulos, 2004).

SLT has been gaining momentum in organisational studies concerned with learning. However, notwithstanding the recent calls for taking a practice-based approach to entrepreneurial learning (Rae and Carswell, 2001; Rae, 2002, 2004; Taylor and Thorpe, 2004; Thorpe *et al.*, 2005) – with the exception of a few peripheral studies (Theodorakopoulos *et al.*, 2005; Theodorakopoulos & Ram, 2006) - SLT has not been used systematically to any length in this context. The chapter relates to this literature and adds value by considering how new knowledge underpinning entrepreneurial action in growth-oriented, small firms is created in the context open innovation. This entails how such knowledge is generated in shared spaces where multiple actors operate and how then it is internalised and used to develop new products in an open innovation context. This is an important line of inquiry as, notably, little attention has been paid to how small firms engage in open innovation abilities (Spithoven *et al.*, 2010).

Following from this, SLT is applied as the major analytical lens in this study to investigate the learning processes relating to entrepreneurial action in technology-based, growth-oriented small firms, which engage in open innovation. The components highlighted by Brown and Duguid (1998) and Wenger (1998, 2000) form the crux of the conceptualisation used to address learning as entrepreneurial action in open innovation context. The rest of the chapter is organised as follows. The next section provides the theoretical background, delineating SLT and its constituent elements. This is followed by a section outlining the research design of the study this chapter is based upon. Following from this, the ensuing section presents an instrumental case study (Stake, 1995). The penultimate section discusses and concludes the chapter, followed by the final section, which indicates avenues for further research.

THEORETICAL BACKGROUND

The point of departure is the contention that organising for entrepreneurial action in the context of open innovation entails the construction of an organisational community – "a community of employees rather than a company of employees constrained by traditional hierarchical configurations" Ireland and Hitt (2005, p.67). The citizens of such organisational communities are seen as boundary spanners, through which reciprocal knowledge sharing and learning takes place (e.g. Brown & Duguid, 1991, 1998; Wenger, 1998). Situated learning theory (SLT) aligns well with this view and offers a potent analytical lens for examining the role of organising for effective learning processes underlying entrepreneurial action in growth-oriented, innovatory small firms.

This theory has gained momentum during the last decade, providing an alternative to conventional organisational learning approaches. Its primary tenet is that learning is a fundamentally a social phenomenon reflecting the social nature of human beings capable of knowing and 'it is understood as the development of a new identify based on participation in the system of situated practices' (Gherardi, *et al.*, 1998, p.276). The central construct within situated learning theory is the notion of 'community of practice' (CoP). 'Communities of practice are groups of people who share a concern, a set of problems, or a passion about a topic, and who deepen their knowledge and expertise in this area by interacting on an ongoing basis' (Wenger *et al.*, 2002, p.4). For Wenger, the construct CoP constitutes a point of entry into a broader conceptual framework, which underscores the importance of community, practice, learning, meaning and identity as elements that '...are deeply interconnected and mutually defining' (Wenger, 1998, p.5). These components illuminate the learning process, pointing out what matters about innovating as learning, as well as the nature of knowledge and knowing.

Situated Learning theorists (Brown & Duguid, 1998; Wenger, 2000; Swan *et al.*, 2002; Wenger *et al.*, 2002; Scarbrough *et al.*, 2004) maintain that learning and innovation (or innovation as learning) takes place in the interface between members of interrelated CoPs. Constellations of interrelated CoPs form a 'social learning system' (Wenger, 2000) that produces an 'ecology of knowledge' (Brown & Duguid, 1998). Employees, beyond being members of their organisational CoP, also belong to other broader CoPs, owing to their professional networks and specialisation (Brown & Duguid, 2001a; Swan *et al.*, 2002). For instance, knowledge workers of software development companies such as Alpha-Tech, including engineers, technical architects, consultants and salespeople, all possess specialist skills and knowledge to variant types and degrees, which gives them access to their wider 'professional' CoPs, beyond their organisation. Connected in this way, these boundary spanners "can rely on a complex system of overlapping communities, common backgrounds, and personal relationships to help evaluate and propagate knowledge" (Brown & Duguid, 1998, p.102). Hence, knowledge generated inter-organisationally, i.e. in the wider, professional CoP, can be transferred into the organisation and vice versa.

Conversely, knowledge that may not be shared within an organisation because of epistemic differences between employees, may flow out easily to other firms whose employees work for competitors but are members of the same wider, professional CoP and share the same epistemic principles (Brown & Duguid, 1998, 2001a). Brown and Duguid refer to this phenomenon as the simultaneous stickiness and leakiness of knowledge. Such simultaneously sticky and leaky knowledge can be market or technology-specific and is pertinent in identifying and exploiting entrepreneurial opportunities. Therefore, building a strong organisational CoP and creating cohesion between employees belonging to different professional CoPs becomes important for enabling inflows and deterring outflows of knowledge. Whilst it is difficult to achieve this in large organisations due to specialisation of knowledge and expertise resulting from division of labour, the small firm offers a fertile ground. The small firm is typically characterised by a flexible and multitasking workforce, where employees have a better understanding of each others' work, oftentimes having to perform various functions, in different areas.

Wenger's (1998, 2000) conceptualisation highlights the importance of three modes of belonging to a community of practice: engagement, imagination and alignment. Engagement refers to social participation of members locally, in the organisational activities. Imagination relates to being able to think of oneself as a member of a wider community beyond local practice, while alignment refers to coordinating perspectives, interpretations and actions so that higher goals are realised. Wenger (*ibid*) argues that enhancing these three modes of belonging improves three key characteristics of a CoP: its quality (higher learning energy), its boundary experiences (i.e. more generative interfaces with other CoPs) and the identity of those belonging to a CoP (e.g. identities become more receptive to new ideas and possibilities). In turn, a stronger CoP with more productive boundaries and healthier identities underpin the ability to create new meaning. That is, an organisation as a potent CoP is more capable of identifying and exploiting entrepreneurial opportunities, which in turn has positive impact on business performance.

With regard to building a potent organisational CoP to underpin innovatory learning and entrepreneurial action, Wenger (1998, p.22) argues that 'if learning occurs naturally then what is needed is not to create learning, but rather to create the circumstances that make learning empowering and productive'. In this respect, organising for learning pertaining to entrepreneurial action in an open innovation context alludes to paying due regard to factors that can enhance communal learning. Boundary events, brokers and translators, boundary objects, membership structures and inclusive participation are highlighted as crucial organisational design elements in this respect (Brown & Duguid, 1998; Wenger, 1998; 2000). Prescriptive recommendations relating to organising interventions concerned with nurturing and shaping CoPs to enhance learning capabilities and performance bring to fore the aforementioned design elements (Lesser & Everest, 2001; Wenger *et al.*, 2002; Plaskoff, 2003).

Although such prescriptive approaches provide a compass useful for practitioners, such popularised versions of SLT tend to disregard or suppress Lave and Wenger's (1991, p.42) initial position that learning processes are integral to the exercise of power and control, which can result in "alienation from full participation in learning… Hegemonic relations truncate possibilities for identities of mastery within CoPs". In popularising SLT (e.g. Wenger, 1998; Brown and Duguid, 2001a, b; Wenger *et al.*, 2002) some radical, key elements relating to power relations and politics have been marginalised or disregarded. Contu and Willmott (2000, 2003) in their discourse bemoan this shortfall and are particularly critical of the way by which 'situated learning' and the concept of CoPs has been promoted within a managerialist/functionalist remit as a medium or even as a technology of consensus and stability. Contu and Willmott (2003, p.284) assert that "challenging and innovative elements [of SLT], such as the idea that learning practices are shaped, enabled and constrained by relations of power are dimly recognized or discarded". As they succinctly put it in an earlier article, "In Communities of Practice and Social Learning Systems, references to alienation and blocked mobility have completely disappeared to be replaced with concepts about specifying the learning technologies that will facilitate the realisation of a seemingly consensual vision of progress" (Contu & Willmott, 2000, p.271).

In the same vein, Fox (2000) lends support to the argument that the popularisation of Lave and Wenger's (1991) conceptualisation of learning situated in CoPs has selectively appropriated notions that are congruent with a managerialist mindset – discarding subversive insights relating to power and conflict. He critically notes the shift from earlier participation in an analytic community engaged in practices related to enhancing understanding for emancipation purposes to an analytic community concerned with prediction and control for improving performance. In this shift, "wider issues of power and conflict are safely tucked away" (Fox, 2000, p.403). The fact that Brown and Duguid (2001a, p. 210) mention only in the footnotes of their discourse that "in most organizations the power to negotiate is not equally distributed…parties negotiating around boundary objects rarely negotiate on equal terms" lends support to Fox's (2000) contention. Therefore, it is postulated that a more central consideration of power is needed to extend the contribution of CoPs/SLT (Contu & Willmott, 2000, 2003; Fox, 2000; Marshall & Rollinson, 2004). Applying SLT as the major analytical lens without paying due regard to critical issues relating to power relations and political activity can obscure elements that may be pivotal in enhancing understanding of the learning and knowing processes underlying entrepreneurial action in growth-oriented small firms. The next section outlines the research design of the study this chapter is based upon.

RESEARCH DESIGN

The empirical part of the paper grounds SLT on a longitudinal case study. This entails a growth-oriented small firm operating in the Computer Software and Services sector. The latter constitutes a new economy, growing sector, representative of high knowledge-content services, where open innovation is most prominent. Firms operating in this sector are of paramount socioeconomic significance, as they influence considerably productivity in almost all industrial sectors and enable small firms to compete more effectively with large firms (OECD, 2003; EITO, 2006).

More than this, professional/knowledge-based service firms, such as the Computer Software and Services sector, are often considered as model firms with regard to entrepreneurial behaviour and knowledge creation for firms in other industries (e.g. Alvesson, 2000; Empson, 2001; Løwendahl *et al.*, 2001). In the same line, logic suggests that such firms afford better opportunities to gain insight into the dynamics of organising to learn in an open innovation context, than those by traditional firms. Hence, the case-firm examined in this study is purposively selected on theoretical grounds, aiming at analytical generalisation to theory - i.e. not statistical generalisation (Yin, 2003). AlphaTech as a knowledge-based, growth-oriented, innovatory small firm constitutes an instrumental case (Stake, 1995).

Rapport with the firm was developed over eight months, involving in-depth personal interviews with key decision makers/key agents, knowledgeable of growth-related entrepreneurial action. These key informants had experienced directly and contributed significantly to entrepreneurial growth attributable to new product development, in different capacities; from process manager, to technical architect, to project manager, to sales manager, and to Managing Director. During the interviews participants discussed a learning episode of entrepreneurial action in detail (i.e. learning is seen as involving both cognitive change and action in accord with Vera and Crossan, 2004). The learning episode technique is a variation of the 'critical incident' or 'critical event method', which has been applied to entrepreneurship studies (Deakins & Freel, 1998; Cope & Watts, 2000; Cope, 2003; Chell, 2004; Zhang *et al.*, 2006). Focus was on the nature of organising for entrepreneurial action relating to new product development, which took place over a period of eight months approximately and contributed significantly to the growth of the company.

The interviews aimed at encouraging the participants to expand on the process that led to the episode, what caused it, how it was resolved, who was involved in that process, what was learnt, how and when. The purpose was to obtain an insight on how issues evolved over time and where and how ideas were developed and integrated, in order to understand how the process of entrepreneurial action was effected in an open innovation context.

The main data collection method was formal interviews with key informants that had contributed significantly and experienced directly the entrepreneurial action of Alpha-Tech, in different capacities. Formal interviews were supplemented by extensive informal discussions during lunch breaks and casual encounters outside the premises of Alpha-Tech, as well as non-participant observation and examination of relevant documentation and archival data, where available. Concerted efforts were made to triangulate sources to address the problems associated with retrospection, mainly lapses in memory and *ex post* rationalisation (Easterby-Smith *et al.*, 2002), In congruence with Weick (2002), studying the phenomenon retrospectively provided the opportunity to capture the respondents' reflection on what happened during this episode of entrepreneurial action not as an isolated incident, but as an indicator of wider tendencies (Cope & Watts, 2000; Cope, 2003) in organising for acting entrepreneurially in an open-innovation context.

An analytic description of the phenomenon studied in Alpha-Tech supported extension logic (Yin, 2003), following Miles and Huberman's (1994) procedure. The QSR NVivo software was used to code and retrieve, which proved quite useful in handling large volumes of data and managing complexity. Notably, the package was also useful in linking ideas, exploring patterns and in creating an audit trail/case-study database so that the findings are trustworthy (Yin, 2003).

ALPHA-TECH AS AN INSTRUMENTAL CASE

Organisational Context

Strategic Orientation and Organisational Structure

Alpha-Tech provides information assurance technology for software applications. It is a small firm based in the UK, with representation in Canada, the US and Australia. Overall, it employs 48 people, of which 70% are technical staff. Founded in 1990, the company specialises in delivering sophisticated messaging and directory solutions to military, national defence and other government departments as well as to organisations in the financial and communications services sectors worldwide. Its two main markets are defence and government. Its customer base comprises a host of large organisations, including the UK Ministry of

Defence (MoD), HM Revenue and Customs, the Australian, Canadian and Luxembourg Departments of Defence, NATO (North Atlantic Treaty Organisation), the National Security Agency, the US Navy, the Foreign and Commonwealth Office, the Government of Canada, and the Queensland Government. Its solutions apply to environments where the sending and receiving of a message is critical. In these niche markets the value of communications is high, the need for secure, reliable and auditable message handling is paramount and the existing infrastructure is typically complex. Alpha-Tech works mainly as a component provider either to systems integrators such as BT, EADS, EDS, EID, Fujitsu, Hewlett Packard, IBM, LogicaCMG, or to the end customer. By participating in such procurement systems Alpha-Tech engages actively in open innovation.

Alpha-Tech's messaging solutions allow organisations to build on their existing infrastructure to address security needs and apply accountability, traceability and control to their messaging/directory environment. The four elements that its solutions include are: security, data labelling, access control and roles-based messaging. The firm provides a full technology and service capability to support secure information assurance programmes. This includes consultancy and support on messaging, directory and security technologies, supply of technology components for system solutions, implementation services to install, configure and integrate Alpha-Tech's technology into the final solution, as well as analysis, design, development and maintenance of new or enhanced functionality.

The Board of Directors consists of the Managing Director, the Finance Director, the Sales and Marketing Director and the Professional and Technical Services Director. Work is organised around the above three areas, with middle level managers being responsible for each department. At its broadest, there are three important kinds of knowledge in Alpha-Tech: management specific, market specific and technology specific. The 'dominant coalition' (Daft and Weick, 1984; March and Simon, 1993), comprising the company's Directors and functional managers, is concerned with the management of strategic and tactical activities. The business development team and the consultants are more concerned with market developments and customer needs, while the technical architects and the software engineers are rather technology-focused and follow closely the evolution of standards and trends. These groups have a sound knowledge in their areas. They are significant for generating, diffusing and exploiting knowledge in Alpha-Tech and perform distinct roles in the company. The structure of Alpha-Tech is relatively flat, enabling rapid communication between members of different functions.

The company's approach to growth is to develop new products for existing markets. In business parlance, Tony, a Senior Technical Architect of Alpha-Tech, commented on the strategic orientation of the firm in this way:

It's very risky to do a new product in a new market, so what we are doing is... the focus we've got is to do more products, new products, but to our existing market. What's crucial is to lead the way, to set the standards, so that others have to develop their products to fit ours – that's key. (Tony, Senior Technical Architect)

Organisational Culture

The management team promotes a culture of continuous development. Central to this are efforts to recognise and build on each individual's talents so that employees find enjoyment and satisfaction in their work. The firm is recognised as an Investor in People (IiP) and is accredited to ISO9001/TickIT [1] and BS7799 [2]. Staff development is regarded as crucial for its competitive performance and all employees are encouraged to pursue training, improve qualifications and further education opportunities. According to the respondents, looking after and considering the views expressed by the staff promotes "engagement"; a culture of enthusiasm and professionalism that strengthens Alpha-Tech's organisational community, with positive effects on the competitiveness of the firm. On this point, Brian maintained:

We're trying to embed that culture [of enthusiasm and professionalism] in everybody. I am not saying we got it right yet. It is a really difficult thing. I think the answer is that people must be confident and they need to be skilled. People here are encouraged to develop themselves, study, take up courses, etc. And they are encouraged to participate actively. This concept of engagement is fundamental, because if people are engaged in the business they will do that extra bit. They will, you know, if they feel part of it, if they feel they can make a difference then people will do...How can we engage them? I think the way you do that is you make them feel that their value to the business is beyond the nine to five. Their opinions count. All right, you know about mushroom management. Keep mushrooms in the dark, open the door, top more shit and then you get mushrooms. If you treat people like that, then they will not engage, but if you go to seek their opinions and their opinions are respected... One of the things that we are doing is every other quarter they get to meet a director that they don't work for. They will sit and just talk through about what they think about the business, what can be improved and that then gets into a matrix and gets presented back to the business as an action plan. (Brian, Managing Director)

Evidence of the importance placed on the above is the post of 'Business Improvement Manager', held by Elaine, which was created in 2002 to streamline the various activities within the business. Elaine, commenting upon her work in Alpha-Tech, referred to the inherent tension in organising and the act of balancing practice and procedure i.e. having procedures in place that maximise efficiency without stifling creativity (Brown and Duguid, (2000a, 2001b). Balancing practice and procedure provides the ability "... to foster invention – by allowing new ideas to spark – and to further it - by implementing those same ideas" (Brown and Duguid, 2000b, p.74).

What we've tried to do is to set a basic structure, procedures for doing work, for efficiency purposes, that are not too restrictive, so that our process and our people are actually balanced....Of course those [procedures] change are not set in stone. What was right for us five years ago isn't right for us now...We take into account how people feel about it, we consult with them and in the past there have been changes to integrate work and enable people to be more proactive, if you like. Too much structure, too many procedures can have the opposite effect. People feel constrained, they become suspicious, they can't be creative... it can kill initiative and it's difficult to integrate work. (Elaine, Business Improvement Manager)

Hitt *et al.* (2001, p.505) maintain that culture consists of a "complex set of ideologies, symbols and core values that is shared throughout the firms and influences the way it conducts its business... and is the central task of strategic leaders". However, Schein (1997) points out that culture cannot be engineered, claiming that, at best, a stage can be set for it to evolve. Setting the stage for nurturing a culture conducive to business performance, Alpha-Tech has introduced a set of core values. Yet, it is noteworthy that in Alpha-Tech culture is regarded as something that can not only be nurtured but also measured. As Elaine pointed out, these values are stated in artefacts such as internal documents and form an integral part of staff assessment and performance management:

We have a set of values. We've worked very hard to make those an intrinsic part of the business. They are in no way, shape or form a rocket science, but they are something that we use extensively throughout the business. So, we have customer orientation, communication flexibility, initiative, accountability. And we use those, as you can see, as a fundamental part of our performance management process. Everyone is aware of them, because they were told about them, they are posted on our intranet as well...you know, there are some things that are very visible. They are referred to in-company meetings and department meetings, they can be, whatever the context they would be put around. Particularly in performance reviews. We actually have three measures, basically exceptional, acceptable and unacceptable and to add some gradation into that, we have plus, equals and minus. So, we actually have a grading of nine places. (Elaine, Business Improvement Manager)

Notably, such an approach militates against the perennial problem of knowledge hording and stickiness in professional firms (Alvesson, 2000, 2001). The next subsection deals with Alpha-Tech's operating environment.

Operating Environment

Computer Software and Services Industry Profile

Ovum (2006) estimates the UK Computer Software and Services industry at £26 billion, accounting for around 7% of the world output. Outsourcing IT operations to Computer Service companies has been a dominant trend since the beginning of the 1990s, which can be attributed to a combination of factors. These include dynamics relating to organisations concentrating on their core competence, the rising of mergers and acquisitions which often require computer systems to be modified or expanded, along with the need for solutions that are sophisticated, robust, reliable, scalable, interoperable and capable of dealing with geographic dispersion (Key Note, 2004; Forrester Research, 2006; Mergent, 2006).

Computer Software and Services industrial structure is characterised by 'dualism', where there is a symbiosis of small-scale niche firms with global multinational companies. Although the sector is dominated by 10 large corporations - IBM, EDS, CSC, Accenture, Fujitsu Services, Hewlett-Packard, BT, CapGemini, Atos Origin and LogicaCMG - there are clear variations as to the numbers of companies operating in the different segments of the industry. This is due to the relatively low entry barriers, particularly in terms of financial and capital requirements in some sub-sectors, which is conducive to the generation, rapid growth and innovative output of new firms within many advanced industrial economies, such as the UK (Key Note, 2004). According to Business Monitor National Statistics data, in 2003 there were approximately 92,000 businesses in the UK engaged in the Computer Software and Services sector. Table 1 below portrays a break-down of the sector by turnover size-band.

The industry consists of five sub-sectors: Custom/Bespoke Software Development; Systems Building; Support Services; Operational Services; and Planning and Design. Systems Building is the largest sub-sector of computer services. Although Alpha-Tech operates predominantly as a bespoke software developer and support provider, it undertakes activities related to all sub-sectors.

Environmental Dynamism and Complexity

The Computer Software and Services sector has entered a mature phase of development marked by fierce competition and consolidation, especially since 2001. To deal with competitive pressures and improve scale and margins, major multinationals have been expanding, gaining market share and strengthening product and service portfolios through merger and acquisition activities. Since 2001, a number of major mergers and acquisitions

Table 1. Number of UK VAT-based enterprises in the computer software and services industry by size-band – 2003

Turnover Size-band (£000)	Number of firms	% of Total
1-49	23,760	25.8
50-99	35,305	38.3
100-249	20,380	22.1
250-499	4,965	5.4
500-999	3,215	3.5
1,000-4,999	3,430	3.7
5,000+	1,110	1.2
Total	**92,165**	**100.0**

Source: Key Note (2004) - National Statistics/Business Monitor PA1003

have taken place, which has had a significant impact on the sector. Hewlett-Packard and Compaq, IBM and PricewaterhouseCoopers, Atos Origin and the IT operations of KPMG, Logica and CMG, Cap Gemini and Ernst & Young constitute conspicuous examples (Key Note, 2004). Moreover, acquisition activities include the takeover of previously independent firms and therefore small firms remain vulnerable (Mergent, 2006). In discussions over the dynamism of the external operating environment, respondents of Alpha-Tech commented upon this process in the industry as a major force of change that can be challenging:

I would say that the environment is dynamic, not so much in terms of technological change, although new standards are being developed, but mostly in terms of business developments, like mergers and take-overs. So as large system integrators consolidate, you may find yourself in a position where now you are competing with a customer because the company they bought makes what we make...In that regard, the environment is dynamic and changes can have serious implications as to who you compete with and who you collaborate with in producing something new, where the new ideas for new products and innovative features come from if you like... (Brian, Managing Director, Alpha-Tech)

Given security concerns related to the surge of terrorism and other international crime threats, there has been an increasing demand for ICT security services (AIIM International, 2006). Regarding the former, the recent transformation of the military, national security and policing doctrines influences the design and deployment of command, intelligence and communication chains. This transformation highlights the need to improve the interoperability of systems used by different government departments, agencies and armed forces (MoD, 2004). Hence, the ability to respond rapidly against time-critical targets such as those related to national security is essential and requires strategic ICT systems that can support interoperability and mobility.

This has led to innovations that offer wireless technologies, multi-functionality and cross-platform compatibility, i.e. the ability to perform a variety of tasks across a wide range of devices. Forecast International (2006) projects sales of communications systems for the U.S. military (the leading market) to reach $2 billion over the next 10 years. Over the same period, nearly $0.7 billion will be spent on communications research and development projects which aim to develop rapidly reconfigurable, affordable, robust, interoperable, and evolvable communications technologies that support network-centric warfare. These developments favour firms like Alpha-Tech who operate in the forefront of this niche-market.

However, a major challenge compounding problems related to delivery and support in the sector is complexity, as the software (and hardware) must be effectively integrated right from the outset (Gartner, 2006). This in turn fosters further consolidation, but also collaboration between firms providing complementary services. Therefore, the industry is characterised by complex supply chains, which further increases environmental uncertainty (Duncan, 1974). Firms within the sector act as suppliers/customers to and often compete with each other (Ovum, 2006). A software developer can act as an outside supplier to a computer services firm, while software publishers may buy in the services of a service provider. Multinationals like IBM, Hewlett-Packard and Oracle operate service divisions alongside product divisions (either software and/or hardware) and compete with their suppliers in certain areas. In conversations on the complexity of the operating environment, respondents of Alpha-Tech referred to complex relations in their supply chains in this way:

It's quite complex. Large system integrators like Fujitsu or EDS can be customers or competitors. System integrators subcontract work to us because they do not have the resources to do it – you see, it's quite niche our area. Other times they may develop something in-house so they are in competition with us. And sometimes we work with competitors, other firms like us, as partners on a given project led by the systems integrator. (Elaine, Business Improvement Manager, Alpha-Tech)

We often have to work together with different vendors and make our products work with theirs. That can be very tricky, depending on how the other party behaves, i.e. how collaborative they are and often requires an awful lot of effort. (Tony, Senior Technical Architect, Alpha-Tech)

It is noteworthy that computer services companies are amongst the largest clients of computer services in the UK (Key Note, 2004). Most computer services companies - especially those involved in systems integration services, like Alpha-Tech develop - strategic relationships with key suppliers/customers (UN, 2002). In this respect, the adoption of more widespread, international software standards and certification procedures in safety-critical areas has allowed users to feel more comfortable about using non-indigenous suppliers (OECD, 2003). A number of these standards have specifically been set at an international level by bodies such as the European Space Agency. Notably, this in turn has helped Computer Software and Services firms like Alpha-Tech standardise their products for a wider market and internationalise their activities. Against this backcloth, the competitiveness of firms operating in the UK Computer Software and Services industry, and more specifically within the defence market, is determined by their ability to capitalise on a knowledge-base, which to a large extent is co-created by different actors, who at times act as competitors. The capability to partner with organisations that offer complementary skills and competencies in one project and compete with them in another is crucial.

Since 1990, having a close relationship with standards organisations such as the International Telegraphic Union (ITU) and the Internet Engineering Task Force (IETF), Alpha-Tech has been contributing to the development of the technology that is used today in information assurance. Through this process it gained technical competencies related to developing and applying secure messaging software. Based on these competencies and expertise, Alpha-Tech entered the defence market and subsequently developed a significant compression technology. The latter constitutes a learning episode of entrepreneurial action in Alpha-Tech and is dealt with in the next section

Entrepreneurial Action in Open Innovation Context

A Significant Learning Episode of Entrepreneurial Action - Developing a Compression Technology

There has been a significant increase in deployed military operations, where the requirements of mobility and interoperability of communications is ever-significant. In mobile operations, of particular importance is the ability to communicate across low bandwidth environments, between strategic and tactical domains, where networks are rather unstable and throughput capacity is limited. This creates a need for compression techniques that can address these requirements. NATO has developed STANAG 4406 Annex E, a draft standard which describes a tactical protocol to be used for the interconnection of MMHS over networks with reduced throughput capacity. This presented Alpha-Tech with an opportunity for engineering a product that could support this standard. Building on its capabilities of providing MMHS, from November 2004 until June 2005 Alpha-Tech developed a compression technology that conforms to the NATO standard in question. The product is a 'message transfer agent' that enables mobile and tactical messaging in support of land, sea and air operations. It ensures that all elements of message handling flow smoothly and can be shared across deployed forces in low-bandwidth, mobile and tactical environments and enables existing messaging infrastructures to be used in conjunction.

There were two main reasons for developing the new compression technology. The first refers to the need to gain an edge over competitors. The second relates to the wish to provide comprehensive solutions and compete at a different level. On this point, the respondents maintained:

It was initially called the CTE, for the tactical environment where you have less communication or less stable network... We are the first to develop such a product and I think this strengthens our position in the defence market. You see, we needed to develop something superior and also, most important, to set the rules of the game if you like.(Tony, Senior Technical Architect)

So there is, you know, there is host of different issues for us in this. Beyond the fact that we set the pace with this technology, so far we've been a software house, but this helps us become a solutions house... which means bundling products, cross-selling, better margins and competing at a different level. (Brian, Managing Director)

Beyond solidifying its position as a software house, Alpha-Tech aims at enhancing the range of services offered. Its new product offer, the CTE, integrates different components developed by Alpha-Tech and affords new possibilities. Alpha-Tech can provide variants of the new product for different users with different requirements, such as the Army or the Navy, delivered as a component or integrated into a system. This enables Alpha-Tech to operate beyond developing software to designing solutions and integrating products into a system. Crucially, it manifests a shift in strategic orientation from a software house to a provider of complete solutions. The latter implies much better profit margins but also more challenges, as providing both hardware and software and offering complete solutions relates to more complex and demanding activities.

The next sections deals with the learning mechanism at play in the indicative learning episodes of entrepreneurial action presented above. Such learning lies at the crux of open innovation.

Accessing Interorganisational Knowledge

Participation in Boundary Events and Use of Boundary Objects

With regard to effecting coordination between companies contributing to a given defence project, CWID (Coalition Warrior Interoperability Demonstration) exercises constitute major boundary events (Brown & Duguid, 1998; Wenger, 1998, 2000; Wenger *et al.*, 2002) in the sector Alpha-Tech is operating. They bring together members of a constellation of CoPs, which encompass military procurement specialists, opinion leaders such as consultants, military organisations and departments, system integrators and software developers. Prototypes such as the compression technology featuring STANAG 4406 Annex E, the new product relating to the learning episode of entrepreneurial action considered, are showcased in a real life environment to prospective buyers such as MoDs and NATO armed forces and constitute key boundary objects (*ibid*).

The importance of the interfaces between boundary spanners of this social learning system during CWID exercises, as intercommunal, sensemaking boundary processes, was highlighted by all respondents. Boundary spanners of these communities examine how effectively new technologies, in conjunction with existing systems in place, enable secure communication between units stationed in different countries. It is via negotiability of meaning (Wenger, 1998, 2000) that software developers, systems integrators, experts advising the end-customer and end-users make sense of what is available, what needs to be developed and then work out related challenges in praxis (Weick & Wesley, 1996). Brian and Tony, departing from a discussion on the learning episode, elaborated upon the role these events play in making sense of entrepreneurial opportunities:

The CWIDs are major events because all the players are there: customers, competitors are there, advisors and decision makers; we talk to all these people. By talking to them we get ideas. I mean, from informal discussions, because we are specifically focused on what we are doing in the CWID interoperability exercises. (Brian, Managing Director)

...And there may be new concepts that we come back with, new ideas for things that we should be looking at addressing and there might be a particular need and they say, we need to have this, sort of, functionality, no one has solved this problem; that needs solving. Or it may be something that we picked up, something they haven't bluntly said but we picked that up based on what others supply. (Tony, Senior Architect)

NATO standards, project plans and demonstrated technologies constitute significant boundary objects (Star & Griesemer, 1989; Brown & Duguid, 1998; Wenger, 1998), which support the technology development process through negotiability of meaning between the various actors involved in a supply chain. On this point Elaine remarked:

So you know, as we're working together according to the project plan, you find that they may be doing something wrong or we are doing something wrong that we have to resolve. There could be crucial issues to be resolved but some are just minor things where we have different interpretations of the standard specifics, so there might be a slight disconnection; you have to just agree what you need to fix. (Elaine, Business Improvement Manager)

As the respondents of Alpha-Tech made clear during the discussion about these events, CWID demonstrations are instructive forums in acquiring valuable information/knowledge, not only on technological developments, but often on wider areas of the business context. Software developers such as Alpha-Tech, while competing in areas where they supply similar products, also have to cooperate on various occasions, in order to make their products compatible in a configuration set up by a systems integrator. Their workers are "Practitioners [who] have a lot in common: their training, their institutional frameworks, their interests and in particular the warrants with which they evaluate what is important and what is not" (Brown & Duguid, 1998, p.102). As members of a CoP or a constellation of CoPs concerned with secure messaging systems, they are able to enhance their knowledge in the field through casual boundary interactions. Social participation in praxis, informal discussions and story-telling help the circulation of tacit knowledge amongst such professionals, who may be working for competing firms (Brown & Duguid, 2001a; Tsoukas, 2002; Wenger, 2000; Wenger *et al.*, 2002; Scarbrough *et al.*, 2004). The respondents acknowledged that the lines that let knowledge leak out, also let if flow in; trying to block leaks risk cutting the firm off from essential knowledge (Brown and Duguid, 2000a, 2001a). Tony's quote on this point is illustrative:

The CWID is quite useful. So, because we are getting their systems to work with our systems, we are getting our systems to interoperate as part of the demonstration, we will talk to competitors. I mean, you will see their products, if they have any problem then you will go over and help them and you may sit with them while they configure and solve the problem and so you can see what's in their stuff which is good. It's pretty casual with the competitors...It's normally the consultants and engineers who are down there getting it to work. I mean, because of the people who go to these events and certainly the people who are doing it, there is a sort of camaraderie between them - they all try to deliver; rather than being there as competitors. They are fellow developers or fellow consultants. So, we do talk during the event and sometimes later on over a drink. Yeah, and I think we do offer advice in some areas where it's possible – you hear all sort of stories about how you fix something or what the X and Y do to solve a problem they have, or generally what's happening out there. So, it's sort of a complementary position or not competing position if we do a combined product. So, depending on what you think, there is going to be a risk [of knowledge leak] but I suppose it depends on what you make of it. (Tony, Senior Architect)

Moreover, it is noteworthy that through social participation, use can give rise to generation and dissemination, as new meaning can be formed during trials, which underscores the cyclicality, complexity and 'messiness' of the strategic entrepreneurial learning process.

Establishing Intra-Organisational Mechanisms for Knowledge Internalisation and Entrepreneurial Action

There are various channels through which different market and technology knowledge pertinent to entrepreneurial opportunities is diffused and merge to produce and/or refine new meaning at a group/organisational level in Alpha-Tech. Quarterly company meetings, monthly departmental meetings, most importantly casual interactions with colleagues in the office and in the field, all contribute to shaping the intracommunal identity of Alpha-Tech and 'learning what matters and what not in practice' (Wenger, 1998).

However, a distinct mechanism established by Alpha-Tech's management is the so-called Product Strategy Forum (PSF), which relates to strategy formation and new product development. The role that PSF plays in making sense of entrepreneurial opportunities is exemplified in the learning episode of entrepreneurial action considered. In 2004 the members of the PSF, acting on market signals and input from the advisory groups and after making sense of an opportunity to develop a compression technology at individual and group levels, diffused related knowledge at an organisational level. Elaine, referring to the interface between members of PSF, explained that the above constitutes a strategic mechanism:

You know, our technical architects go out there and find something wonderful, this is what we think is going happen in the industry. They would bring it in and they would do a business case to say why we should invest in it. At this point it would actually go through a review to see its commercial viability, where the Business Development team say if it's something that we want to do, if it supports what we already have in place etc. And we would analyse it to see if it meets high level requirements. How much asset is it going to take? How many people will be required to develop it? Is it going to be commercial? You know, what return are we going to get on the investment? (Elaine, Business Improvement Manager)

The importance of sociality and informality in disseminating information/knowledge relating to innovatory activities through casual conversations and exchange of 'war stories' (e.g. Orr, 1996; Brown and Duguid 2000b, 2001; Orlikowski, 2002) cannot be overlooked. Yet, it is noteworthy that the management in Alpha-Tech makes an earnest effort to organise and systematise the dissemination of information and knowledge within the company. Members of the PSF meet regularly to discuss business opportunities and make sense of what is required to exploit effectively such opportunities. Such meetings are seen as vital occurrences. They help the flow of information/knowledge and refinement of new meaning, generated by boundary spanners such as members of the PSF, which is necessary for identifying and exploiting successfully entrepreneurial opportunities.

Minding the Periphery, Widening Participation and Optimising Knowledge

During one of the quarterly feedback sessions mentioned by Brian earlier, it became apparent that amongst other employees, some Engineers and Consultants in particular were feeling excluded from product strategy activities that the then PSG was undertaking and that they could not contribute to Alpha-Tech's future development. In Wenger's (1998, p.197) parlance, these employees felt that it was just the board and the people comprising the PSG (Business Development and Technical architects) that had "ownership of meaning" within Alpha-Tech's structural relations that determine "economies of meaning". Therefore, these employees viewed their position within the company's social configuration as marginal, their negotiability restricted and currency impoverished. Brian explained that this became clear from the feedback process that Alpha-Tech undertakes periodically, elaborating in this way:

Part of the stuff that came out from the last feedback was that people were not feeling they really had the power to influence what the PSG says the company should do. They felt that the PSG did what they did without listening to the people... they said: 'we are not involved enough. We are not getting our opinions into the product strategy'. Yeah, it came from a whole range of people, particularly from developers [engineers] and consultants. (Brian, Managing Director)

Wenger (1998) cautions that "...this situation when it persists results in a mutually reinforcing condition of both marginality and inability to learn". However, Alpha-Tech's management team acknowledged the need to widen participation in the development process, 'mining the periphery' of the organisational CoP (Brown, 2004). Conceivably, in doing so, it amended structural relationships of "ownership of meaning" and strengthened the negotiability of sub-communities that felt marginalised. It responded to negative sentiments by making more transparent and inclusive the workings of the group, tapping into the periphery of the organisational CoP and addressing power asymmetries (Yanow, 2004). Tony's comment resonates with Brown and Duguid (2000b) contention that democratising the development process and making it more participative without creating a stifling bureaucracy is a balancing act:

And so, there were various moves around, trying to make it open, without being run by a committee - a committee is at the other end of the spectrum - so there is a balancing act. And as part of this, the way it has been changed is that now the speaking is no longer restricted to the PSG, or the PSF as is now called. Its goal is particularly to be more open and involve more people than rather the Business Development and the Technical Architects as we need to pull in other people's experiences. We need to pull their input into the discussions of the PSF to better product development. (Tony, Senior Technical Architect)

Moreover, the ATP work protocol is worth mentioning an example of a tool created to facilitate product development in Alpha-Tech. This aligns well with the broader messaging software CoP and reflects the input of other sub-communities, apart from Business Development and Technical Architects, such as Consultants and Engineers. As a boundary object, it helps aligning different sub-CoPs in the firm towards higher organisational objectives (Wenger, 1998). In conjunction with a culture that fosters knowledge sharing, as pointed out by Elaine and Tony earlier, this approach constitutes a response to the perennial challenge of fostering identification in professional firms and combating stickiness/leakiness of knowledge, echoing Alvesson's (2000, 2001) observations. Notably, such organising reinforces positive behaviour through work procedures and a culture conducive to capturing, diffusing and internalising new knowledge that enables entrepreneurial action in an open innovation context.

DISCUSSION AND CONCLUSION

UK small firms operating in the Computer Software and Services sector have experienced an increasingly dynamic and complex external environment over the last two decades. The main challenges facing these firms relate to intensified rivalry, moderate industry growth rates, susceptibility to economic downturns and rapid technological change. Despite the mature status of the overall industry, there are still opportunities in key segments that make up the Computer Software and Services sector. The market for secure messaging systems in the private, public and national defence sectors has been growing in the EU and the USA - especially since the end of the 1990s - creating opportunities for software firms specialising in secure messaging like Alpha-Tech. In the final analysis, fitting their customers' supply chains requires capabilities relating to quality of service, speed, dependability, flexibility, cost effectiveness and innovativeness. With regard to the latter, firms operating in this sector have to be able to absorb, modify and utilise knowledge created in the open innovation space, in order to identify and exploit successfully entrepreneurial opportunities.

Against this backdrop, Alpha-Tech as an exemplar, growth-oriented, innovatory small firm was scrutinised. A major learning episode of entrepreneurial action was considered, revealing wider tendencies, in accessing external knowledge and internalising it. Alpha-Tech underwent this episode in order to gain competitive advantage and address effectively challenges stemming from its operating environment. SLT elucidated the learning mechanism underlying its entrepreneurial action. From this perspective, drawing attention to intra- and interorganisational/intercommunal power relations and political activity revealed that such phenomena condition situated learning and, ultimately, entrepreneurial capabilities and organisational performance. Notably, the episode of developing new compression technology considered in this chapter exemplifies broader tendencies, which potentially characterise the entrepreneurial learning processes underlying innovation in this kind of firms. That is to say, these tendencies transcend the particular episode and relate to entrepreneurial learning indicatively.

From the study of Alpha-tech as an instrumental case (Stake, 1995) and not as a sample of one, it is evident that the learning mechanism underlying entrepreneurial action (i.e. the identification and exploitation of entrepreneurial opportunities is enabled by key actors' legitimate social participation at intra- and interorganisational/intercommunal levels. Their sensemaking, learning, and applying of skills and knowledge is linked with exploration and exploitation of pertinent knowledge. These underpinnings of entrepreneurial action are inextricably linked with social participation in a web of learning systems and ecologies of knowledge (Brown & Duguid, 1998; Wenger, 2000; Brown & Duguid, 2001a). They take place in the interplay between social competence and personal experience at individual, group, organisational and interorganisational levels (Wenger *et al.*, 2002; Richter, 2004). Hence, such learning constitutes a strategic resource crucial to competitive advantage, echoing Deakins and Freel (1998, 2003), Cope (2003) and Harrison and Leitch (2005).

Alpha-Tech exemplifies the importance of systematising absorptive and transformative activities related to assessing, assimilating and exploiting knowledge relevant to the identification and exploitation of entrepreneurial opportunities, lending support to Jones and Craven's (2001) work. Perhaps a distinctive feature of Alpha-Tech is its relational capacity - i.e. its ability to build and exploit intra- and interorganisational/intercommunal relations systematically, supporting both formal and informal processes to this end (Keogh, 1999; Keogh *et al.*, 2000) and addressing power asymmetries (Yanow, 2004). In congruence with Thorpe *et al.* (2005), the findings of this study suggest that the more active the management team is in encouraging participation in different communities – i.e. in promoting multimembership in Wenger's (1998, 2000) parlance - the more likely such interactions are to provide resources for reflecting on current activities and innovating.

It is evident that identifying and exploiting entrepreneurial opportunities hinges on displaying a learning capability defined in social communities (Wenger, 1998, 2000; Wenger *et al.*, 2002). It is an outcome of a process of contextual learning, where social and intuitive understanding of problems, needs and solutions is developed through situated immersion and social participation in industry or other community networks (Rae, 2004; Taylor and Thorpe, 2004). Such a learning capability constitutes an entrepreneurial capability (Alvarez and Barney, 2000), which would appear to be a function of the intra- and intercommunal capacities of the organisation. The term 'intercommunal capacity' relates to the competency to gather information and create knowledge pertinent to the identification and exploitation of entrepreneurial opportunities through interaction with other organisations/CoPs. Conversely, 'intracommunal capacity' refers to the competency to absorb, develop, operationalise and institutionalise such knowledge within the organisational CoP. Intra- and intercommunal capacities confer a learning capability that can be classed as dynamic in its own right. This underpins other dynamic capabilities, which give rise to organisational core competencies (Zahra & George, 2002; Zollo & Winter, 2002). Therefore, firms that wish to grow should seek actively to nurture their inter- and intracommunal capacity. This could be achieved by focusing the firm's energy on the design elements put forth by SLT. The latter include developing a conducive culture, participating in generative boundary events, using effective boundary objects, minding the periphery and widening participation in the development of the firm.

FUTURE RESEARCH DIRECTIONS

Application and refinement of SLT in this chapter contributes to a better understanding of the learning mechanism relating to entrepreneurial action in growth-oriented and innovatory small firms. It constitutes an advancement of a substantive theory of entrepreneurial learning in the context of open innovation. However, it has to be born in mind that transferability of findings is space and context dependent. The conceptualisation produced could be advanced, incrementally, to middle-range theory through further research involving multiple-case studies of firms that differ in size and age, operating in diverse sectors and possibly by considering different dimensions (see Eisenhardt, 1989 & Eisenhardt & Graebner, 2007).

Finally, another issue relating to the scope of the project concerns the level of analysis in studies of entrepreneurship (Davidsson & Wiklund, 2001) and CoPs (Roberts, 2006). This study scrutinised the learning processes relating to entrepreneurial action in growth-oriented, innovatory small firms, at a firm level. Examining this phenomenon at a meso or macro level - for instance at a constellation of interrelated CoPs level - from a SLT perspective would be worthwhile, as it could draw attention to large-scale properties of composite social learning systems.

REFERENCES

AIIM International. (2006). *The state of the ECM industry: Moving from 'why' to 'how.* Silver Spring, MD: AIIM International.

Alvarez, S., & Barney, J. (2000). Entrepreneurial capabilities: A resource based view. In Meyer, G. D., & Heppard, K. A. (Eds.), *Entrepreneurship as strategy: Competing on the entrepreneurial edge*. Thousand Oaks, CA: Sage.

Alvesson, M. (2000). Social identity and the problem of loyalty in knowledge intensive companies. *Journal of Management Studies, 37*(8), 1101–1123. doi:10.1111/1467-6486.00218

Alvesson, M. (2001). Knowledge work: Ambiguity, image and identity. *Human Relations, 54*(7), 863–886. doi:10.1177/0018726701547004

Bröring, S., & Herzog, P. (2008). Organising new business development: Open innovation at Degussa. *European Journal of Innovation Management, 11*(3), 330–348. doi:10.1108/14601060810888991

Brown, S. J. (2004). Minding and mining the periphery. *Long Range Planning, 37*(2), 143–151. doi:10.1016/j.lrp.2004.01.001

Brown, S. J., & Duguid, P. (1998). Organizing knowledge. *California Management Review, 40*(3), 90–111.

Brown, S. J., & Duguid, P. (2000a). *The social life of information*. Cambridge, MA: Harvard Business School Press.

Brown, S. J., & Duguid, P. (2000b). Balancing act: How to capture knowledge without killing it. *Harvard Business Review, 78*(3), 73–80.

Brown, S. J., & Duguid, P. (2001a). Knowledge and organisation: A social practice perspective. *Organization Science, 12*(2), 198–213. doi:10.1287/orsc.12.2.198.10116

Brown, S. J., & Duguid, P. (2001b). Creativity versus structure: A useful tension. *MIT Sloan Management Review, 42*(4), 93–94.

Chell, E. (2004). Critical incident technique. In Cassell, C., & Symon, G. (Eds.), *Essential guide to qualitative methods in organisation studies* (pp. 45–60). London, UK: Sage.

Chesbrough, H. (2003). *Open innovation: The new imperative for creating and profiting from technology*. Boston, MA: Harvard Business School.

Chesbrough, H. (2006). *Open innovation: Researching a new paradigm*. Oxford, UK: Oxford University Press.

Chesbrough, H., & Crowther, A. K. (2006). Beyond high tech: Early adopters of open innovation in other industries. *R & D Management, 36*, 229–236. doi:10.1111/j.1467-9310.2006.00428.x

Cohen, W. M., & Levinthal, D. A. (1990). Absorptive capacity: A new perspective on learning and innovation. *Administrative Science Quarterly, 35*(1), 128–152. doi:10.2307/2393553

Contu, A., & Willmott, H. (2000). Comment on Wenger and Yanow. Knowing in practice: A delicate flower in the organizational learning field. *Organization, 7*(2), 269–276. doi:10.1177/135050840072004

Contu, A., & Willmott, H. (2003). Re-embedding situatedness: The importance of power relations in learning theory. *Organization Science, 14*(3), 283–296. doi:10.1287/orsc.14.3.283.15167

Cope, J. (2003). Entrepreneurial learning and critical reflection: Discontinuous events as triggers for higher-level learning. *Management Learning, 34*(4), 429–450. doi:10.1177/1350507603039067

Cope, J., & Watts, G. (2000). Learning by doing: An exploration of experience, critical incidents and reflection in entrepreneurial learning. *International Journal of Entrepreneurial Behaviour and Research, 6*(3), 104–124. doi:10.1108/13552550010346208

Daft, R. L., & Weick, K. E. (1984). Towards a model of organisations as interpretation systems. *Academy of Management Review, 9*(2), 284–295.

Davidsson, P., Delmar, F., & Wiklund, J. (2002). Entrepreneurship as growth: Growth as entrepreneurship. In Hitt, M. A., Ireland, R. D., Camp, M. S., & Sexton, D. L. (Eds.), *Strategic entrepreneurship: Creating a new mindset* (pp. 329–342). Oxford, UK: Blackwell.

Deakins, D., & Freel, M. (1998). Entrepreneurial learning and the growth process in SMEs. *The Learning Organization, 5*(3), 144–155. doi:10.1108/09696479810223428

Deakins, D., & Freel, M. (2003). *Entrepreneurship and small firms* (3rd ed.). London, UK: McGraw-Hill.

Duncan, R. (1974). Characteristics of organisational environments and perceived environmental uncertainly. *Administrative Science Quarterly, 17*(3), 313–327. doi:10.2307/2392145

Easterby-Smith, M., Thorpe, R., & Lowe, A. (2002). *Management research: An introduction* (2nd ed.). London, UK: Sage.

Eisenhardt, K. M., & Graebner, M. E. (2007). Theory building from cases: Opportunities and challenges. *Academy of Management Journal, 50*(1), 25–32. doi:10.5465/AMJ.2007.24160888

EITO. (2006). *The EU ICT industry*. EITO. Retrieved from http://www.eito.com

Empson, L. (2001). Introduction: Knowledge management in professional service firms. *Human Relations, 54*(7), 811–817. doi:10.1177/0018726701547001

Forecast International. (2006). *The market for U.S. military communications systems*. Newtown, CT: FI.

Forrester Research. (2006). *Protecting communication in the emerging information workplace*. New York, NY: Forester.

Fox, S. (2000). Communities of practice, Foucault and actor network theory. *Journal of Management Studies, 37*(6), 853–867. doi:10.1111/1467-6486.00207

Gartner, W. B., Carter, N. M., & Hills, G. E. (2003). The language of opportunity. In Steyaert, C., & Hjorth, D. (Eds.), *New movements in entrepreneurship* (pp. 103–124). Cheltenham, UK: Edward Elgar.

Gartner. (2006). *Forecast: Security software, worldwide, 2005-2010*. New York, NY: Author.

Gherardi, S., Nicolini, D., & Odella, F. (1998). Toward a social understanding of how people learn in organizations. *Management Learning, 29*(3), 273–297. doi:10.1177/1350507698293002

Gibbons, M., Limoges, C., Nowotny, H., Schwrtzman, S., Scott, P., & Trow, M. (1994). *The new production of knowledge: The dynamics of science and research in contemporary societies*. London, UK: Sage.

Harrison, R. T., & Leitch, C. M. (2005). Entrepreneurial learning: Researching the interface between learning and the entrepreneurial context. *Entrepreneurship Theory and Practice, 29*(4), 351–371. doi:10.1111/j.1540-6520.2005.00089.x

Hitt, M. A., Ireland, R. D., Camp, M. S., & Sexton, D. L. (2001). Guest editors' introduction to the special issue strategic entrepreneurship: Entrepreneurial strategies for wealth creation. *Strategic Management Journal, 22*, 479–505. doi:10.1002/smj.196

Huysman, M. (2004). Communities of practice: Facilitating social learning while frustrating organizational learning. In Tsoukas, H., & Milonopoulos, N. (Eds.), *Organizations as knowledge systems: Knowledge, learning and dynamic capabilities* (pp. 67–85). Basingstoke, UK: Palgrave MacMillan.

Ireland, R. D., & Hitt, M. A. (2005). Achieving and maintaining strategic competitiveness in the 21st century: The role of strategic leadership. *The Academy of Management Executive, 19*(4), 63–77. doi:10.5465/AME.2005.19417908

Jones, O., & Craven, M. (2001). Expanding capabilities in a mature manufacturing firm: Absorptive capacity and the TCS. *International Small Business Journal, 19*(3), 39–55. doi:10.1177/0266242601193003

Keogh, W. (1999). Understanding processes and adding value within innovative small firms. *Knowledge and Process Management, 6*(2), 114–125. doi:10.1002/(SICI)1099-1441(199906)6:2<114::AID-KPM56>3.0.CO;2-A

Keogh, W., Stewart, V., & Taylor, J. (2000). Science and technology based SMEs: Learning from the marketplace. *International Journal of Entrepreneurial Behaviour and Research, 6*(4), 187–203. doi:10.1108/13552550010354894

Key Note. (2004). *Market report: Computer services* (7th ed.). London, UK: Author.

Lave, J., & Wenger, E. (1991). *Situated learning: Legitimate peripheral participation*. Cambridge, UK: Cambridge University Press.

Lesser, E., & Everest, K. (2001). Using communities of practice to manage intellectual capital. *Ivey Business Journal, March/April*, 37-41.

Løwendahl, B. R., Revang, Ø., & Fosstenløkken, S. M. (2001). Knowledge and value creation in professional service firms: A framework for analysis. *Human Relations, 54*(7), 911–931. doi:10.1177/0018726701547006

March, J. G., & Simon, H. A. (1993). *Organisations* (2nd ed.). Cambridge, MA: Blackwell.

Marshall, N., & Rollinson, J. (2004). Maybe Bacon had a point: The politics of interpretation in collective sensemaking. *British Journal of Management, 15*(S1), S71–S86. doi:10.1111/j.1467-8551.2004.00407.x

Mergent (2006). *IT & high technology: Europe*. New York, NY: Mergent

Miles, M. B., & Huberman, A. M. (1994). *Qualitative data analysis* (2nd ed.). Thousand Oaks, CA: Sage.

MoD. (2004). *UK Defence command paper: Delivering security in a changing world*. London, UK: MoD.

OECD. (2003). *ICT and economic growth: Evidence from OECD countries, industries and firms*. Paris, France: OECD.

Orlikowski, W. J. (2002). Knowing in practice: Enacting a collective capability in distributed organizing. *Organization Science*, *13*(3), 249–273. doi:10.1287/orsc.13.3.249.2776

Orr, J. (1996). *Talking about machines: An ethnography of a modern job*. Ithaca, NY: IRL Press.

Ovum. (2006). *Market trends: The outlook for software and IT services to 2010*. London, NY: Author.

Plaskoff, J. (2003). Intersubjectivity and community building: Learning to learn organizationally. In Easterby-Smith, M., & Lyles, M. A. (Eds.), *The Blackwell handbook of organizational leaning and knowledge management* (pp. 161–183). Oxford, UK: Blackwell.

Rae, D. (2002). Entrepreneurial emergence: A narrative study of entrepreneurial learning in independently owned media business. *International Journal of Entrepreneurship and Innovation*, *3*(1), 53–59. doi:10.5367/000000002101299042

Rae, D. (2004). Practical theories from entrepreneurs' stories: Discursive approaches to entrepreneurial learning. *Journal of Small Business and Enterprise Development*, *11*(2), 195–202. doi:10.1108/14626000410537137

Rae, D., & Carswell, M. (2001). Towards a conceptual understanding of entrepreneurial learning. *Journal of Small Business and Enterprise Development*, *8*(2), 150–158. doi:10.1108/EUM0000000006816

Richter, I. (2004). Individual and organizational learning at the executive level. In Grey, C., & Antonacopoulou, E. (Eds.), *Essential readings in management learning* (pp. 130–148). London, UK: Sage.

Scarbrough, H., Swan, J., Laurent, S., Bresnen, M., Edelman, L., & Newell, S. (2004). Project-based learning and the role of learning boundaries. *Organization Studies*, *25*(9), 1579–1600. doi:10.1177/0170840604048001

Schein, E. H. (1997). *Organizational culture and leadership* (2nd ed.). San Francisco, CA: Jossey-Bass.

Sexton, D. L., & Smilor, R. W. (1997). *Entrepreneurship 2000*. Chicago, IL: Upstart.

Spithoven, A., Clarysse, B., & Knockaert, M. (2010). Building absorptive capacity to organise inbound open innovation in traditional industries. *Technovation*, *30*, 130–141. doi:10.1016/j.technovation.2009.08.004

Stake, R. E. (1995). *The art of case study research*. London, UK: Sage.

Star, S. L., & Griesemer, J. R. (1989). Institutional ecology, "translations", and boundary objects: Amateurs and professionals in Berkeley's Museum of Vertebrate Zoology, 1907-1939. *Social Studies of Science*, *19*(3), 387–420. doi:10.1177/030631289019003001

Swan, J., Scarbrough, H., & Robertson, M. (2002). The construction of communities of practice in the management of innovation. *Management Learning*, *33*(4), 477–496. doi:10.1177/1350507602334005

Taylor, D. W., & Thorpe, R. (2004). Entrepreneurial learning: A process of co-participation. *Journal of Small Business and Enterprise Development*, *11*(2), 203–211. doi:10.1108/14626000410537146

Theodorakopoulos, N., & Ram, M. (2006). Supplier diversity intermediation: Nurturing and bridging communities of practice. *Proceedings of 29th ISBE Conference*, October 31 – November 2, Cardiff, Wales.

Theodorakopoulos, N., Ram, M., & Worthington, I. (2005). Supply chain learning: Experimenting with supplier diversity. *International Journal of Entrepreneurship and Management, 1*(4), 461–478. doi:10.1007/s11365-005-4773-8

Thorpe, R., Gold, J., Holt, R., & Clarke, J. (2006). Immaturity: The constraining of entrepreneurship. *International Small Business Journal, 24*(3), 232–252. doi:10.1177/0266242606063431

Thorpe, R., Holt, R., Macpherson, A., & Pittaway, L. (2005). Using knowledge within small and medium-sized firms: A systematic review of the evidence. *International Journal of Management Reviews, 7*(4), 257–281. doi:10.1111/j.1468-2370.2005.00116.x

Tsoukas, H. (2002). Knowledge-based perspectives on organizations: Situated knowledge, novelty and communities of practice. *Management Learning, 33*(4), 419–426. doi:10.1177/1350507602334001

Tsoukas, H., & Mylonopoulos, N. (2004). Introduction: What does it mean to view organizations as knowledge systems? In Tsoukas, H., & Milonopoulos, N. (Eds.), *Organizations as knowledge systems: Knowledge, learning and dynamic capabilities* (pp. 1–26). Basingstoke, UK: Palgrave MacMillan.

UN. (2002). *Changing dynamics of global computer software and services industry: Implications for developing countries*. United Nations Conference on Trade and Development, Technology for Development Series. New York, NY: United Nations Publications.

Vera, D., & Crossan, M. (2004). Strategic leadership and organizational learning. *Academy of Management Review, 29*(2), 222–240.

Weick, K. E. (2002). Essai: Real-time reflexivity - Prods to reflection. *Organization Studies, 23*(6), 893–898. doi:10.1177/0170840602236011

Weick, K. E., & Wesley, F. (1996). Organizational learning: Affirming an oxymoron. In Clegg, S. R., & Nord, W. R. (Eds.), *Handbook of organization studies* (pp. 440–458). London, UK: Sage.

Wenger, E. (1998). *Learning, meaning and identity*. New York, NY: Cambridge University Press.

Wenger, E. (2000). Communities of practice and social learning systems. *Organization, 7*(2), 225–246. doi:10.1177/135050840072002

Wenger, E., McDermott, R., & Snyder, W. M. (2002). *Cultivating communities of practice*. Boston, MA: Harvard Business School Press.

Yin, R. (2003). *Case study research: Design and methods* (3rd ed.). London, UK: Sage.

Zahra, S. A., & George, G. (2002). Absorptive capacity: A review, reconceptualization, and extension. *Academy of Management Review, 27*(2), 185–203.

Zhang, M., Macpherson, A., & Jones, O. (2006). Conceptualising the learning process in SMEs: Improving innovation through external orientation. *International Small Business Journal, 24*(3), 299–323. doi:10.1177/0266242606063434

Zollo, M., & Winter, S. G. (2002). Deliberate learning and the evolution of dynamic capabilities. *Organization Science, 13*(3), 339–351. doi:10.1287/orsc.13.3.339.2780

ADDITIONAL READING

Alvesson, M. (2000). Social Identity and the Problem of Loyalty in Knowledge Intensive Companies. *Journal of Management Studies, 37*(8), 1101–1123. doi:10.1111/1467-6486.00218

Bröring, S., & Herzog, P. (2008). Organising New Business Development: Open Innovation at Degussa. *European Journal of Innovation Management*, *11*(3), 330–348. doi:10.1108/14601060810888991

Brown, S. J., & Duguid, P. (2000). *The Social Life of Information*. Cambridge, MA: Harvard Business School Press.

Chesbrough, H. (2003). *Open Innovation: The New Imperative for Creating and Profiting from Technology*. Boston, MA: Harvard Business School.

Contu, A., & Willmott, H. (2003). Re-embedding Situatedness: The Importance of Power Relations in Learning Theory. *Organization Science*, *14*(3), 283–296. doi:10.1287/orsc.14.3.283.15167

Cope, J. (2003). Entrepreneurial Learning and Critical Reflection: Discontinuous Events as Triggers for Higher-level Learning. *Management Learning*, *34*(4), 429–450. doi:10.1177/1350507603039067

Davidsson, P., Delmar, F., & Wiklund, J. (2002). Entrepreneurship as Growth: Growth as Entrepreneurship. In Hitt, M. A., Ireland, R. D., Camp, M. S., & Sexton, D. L. (Eds.), *Strategic Entrepreneurship: Creating a New Mindset* (pp. 329–342). Oxford: Blackwell.

Empson, L. (2001). Introduction: Knowledge Management in Professional Service Firms. *Human Relations*, *54*(7), 811–817. doi:10.1177/0018726701547001

Gherardi, S., Nicolini, D., & Odella, F. (1998). Toward a Social Understanding of How People Learn in Organizations. *Management Learning*, *29*(3), 273–297. doi:10.1177/1350507698293002

Gibbons, M., Limoges, C., Nowotny, H., Schwrtzman, S., Scott, P., & Trow, M. (1994). *The New Production of Knowledge: The Dynamics of Science and Research in Contemporary Societies*. London: Sage.

Huysman, M. (2004). Communities of Practice: Facilitating Social Learning While Frustrating Organizational Learning. In Tsoukas, H., & Milonopoulos, N. (Eds.), *Organizations as Knowledge Systems: Knowledge, Learning and Dynamic Capabilities* (pp. 67–85). Basingstoke: Palgrave MacMillan.

Ireland, R. D., & Hitt, M. A. (2005). Achieving and Maintaining Strategic Competitiveness in the 21st Century: The Role of Strategic Leadership. *The Academy of Management Executive*, *19*(4), 63–77. doi:10.5465/AME.2005.19417908

Keogh, W., Stewart, V., & Taylor, J. (2000). 'Science and Technology Based SMEs: Learning from the Market Place. *International Journal of Entrepreneurial Behaviour and Research*, *6*(4), 187–203. doi:10.1108/13552550010354894

Lesser, E., & Everest, K. (2001). Using Communities of Practice to Manage Intellectual Capital. *Ivey Business Journal*, March/April: 37-41.

Løwendahl, B. R., Revang, Ø., & Fosstenløkken, S. M. (2001). Knowledge and Value Creation in Professional Service Firms: A Framework for Analysis. *Human Relations*, *54*(7), 911–931. doi:10.1177/0018726701547006

Plaskoff, J. (2003). Intersubjectivity and Community Building: Learning to Learn Organizationally. In Easterby-Smith, M., & Lyles, M. A. (Eds.), *The Blackwell Handbook of Organizational Leaning and Knowledge Management* (pp. 161–183). Oxford: Blackwell.

Scarbrough, H., Swan, J., Laurent, S., Bresnen, M., Edelman, L., & Newell, S. (2004). Project-Based Learning and the Role of Learning Boundaries. *Organization Studies*, *25*(9), 1579–1600. doi:10.1177/0170840604048001

Stake, R. E. (1995). *The Art of Case Study Research*. London: Sage.

Swan, J., Scarbrough, H., & Robertson, M. (2002). The Construction of Communities of Practice in the Management of Innovation. *Management Learning, 33*(4), 477–496. doi:10.1177/1350507602334005

Taylor, D. W., & Thorpe, R. (2004). Entrepreneurial Learning: A Process of Co-participation. *Journal of Small Business and Enterprise Development, 11*(2), 203–211. doi:10.1108/14626000410537146

Theodorakopoulos, N., Ram, M., & Worthington, I. (2005). Supply Chain Learning: Experimenting with Supplier Diversity. *International Journal of Entrepreneurship and Management, 1*(4), 461–478. doi:10.1007/s11365-005-4773-8

Tsoukas, H. (2002). Knowledge-based Perspectives on Organizations: Situated Knowledge, Novelty and Communities of Practice. *Management Learning, 33*(4), 419–426. doi:10.1177/1350507602334001

Wenger, E. (1998). *Learning, Meaning and Identity*. New York: Cambridge University Press.

Wenger, E. (2000). Communities of Practice and Social Learning Systems. *Organization, 7*(2), 225–246. doi:10.1177/135050840072002

Wenger, E., McDermott, R., & Snyder, W. M. (2002). *Cultivating Communities of Practice*. Boston, MA: Harvard Business School Press.

Yanow, D. (2004). Translating Local Knowledge at Organizational Peripheries. *British Journal of Management, 15*(S1), S9–S25. doi:10.1111/j.1467-8551.2004.t01-1-00403.x

Yin, R. (2003). *Case Study Research: Design and Methods* (3rd ed.). London: Sage.

Zhang, M., Macpherson, A., & Jones, O. (2006). Conceptualising the Learning Process in SMEs: Improving Innovation through External Orientation. *International Small Business Journal, 24*(3), 299–323. doi:10.1177/0266242606063434

KEY TERMS AND DEFINITIONS

Communities of Practice: A group of like-minded individuals pursuing a common enterprise, who share interests, values, beliefs and norms to a certain extent as a result and deepen their knowledge and expertise in this area by interacting on an ongoing basis.

Dynamic Learning: Renewable learning linked with entrepreneurial action, constituting a dynamic capability.

Entrepreneurial Action: Identification and exploitation of entrepreneurial opportunities, linked with innovation (i.e. the creation of new products and/or organising processes) and growth in employment and/or turnover.

Entrepreneurial Learning: Learning that is manifested in changes in cognition and behaviour at a dominant coalition/firm level, relating to entrepreneurial action.

Growth-Oriented Small Firm: A firm employing less than 50 workers, committed to growth in terms of employment and/or turnover, above the average industry rate.

Instrumental Case: A firm that serves as an exemplar, indicating wider tendencies that transcend local context and providing lessons that are applicable to cases that are similar in key dimensions.

Learning Episode of Entrepreneurial Action: A bracketed period of time referring to the identification and exploitation of entrepreneurial opportunities, which is inextricably linked to innovatory activity i.e. the creation of new products and or services.

Open Innovation: In contrast to closed innovation, ideas and knowledge are developed outside the firm, often through collaboration between a host of actors, representing different constituencies.

ENDNOTES

1 The ISO 9000 series of standards is published by ISO (International Organization for Standardization). These standards relate to quality management systems designed to enable organizations to meet their customer needs consistently.

2 The BS7799 (British Standard 7799) is an internationally recognised security standard. In 2007 it was incorporated into the ISO 27000 series of standards as ISO/IEC 27002, which includes many enhancements and improvements on previous versions.

Chapter 11
Open Innovation Strategies in SMEs:
Development of a Business Model

Hakikur Rahman
University of Minho, Portugal

Isabel Ramos
University of Minho, Portugal

ABSTRACT

Open innovation in entrepreneurships already finds its acceptance at all levels of the business industry for adding value to the business. The value could be in the form of economic gain or enhancement of knowledge leading to a sustained financial base. Open innovation adopts various strategies to accomplish the task for enhancing the value gain. Varying by size, nature, pattern, or characteristics of the firm various strategies are being adopted by enterprises. Though largely known to be familiar in corporate business houses, in recent years open innovation is also becoming increasingly familiar in small and medium enterprises (SMEs) and the trend is rapidly increasing. However, despite the potency of open innovation strategies, most of the enterprises are yet to find a sustained business model, especially for the SMEs working at the periphery of that value chain. This forms the basis of the current study. This chapter is trying to formulate a business model incorporating partnership approach from academia, research houses, intermediaries, and other stakeholders.

INTRODUCTION

Open innovation has emerged as a new paradigm where firms commercialize external and internal ideas, knowledge, expertise and or technologies and use both external and internal resources. In an open innovation process, models can be launched by taking ideas from internal or external sources and new technology can enter at various stages of the development process. Models can also go to the market in many ways, such as out-licensing of intellectual property or a spin-off venture in addition to traditional sales channels. Usually, open innovation combines internal and external ideas

DOI: 10.4018/978-1-61350-519-9.ch011

into architectures and systems whose requirements are typically defined by a business model. Eventually, the business model utilizes both external and internal ideas to generate business value, while defining internal mechanisms to claim some portion of that value (Chesbrough, 2003a: xxiv; West, Vanhaverbeke & Chesbrough, 2006). This notion of open innovation, was initially proposed by Chesbrough (2003a; b) and has rapidly gained the interest of both researchers and practitioners, illustrated by a number of special issue publications, dedicated conferences and a fast growing body of literature (Fredberg, Elmquist & Ollila, 2008). This new paradigm inspires enterprises to find the most appropriate business model to commercialize novel products or services, regardless of the existence of any model within the enterprise or must be sought externally (De Jong, Vanhaverbeke, Kalvet & Chesbrough, 2008).

Depending on the demand and reality, open innovation is based on various principles. Among them, the prime ideology could be the necessity to tap external knowledge to enrich the knowledge of the internal experts; the second could be the essentiality of external R&D for an enterprise to generate significant value; the next could be the trend of research that may not be completely dependent on internal resources and profit would be the main aim of entrepreneurship, irrespective of the ideas; and the important one is the acquisition of strategies that are built on a sustained business model, before reaching to the market. (Chesbrough, 2003b; 2003c; De Jong, Vanhaverbeke, Kalvet & Chesbrough, 2008). Apart from other functions, the business model creates value within the value chain and captures a part of it for the focal firm (Chesbrough, 2006b).

However, as this study has found, open innovation business model for small and medium scale enterprises (SMEs) is scant. This could be due to the scarce resources of the SMEs community, despite their major contribution in economy and job market, especially in Europe and other developed economies of the world. Furthermore, due to any acceptable and experimentally verified business model most of the developing and transitional economies are also not in a befitting situation in adopting open innovation strategies in their business model.

This chapter intends to establish an open innovation business model synthesizing various other business models that are being developed by contemporary researchers and being adopted and accepted by leading enterprises, though hardly they could be designated as SMEs. But, it is expected that with initial support from the parent entities, the spin-off firms can take significant leaps in strengthening their business engines. Along this perspective, the chapter has developed a business model incorporating crucial supports from all stakeholders in a two tier process. Later on the chapter has focused on a few research and practice challenges and hints on future research aspects before the conclusion.

BACKGROUND

Innovation can be termed as the outcome of a set of activities that utilizes knowledge to create new value to those benefiting from its use (De Sousa, 2006). It is clear that successful innovation under complexity, uncertainty and transformation can only be achieved through collaborative approaches that integrate knowledge inside and outside the enterprise. This model, in turn more popularly known as open innovation business model, is being increasingly used by enterprises throughout the world. It has been observed that this format of open innovation business model is a must for SMEs, which usually lack knowledge of competiveness to fully comply with the innovation process. It has also been observed that larger organizations are increasingly moving away from their traditional R&D approach to a more collaborative approach, varying in nature and context (De Sousa, 2008).

Open innovation reflects the ability of firms to gain access to not only external resources but also utilizing internal resources of innovations using a defined business model to acquire the business value from such innovations (West, 2006). However, for firms seeking to gain additional revenues through open innovation predominantly depends on two factors. Firstly, in strict sense, technology licensing depends on the firm's intellectual property (IP) strategy, which defines the role of the IP for the innovator and for the licensee(s). Secondly, in specific sense, the innovator must develop a business model consistent to the value of the IP and the innovator's positioning in the value network (Chesbrough & Rosenbloom, 2002; West, 2006). Therefore, a firm's business model depends not only on its IP and the value proposition (explicit factors) (Chesbrough & Rosenbloom, 2002; Chesbrough, 2003a), but also on the corresponding business models of the suppliers, customers, competitors and intermediaries (implicit factors) within the value network (West, 2006).

The open innovation business model entails that enterprises can and should use both internal and external ideas and knowledge to advance their innovation processes. It also demands that internal ideas can be taken to the market through external channels (for example, spin-offs, external licensing of intellectual property, or other forms of technology exploitation) to generate additional business value. This new paradigm also inspires enterprises to find the most appropriate business model to commercialize new products or services, regardless whether that model exists within the enterprise or must be taken from external sources (De Jong, Vanhaverbeke, Kalvet & Chesbrough, 2008).

However, this newly evolved business trend is driven by two major factors. Firstly, market democratization is increasing the micro-segmentation and demanding more customized solutions in unexpected, challenging and unpredictable circumstances, which require a higher level of interactions and participations of customers, suppliers and other stakeholders. Secondly, the increasing competitive pressure around the globe and shortened product life-cycles is demanding quicker innovation processes at a reduced cost for the existence and survival of entities of small enterprises.

Examples of innovations developed through open business models proliferate in the livelihood of people at large, ranging from the new mobile phone application, gaming products to new lines of potato chips. This desegregation of the traditional innovation process, requiring internal and external knowledge, has opened space for the emergence of new business models and types of entrepreneurships. Some of these new companies act as 'knowledge brokers' or 'intermediaries', providing the links, knowledge sources and even technical knowledge and expertise, so that firms can accelerate and amplify the effectiveness of their innovation processes (De Sousa, 2008).

The next section, which is the main trust of the chapter, is focusing on the concept of open innovation business model, discussing about the basis for a business model and developing an open innovation business model using a flexible partnership approach.

BUSINESS MODEL

Basis for a Business Model

As a new way of conceptualizing innovation, open innovation business model relaxes many of the assumptions presumed in the Chandlerian model (Chandler, 1990), both in the external supply of innovation to be incorporated into a firm's offerings, as well as the potential demand from outside the firm for its internal innovations. However, this does not mean that any innovation model is feasible, any more than the rise of the Internet and WWW, which meant that any 'e-strategy' was profitable. Moreover, experimentation within the open innovation paradigm has the constraint of establishing a business model for creating or using an innovation strategy, a constraint that may have been obscured by the cross-subsidies often seen with vertical integration (West, Vanhaverbeke & Chesbrough, 2006).

Open innovation as a business management model is currently gaining grounds in many industries. Among them pharmaceutical, consumer products, software, information technology, and a few service sectors are leading the global business. In this aspect, balancing the right size and structure of R&D has become one of main objectives in R&D management today (Chesbrough, 2003a; Gassmann & Reepmeyer, 2005; Gaule, 2006). Based on his own consulting model for open innovation, Gaule (2006) builds heavily on Chesbrough (2003a) to analyze the impact of open innovation on several parts of the organization. He further provides a series of short case studies, for example on Procter & Gamble. Motzek (2007) points to the motivation factors for companies to engage in open innovation. His empirical material is based on two organizations that are based on the ideas of open innovation. This sample is most likely different in character from already established companies that change to embrace open innovation. His motivation factors are rather similar to the general knowledge on motivation factors for entrepreneurs (Fredberg, Elmquist & Ollila, 2008).

Chesbrough (2007) argues that a business model has two functions, it creates value and it captures a portion of that value, and that open business models enable enterprises to be more competent in creating and capturing value. He also argues that companies need to adapt their business models to open innovation, as it is a way of generating value from their intellectual property (Chesbrough, 2003c). Chesbrough (2007) further argues that the rising costs of technology development and the shortening product life cycles make it harder for enterprises to justify innovation investments. Using open business models, a firm can beat the cost side through leveraging on external R&D resources to save time and money and at the same time, beat the revenue side by licensing out internal technologies. He also argues that enterprises need to develop their capabilities to experiment with their business models, for instance through alternative brands or spinoffs (to reduce risks) and fundamental changes that require top commitment and support (Chesbrough, 2007; Fredberg, Elmquist & Ollila, 2008).

According to Chesbrough and Schwartz (2007), co-development partnerships are increasingly important in open innovation models. This is in line with Chiaromonte (2006), who argues that the difference of open innovation to traditional outsourcing of innovative capacity is that the outside partners are not seen as suppliers but as peers. Chesbrough and Schwartz (2007) also point out that the use of partners can create business models that reduce R&D expenses, expand innovation output and open up new markets (Fredberg, Elmquist & Ollila, 2008).

For a business model, using partnership approach as mentioned earlier, Fredberg, Elmquist and Ollila (2008) underline the following flow chart, illustrating the partnership approach in an enterprise (see Figure 1).

Based on the above arguments and arguments made by Chesbrough and Rosenbloom (2002), authors like to introduce another flow chart (see Figure 2) illustrating the formulation of innovation strategy in an open innovation business model:

Development of the Business Model

Till to date, most of the studies on open innovation business model have examined open innovation at the organizational level, basically for two reasons. Firstly, innovation is traditionally being conceived as the outcome of deliberate actions of a single entrepreneur, and thus R&D competition has also been styled as an innovation race between two or more entrepreneurs. Secondly, the value of a technological invention is realized only through a business model of the enterprise (Chesbrough & Rosenbloom, 2002). While it is very natural that the open innovation business model may easily

Figure 1. Illustrating the partnership approach in an entrepreneurship

At first, define the business objectives for partnering

Then, classify the R&D capabilities of the firm, focusing their core capacity (key source of advantage); critical capacity (vital to success but not key); and contextual capacity (needed to complete the offering, but not differentiator), and

Finally, align the business models of the two firms.

Figure 2. Shows the formulation of innovation strategies

Articulate a value proposition

Indentify a market segment and its revenue potential

Define the structure of the value chain

Estimate the cost and profit potential

Describe the position within the value network

Formulate the innovator's competitive strategy.

spill over the boundaries of an enterprise or even an industry (West, Vanhaverbeke, & Chesbrough, 2006).

As mentioned earlier, a successful open innovation strategy for SMEs should find creative ways to exploit internal innovation by incorporating external innovation into internal development, and able to motivate external actors to sustain an ongoing stream of external innovations. However, while approaching the global market, it can be observed that large industries, especially the high tech ones, are already involved in open innovation scheme, and competing each other leading the global market. In terms of SMEs with high tech facilities, often generated from universities or research center, they would like to maintain good links to larger organizations, rather than reaching out to the grass roots. On the contrary, SMEs at the local level, equipped with low tech facilities are loosely motivated to promote research, and may not have sufficient resources to identify their needs for innovation through appropriate market analysis, and act accordingly to a defined innovation scheme. The situation becomes more challenging without an appropriate and adaptive

Figure 3. Relationship with the stakeholders in a business model

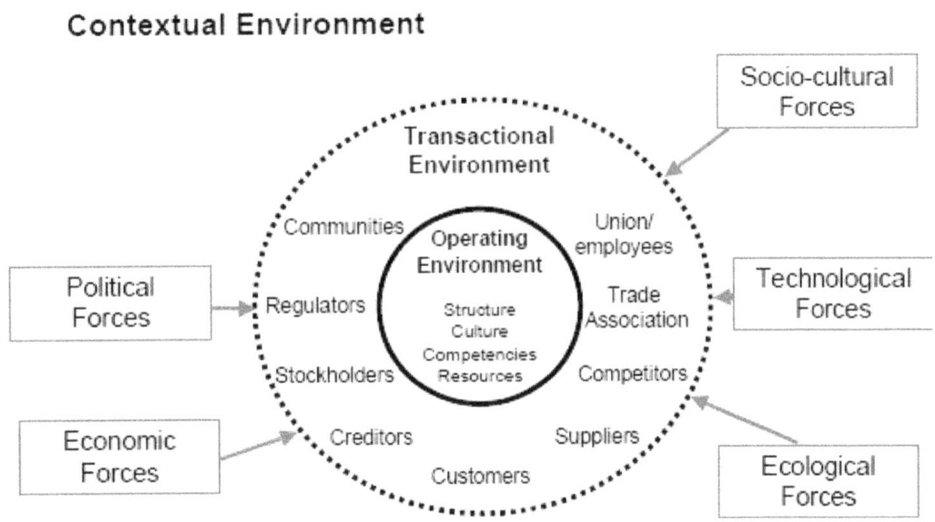

business model (West & Gallagher, 2006; Rahman & Ramos, 2010).

An appropriate and adaptive business model may be seen as the totality of how a company selects its clients, defines and differentiates its responses; classifies those tasks it will perform itself and those it will outsource; configures its resources, goes to market, creates utility for clients; and get hold of profits. It is the entire system for delivering utility to clients and gaining a profit from that activity (Pourdehnad, 2007). Figure 3 shows a relationship diagram with the various actors or stakeholders involved in a business model. This evidently envisages the clear bonding among visible groups of stakeholders among the business communities (a partnership approach).

Triple Helix Model (see Figure 4) is another highly discussing model in this arena. According to this model, a spiral of innovation involves government, university, and industry in multiple reciprocal relationships, to create a flexible overlapping innovation system (another partnership approach) (CSR Europe, 2008).

This research would like to point out another business model that may be utilized in SMEs OI process, which has been developed incorporating mixed approach (see Shorthouse, 2008). Figure 5 shows a mixed approach (amalgamated closed and open innovation approaches) business model. Shorthouse (2008) has suggested this as a joint

Figure 4. The triple helix model (Adopted from CSR Europe, 2008b)

Figure 5. A mixed approach business model (Adopted from Shorthouse, 2008)

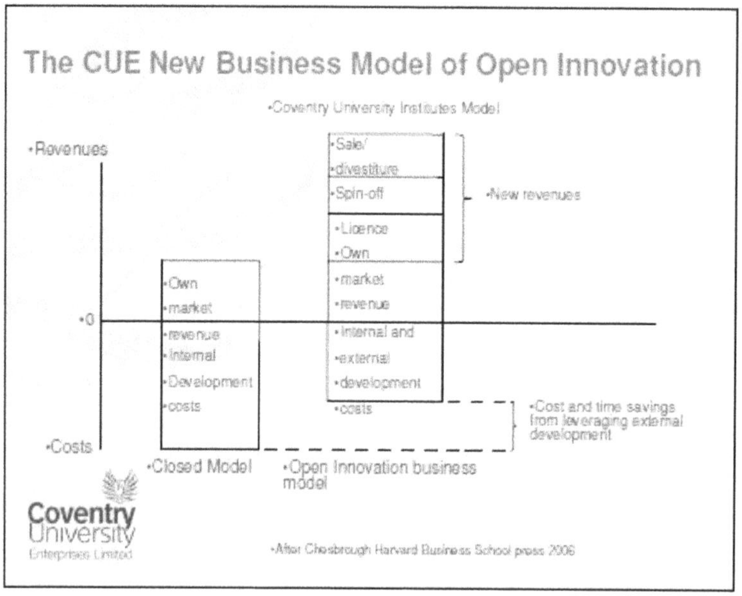

effort to reach the niche market through using both closed and open innovation business model.

However, this study envisages, attaining a sustainable business model in SMEs exposed to open innovation, one need to follow the next two flow charts (see Figure 6, to find out how to prepare SMEs, and Figure 7, to learn about better positioning them) indicated below.

Furthermore, a sustainable business model should also follow to the flow chart 4.

This study further suggests, the business model should incorporate SMEs being inclusive in the following dimension: (See Figure 8)

The proposed business model (to be adopted in a few selected SMEs after a multi-phase survey to be carried out in the next phase of this research) emphasizes on two tiers of relationship; among the core partners in the network (indicated as the core area of the business model) and among the peripheral partners in the network (indicated as the broader area of the business model). In the block, both the segments need not to be isolated from each other. Rather they may remain as active member of the entire community. Therefore, for sustained entrepreneurship the entire group must interact to the grass roots for effective dissemination of open innovation strategies promoting economic and value gain (Rahman & Ramos, 2010). However, while implementing open innovation strategies, despite the introduction of the proposed model, due to the intensity of the competition especially SMEs face tremendous shifting of model architectures adapting the open business environments. Hence, it is recommended that core partners within the network interact faster to provide timely feedback to the peripheral partners for obtaining optimum placement of the business model.

Challenges

The market and thus the industry are always changing, depending on the demand and also the supply. Patents may expire in due period or trademarks get updated in due course, exposing huge investment in challenge. Furthermore, open innovation may not be driven by scale or process, but rather by breakthrough science, or a better understanding of human dynamics validating new targets, which may be totally new or unfamiliar to the market.

Figure 6. Preparing SMEs towards open innovation business model

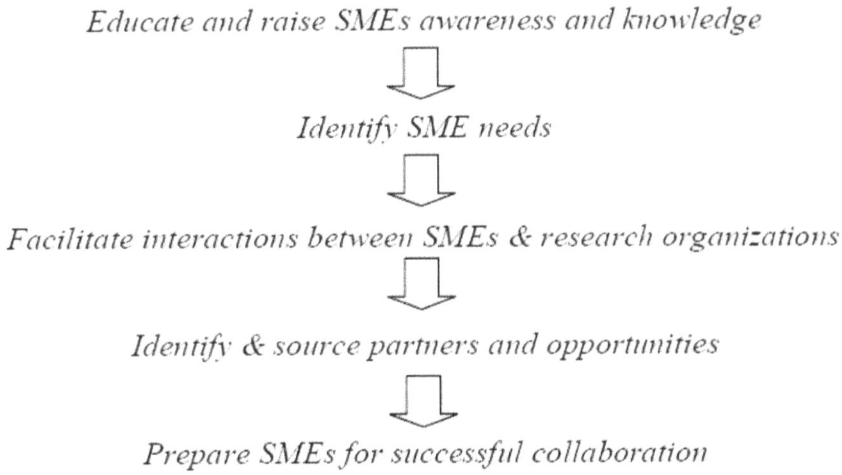

Figure 7. Positioning SMEs towards open innovation business model

Therefore, the open innovation business model and its research practice must be updated dynamically to fit into the new realities (to suit the behavior of the end-user), which is extremely challenging for SMEs (Jaruzelski & Dehoff, 2008).

Open innovation systems (Chesbrough, 2003a) are characterized by the flexible way a firm coordinates a large number of innovation models and assess their value. The system is open, because some models come from outside and some internal models go outside. This approach is consistent with former studies that argue that innovations are increasingly the result of a joint effort of a number of multi-parties (innovation networks, systemic innovations). However, Chesbrough (2003a) argues that in dynamic markets the value of a portfolio of innovation models is difficult to estimate and thereby, that the value of patents –

Figure 8. Building block of the proposed business model (Authors)

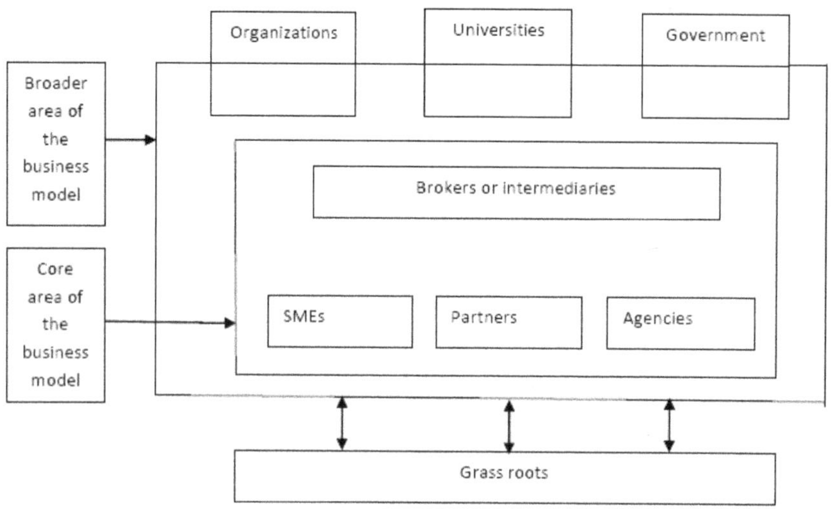

as a specific means to 'secure' an innovation's value – is very indeterminate. Moreover, the value of patents (and or inventions) depends on the specific business model a firm uses to commercialize innovations. Hence, there are different modes of open innovation; they could be spin-offs from research, or out-licensees or technology exploiters. In addition to these, collaboration is not seen as a way out to all innovation problems, but as part of the portfolio in which some if not the majority of models is carried out by the single firm and other models are carried out in partnership with other firms (Dittrich, 2008). This situation exposes SMEs in challenging situations, as they are already exposed to scarce resources and the product life cycles are diminishing.

Therefore, as a newly emerged field of research, open innovation for SME development deserves a prolonged and strategically developed qualitative and quantitative synthesis on available resources, including business models and business practices. Furthermore, while developing an open innovation business model, one should look into the detail of the fundamentals of the innovation process, including the innovation channel, the turn over and also looking into the socio-economic, organizational and cultural differences (Rahman, 2010). Inclusive of above challenges, open innovation for SMEs also deserves further emphasize in resolving issues, like sales, licenses, cash flow, and trust in business relationship; to fulfill the grass roots clientele base by understanding their demand; to develop a transparent business model that add value to it; and to combine appropriate knowledge and technology for building a pipeline of opportunities, enabling a broad-based B2B networking through a sustained organic growth (Rahman & Ramos, 2010).

FUTURE RESEARCH DIRECTIONS

Open innovation is still a moderately new area of research and many of the research questions underline the need to further explore the concept and how enterprises can transform and adapt to this new approach of including external knowledge, resources, and skills in their innovation strategies. The concept is reliant on the organizational and cultural context (Chesbrough, 2007; Fredberg, Elmquist & Ollila, 2008). Similarly, the business model based on the concept is dependent on the strategies and technologies. New technologies will tend to entail new business model. When the technology transforms from one state to another, the value proposition to customers, the value capture by the innovator firm or the relationship of the firm within the value network also transforms (West, 2006; West, Vanhaverbeke & Chesbrough, 2006).

The open innovation model suggests that ideas for innovations can also emerge or go on market from outside the company as well as from inside. The new model assumes that knowledge is spread widely and even the successful innovators with huge R&D resources have to look for the external sources of innovation (Chesbrough, 2006a; Chesbrough, 2006b). The open innovation leverages the role of R&D. Hence, the researchers' job is now, not only to create knowledge, but also to capture it from outside the company. Once a new innovation has taken place, the company can use several business models to bring it to the market. Even if they find that the technology is not suitable for the current business model, it can be licensed or donated to other companies or a new spin-off can be established targeting a new market (Chesbrough 2003a; Chesbrough, 2010). However, the future may reveal that, even if openness in the innovation process is decidedly encouraged there will always be need for some closeness, too (the mixed approach) (Christensen, Olesen & Kjaer, 2005; Viskari, 2006).

CONCLUSION

As observed, open Innovation business models explores the capacity of partners to collaborate each other through technological exchanges, letting ideas flow out of the entrepreneurship in order to find better place for the innovation monetization, and flow into the entrepreneurship as new offerings and new business models (Chesbrough, 2003d; Chesbrough, 2006c). The study finds that, the application of open innovation business models implies high levels of competition due to the opportunity to enter into entirely new markets. At first sight it would not be clear to assume an open innovation business model in a multimarket competition environment, because these schemes rely on different objectives and different assumptions.

This study also finds and supported by another research (see Alcalde, 2009) that open innovation encourages industry competition through trust, confidence and transparent information interchanges. In this context, in order to avoid the competition, multimarket framework, on one hand, may develop threats and retribution mechanism. On the other hand, open innovation presents predominantly technological exchanges as the major driver of the model. Hence, it could be mentioned that traditional mutual self-control mechanisms would not work in open business models. Moreover, the development of technological exchanges among different market players would imply the penetration in entirely new markets. Thus, diversification would be the logical consequence of technological exchanges in open business models. In conclusion, the study suggests by put forwarding the two phase multi-faceted business model, for further explorations and experimentations.

REFERENCES

Alcalde, M. H. (2009). Multimarket contact challenges in open innovation environments. A *Proceeding of the ICERI2009 - International Conference of Education, Research and Innovation, Madrid, Spain, 16-18 November 2009.*

Chandler, A. D. Jr. (1990). *Scale and scope: The dynamics of industrial capitalism*. Cambridge, MA: The Belknap Press of Harvard University.

Chesbrough, H. (2003c). The logic of open innovation: Managing intellectual property. *California Management Review*, *45*(3), 33–58.

Chesbrough, H. (2003d). The governance and performance of Xerox's technology spin-off companies. *Research Policy, 32*, 403–421. doi:10.1016/S0048-7333(02)00017-3

Chesbrough, H., & Schwartz, K. (2007). Innovating business models with co-development partnerships. *Research-Technology Management, 50*(1), 55–59.

Chesbrough, H. W. (2003a). *Open innovation: The new imperative for creating and profiting from technology*. Boston, MA: Harvard Business School Press.

Chesbrough, H. W. (2003b). The era of open innovation. *MIT Sloan Management Review, 44*(3), 35–41.

Chesbrough, H. W. (2006a). New puzzles and new findings. In H. W. Chesbrough, W. Vanhaverbeke & J. West (Eds.), *Open innovation: Researching a new paradigm*, (pp. 15-33). Oxford, UK: Oxford University Press.

Chesbrough, H. W. (2006b). Open innovation: A new paradigm for understanding industrial innovation. In H. W. Chesbrough & W. Vanhaverbeke & J. West (Eds.), *Open innovation: Researching a new paradigm*. Oxford, UK: Oxford University Press.

Chesbrough, H. W. (2006c). *Open innovation: Researching a new paradigm*. Oxford University Press.

Chesbrough, H. W. (2007). Why companies should have open business models. *MIT Sloan Management Review, 48*(2), 22–28.

Chesbrough, H. W. (2010). *Open innovation: A key to achieving socioeconomic evolution, how smaller companies can benefit from open innovation*. Retrieved October 30, 2010, from http://www.jef.or.jp/journal/jef_contents_free.asp?c=3707

Chesbrough, H. W., & Rosenbloom, R. S. (2002). The role of the business model in capturing value from innovation: Evidence from Xerox Corporation's technology spin-off companies. *Industrial and Corporate Change, 11*(3), 529–555. doi:10.1093/icc/11.3.529

Chiaromonte, F. (2006). Open innovation through alliances and partnership: Theory and practice. *International Journal of Technology Management, 33*(2-3), 111–114.

Christensen, J., Olesen, M., & Kjaer, J. (2005). The industrial dynamics of open innovation- Evidence from the transformation of consumer electronics. *Research Policy, 34*(10), 1533–1549. doi:10.1016/j.respol.2005.07.002

De Jong, J. P. J., Vanhaverbeke, W., Kalvet, T., & Chesbrough, H. (2008). *Policies for open innovation: Theory, framework and cases*. Research project funded by VISION Era-Net, Helsinki: Finland.

De Sousa, M. (2006). The sustainable innovation engine. VINE. *The Journal of Information and Knowledge Management Systems, 36*(4).

De Sousa, M. (2008). Open innovation models and the role of knowledge brokers. *Inside Knowledge*, March 2008, 18-22.

Dittrich, K. (2008). Nokia's strategic change by means of alliance networks. A case of adopting the open innovation paradigm? In Sivarajadhanavel, P., & Vellingiri, D. (Eds.), *Open innovation: The networked R and D. Icfai's Professional Reference Book Series*. Chennai, India: Icfai University Press.

CSR Europe. (2008) *R&D open innovation: Networks with SME*. Open Innovation Network, 6 November 2008.

Fredberg, T., Elmquist, M., & Ollila, S. (2008). *Managing open innovation: Present findings and future directions. Report VR 2008:02*. VINNOVA - Verket för Innovationssystem/ Swedish Governmental Agency for Innovation Systems.

Gassmann, O., & Reepmeyer, G. (2005). Organizing pharmaceutical innovation: From science-based knowledge creators to drug-oriented knowledge brokers. *Creativity and Innovation Management*, *14*(3), 233–245. doi:10.1111/j.1467-8691.2005.00344.x

Gaule, A. (2006). *Open innovation in action: How to be strategic in the search for new sources of value*. London, UK: Blackwell.

Jaruzelski, B., & Dehoff, K. (2008). Beyond borders: The global innovation 1000. *strategy+business, 53.*

Motzek, R. (2007). *Motivation in open innovation: An exploratory study on user innovators*. Saarbrücken, Germany: VDM Verlag Dr. Müller.

Pourdehnad, J. (2007). *Idealized design - An open innovation process.* A presentation from the Annual W. Edwards Deming Annual Conference, Purdue University, West Lafayette, Indiana, October 2007.

Rahman, H. (2010). Open innovation: Opportunities and challenges for SMEs. In Cruz-Cunha, M. M., & Varajão, J. (Eds.), *E-business issues, challenges and opportunities for SMEs: Driving competitiveness* (pp. 87–100). Hershey, PA: IGI Global. doi:10.4018/978-1-61692-880-3.ch006

Rahman, H., & Ramos, I. (2010). Open innovation in SMEs: From closed boundaries to networked paradigm. *Issues in Informing Science and Information Technology*, *7*, 471–487.

Shorthouse, S. (2008). *Innovation and technology transfer.* A presentation from the International Conference DISTRICT 2008. International Conference Centre, Dresden, 29 May 2008.

Viskari, S. (2006). *Managing technologies in research organization: Framework for research surplus portfolio. Research Report 176.* Finland: Lappeenranta University of Technology.

West, J. (2006) Does appropriability enable or retard open innovation? In H. Chesbrough, W. Vanhaverbeke & J. West (Eds.), *Open innovation: Researching a new paradigm*. London, UK: Oxford University Press.

West, J., & Gallagher, S. (2006). Challenges of open innovation: the paradox of firm investment in open-source software. *R&D Management Special Issue on Opening-up the Innovation Process,* 36(3), 319-331.

West, J., Vanhaverbeke, W., & Chesbrough, H. (2006). Open innovation: A research agenda. In: H. Chesbrough, W. Vanhaverbeke & J. West (Eds.), *Open innovation: Researching a new paradigm.* London, UK: Oxford University Press.

ADDITIONAL READING

Belussi, F., Sammarra, A., & Sedita, S. R. (2010). Learning at the boundaries in an "Open Regional Innovation System": A focus on firms' innovation strategies in the Emilia Romagna life science industry. *Research Policy*, *39*(6), 710–721. doi:10.1016/j.respol.2010.01.014

Bonner, J., & Walker, O. (2004). Selecting influential business-to-business customers in new product development: Relational embeddedness and knowledge heterogeneity considerations. *Journal of Product Innovation Management*, *21*, 155–169. doi:10.1111/j.0737-6782.2004.00067.x

Brouwers, J., Van Duivenboden, H., & Thaens, M. (2009) The Triple Helix Triangle: Stimulating ICT-driven Innovation at Regional Level, a *Paper from the 2009 Annual Conference of EGPA,* Malta, September 2009

Chesbrough, H. W. (2006). *Open business models: How to thrive in a new innovation landscape*. Boston, MA: Harvard Business School Press.

Chesbrough, H. W. (2007). Why companies should have open business models. *Sloan Management Review*, *48*(2), 22.

De Jong, J.P.J. & von Hippel, E. (2005). User Innovation in SMEs: Incidence and Transfer to Producers, SCALES-initiative (SCientific AnaLysis of Entrepreneurship and SMEs), the Netherlands Ministry of Economic Affairs.

Goduscheit, R. C., Bergenholtz, C., & Jørgensen, J. H. (2009) Open innovation models - a case study of playing poker with chess pieces, A Proceeding of the 10th International CINet Conference, Enhancing the Innovation Environment, Brisbane, Australia, 6-8 September 2009

Huston, L., & Sakkab, N. (2006, March). *P&G's New Innovation Model*, An excerpt from "Connect and Develop: Inside Procter & Gamble's New Model for Innovation,". *Harvard Business Review*, *84*(3),

Lee, S., Park, G., Yoon, B., & Park, J. (2010). Open innovation in SMEs- An intermediated network model. *Research Policy*, *39*(2), 290–300. doi:10.1016/j.respol.2009.12.009

Marais, S. (2010) The definition and development of open innovation models to assist the innovation process, Thesis (MScEng (Industrial Engineering)), University of Stellenbosch, South Africa

Sie, R. L. L. (2010). *Enhancing Open Innovation Through Modelling and Utilizing Knowledge on Coalition Formation (PhD project proposal)*. The Netherlands: Dspace Open University of the Netherlands.

Van der Meer, H. (2007). Open Innovation – The Dutch Treat: Challenges in Thinking in Business Models. *Creativity and Innovation Management*, *16*(2), 192–202. doi:10.1111/j.1467-8691.2007.00433.x

Westergren, U. H., & Holmström, J. (2008) Outsourcing as open innovation: Exploring preconditions for the open innovation model in the process industry, A Proceeding of the 29th International conference on information systems 2008, Paris

Yang, Chia-Han & Shyu, J.Z. (2009) A symbiosis dynamic analysis for collaborative R&D in open innovation, *International Journal of Computational Science and Engineering (IJCSE), Vol. 5, No. 1, 2010* Inderscience Publishers, pp. 1742-7193

Zaletelj, V., Sluga, A., & Butala, P. (2008, March). A Conceptual Framework for the Collaborative Modeling of Networked Manufacturing Systems. *CONCURRENT ENGINEERING: Research and Applications*, *16*(1), 103–114. doi:10.1177/1063293X07084636

KEY TERMS AND DEFINITIONS

Intellectual Property Management (IPM): It is the way of managing intellectual property like; patent renewals, integration and market assessment either manually or using any automated system by the enterprise itself or by an intermediary. IPM involves IP development (planning and implementation of IP in a conducive environment through proper motivation); Market watch (seeking for opportunities by keeping track of other competitors and infringers); IP exchange (acquiring and selling of spin-offs and or joint R&D); and IPR protection (providing legal protection surviving trade secrets and other disciplinary measures).

Open Innovation Business Model: A form of business model involving inbound and outbound ideas defying boundaries of the companies to gain knowledge, expertise and value intended to be beneficial for the entrepreneurships.

Open Innovation Intermediaries: These are companies or agencies that help other companies implement various facets of open innovation, acting as an enabler or guide to explore the market for ideas without getting in over their heads. There are different kinds of open innovation intermediaries, with some acting as agents, thus representing one side of a transaction, while others take on the role of knowledge brokers or market makers.

Open Innovation Strategy: It could be treated as a mandate or a set of clear vision or a well defined process incorporating people (employees, clients, suppliers, other stakeholders) in terms of new ways of working, incentives, fading distinction between work and (social) networks; operations ((e-)processes and (e-)infrastructure) in terms of web access, communities, facilitation of knowledge and creativity sharing; policy (written and unwritten rules) regarding intellectual property, privacy, outside communication; and culture (openness, learning, networking) to create the right (and safe) atmosphere.

Spin-Offs: They may refer to Corporate spin-off, a type of corporate transaction forming a new company or entity; Government spin-off, civilian goods which are the result of military or governmental research; Research spin-off, a company founded on the findings of a member or by members of a research group at a university; Spin-off (media), the process of deriving new radio programs, television programs, video games, or novels from already existing ones; or Spin-off product, a new product which uses the brand name of another product which already has a well-developed image. In case of a spin-off, a parent company distributes shares of a subsidiary to the parent company's shareholders. The shares are usually distributed on a pro rata basis and the subsidiary becomes a separate company. State law and the rules of the stock exchanges determine whether a company must seek shareholder approval for a spin-off.

Chapter 12
Cases Illustrating Risks and Crisis Management

Simona Mihai Yiannaki
European University, Cyprus

ABSTRACT

Undoubtedly, the nature of the relationships between business and risk factors in one country or another does not fit exactly into a "model" nor does it have a pure placebo effect. Yet, models' simplicity may appeal to managers and regulators in understanding important business risks and crisis related phenomena. Backed by this idea, this research underpins a comparative study on SMEs handling risk and crisis management according to a new tailored model of a balance scorecard. This new model of a risk and crisis management aims at improving both SMEs management adaptation and performance across all of crisis' stages, something not attempted so far in the literature. The application of such a 'balanced-scorecard' comes from the author's experience as a banker financing various SMEs industries, as a bank consultant on risk management as well as primarily from the results of a survey performed on a sample of 48 Romanian and Cypriot SMEs, equal-proportionally selected from the area of trading, manufacturing, and services. The results of this case study show, coincidently or not, that there is a significant improvement of the financial performance of the SMEs that employed this model compared to those that did not. The monitoring period: 11/2009-06/2010 was employed as a representative one for the latest global financial crisis which affected the entire European Union region, as well.

INTRODUCTION

The last few years brought severe waves of globalisation and several financial and economic crises, where even large, strong, formerly sound corporations did not cope with. In particular, it seems like in the past 20 years there have been several so-called "once-in-a-lifetime events," such as the floods of Hurricane Katrina or any of the financial crisis, including 1987, 1998, 2000, or 2008, and "no one saw that coming." (Hubbard, 2009) Now it seems the moment to realize the need to protect the most 'numerous' of the firms population: the small and medium size enterprises (SMEs). However, the contribution of SMEs varies considerably across countries. In most economies, the share of enterprises with less than 20 persons engaged is more than 70%, ranging between 67% in Ireland and above 95% in Greece. SMEs also

DOI: 10.4018/978-1-61350-519-9.ch012

account for a smaller share of the total number of employees, ranging between around 11% in the United States and the Czech Republic and more than 35% in Greece (OECD, 2010).

Since SMEs are much more exposed and fragile during and after financial than any other type of companies, whilst they are firmly identified by the EU as the core engine for European economic recovery, growth and innovation,, this chapter looks at a double loop feedback and SME stakeholders' benefit in implementing innovative risk management methods and techniques under a broad Business Socially Responsible strategy.

Subsequently it provides key elements of crisis management resolution not only in times of economic and financial distress but also in times of "black swan" events (Taleb, 2007), such as sudden drops of the National Stock Exchange Index, strong foreign exchange devaluations, sudden and acute raise in unemployment or any other temporary or chronic externality.

Although risk measurement can only be generated by historic data, it has been found that experienced estimators could, using PERT or other well-known models, create usable estimates, which gave usable and effective predictions, with computed standard deviations. One can measure and model risk whatever the risk, static and dynamic, however none can have a crystal ball, as one always have to deal with various levels of uncertainty. This uncertainty is sometimes called error. Any model has also at least 3 inner-errors: model error, parameter error and random error. The task is to minimize them accordingly, but never be able to eliminate them completely. Moreover, the problem of risk management arises further, when the firm doesn't consider any of these.

Others (for example Hubbard, 2010) consider that there are imaginary obstacles that must be overcome, as there are practical measurement solutions that can be applied to any uncertain decision. Hence, one can quantify any uncertainty and then compute the value of reducing that uncertainty by measurement. However one can see during the recent sub-prime crisis that even some of the best quantitative finance modelers at top rating agencies were not able to measure risks of structured products with complicated pay- off's.

Does the picture look though simpler or more effective for the SMEs or not? Except for paying attention at the error and uncertainty part, shouldn't one also look at what is inside the model, before starting measuring?

In pursuing risk measurement, one can learn about the revolutionary thinkers including John von Neumann (inventor of game theory), Harry Markowitz (grandfather of portfolio theory), and the late Fischer Black (Black Scholes option formula), or also find enlightening stories about Fibonacci numbers, chaos theory, the bell curve, regression to the mean, and more (Bernstein, 1996). Yet, despite all the intelligence, computer power, and sophisticated techniques, Bernstein presents us with the growing body of evidence discovered by researchers including the late Amos Tversky's cognitive biases and others who "reveals repeated patterns of irrationality, inconsistency, and incompetence in the ways human beings arrive at decisions and choices when faced with uncertainty." (Bernstein, 1996: pp.271).

In the attempt to model risk, the study employed conditions that can provide indicators of identifiable categories (the five pillars of our proposed model). Once one identifies the patterns indicating the category of risk associated, then one can try to predict when a categorized event might occur. Upon such predictability background, one can attempt to minimize the effects and maximize any recovery efforts. In quantum mechanics, the Heisenberg uncertainty principle states by precise inequalities that certain pairs of physical properties, such as position and momentum, cannot be simultaneously known to arbitrarily high precision. That is, the more precisely one property is measured, the less precisely the other can be measured. Is this paradox trying to teach us a lesson in managing risk and crisis?

The chapter attempts to use though, under a general simplified framework, the key risk evaluation framework which nowadays SME's managers and staff could utilize in reaching their strategic, operational, marketing and financial targets within or not a crisis context.

One big challenge of the study is that we drew under a synthetic five pillar system a balanced-scorecard (herewith BS) which not only can track dawn repeated risk/crisis events and measure their impacts, but it can accommodate further potential risks analysis and mitigation while leaving room for creative improvement at the same time.

In the view of this research, the SMEs are being treated just like any company which needs to support its long-term soundness and profitability in a socially responsible way.

In this light, this study has build a cross-sectoral pool of SMEs from various industries from a particular environment and tested their potential ability to alleviate risk impact and deal with crisis management by utilizing the BS.

BACKGROUND: METHOD ANALYSIS AND CHAPTER FRAMEWORK

Issues, Controversies, Problems: The Balanced Scorecard Concept

This section starts by responding to the question, as why one should utilize the balanced-scorecard (BS) instrument (Kaplan and Norton, 1996) at the base of this model?

Although, there are a few empirical studies linking the use of BS to better decision making or improved financial performance of companies, broadcast surveys of usage have difficulties in this respect, due to the wide variations in definition of 'what a BS is'. Single organization case studies suffer from the 'lack of a control' issue common to any study of organizational change - you don't know what the organization would have achieved if the change had not been made (Lingle & Schieman, 1996), so it is difficult to attribute changes observed over time to a single intervention (the BS), despite obtaining results which demonstrate the usefulness of the BS (Mooraj *et al.*, 1999).

Another criticism among others (Schneiderman, 1999), usually from pundits and consultants, is that the BS does not provide a bottom line score or a unified view with clear recommendations: it is simply a list of metrics (Jensen, 2001). Typically however, the unanswered question relates to things outside the scope of BS itself (such as developing strategies) (Rohm, 2004), which in reality could have a decisive impact in the organization change-over and financial existence. For this reason, the suggested BS is designed to incorporate these aspects, under a flexible quantification.

On the other side of the propellers of the BS, according to Kurtzman (1997), it appears that 64% of companies questioned were measuring performance of companies in a similar way to the BS. Yet, again Kurtzman (1997) sees a strong controversy between two schools of thought. He refers to these schools as *the balance people* and *the numbers people*.

A synthetic table format of these two schools is presented in Table 1.

The balance people promote the use of corporate, or BS. Kurtzman utilizes the so called Analog's scorecard, which includes a number of nonfinancial measures such as number of new products, the rate of on-time deliveries, product development cycle times. He makes his scorecard link these measurements to financial indicators such as the percentage of sales increase due to new-product introductions or gross margins post introduction of new products.

Table 1. The balance people and the numbers people (Kurtzman, 1997)

The Balance People	The Numbers People
The corporate scorecard should be balanced, including nonfinancial soft matters like employee satisfaction.	The balanced scorecard confuses the issue. Measurements should be purely financial.

Kurtzman uses Shell Oil as an example in the numbers people category, which seems to be a good example of a high potential pooled-risks affected corporation. Shell's business model is based on a matrix with four financial measurements as indicated in Table 2.

Shell management's view is that a balanced group of measurements with a number of 'soft categories' leads to confusions. This company's view is an explicit result driven one, where each activity is measured in terms of financial benefit, which in reality is neither a transparent way of sharing the company strategy nor it can differentiate between core activities and corporate social responsibility ones, for example.

At the same time, choosing to calculate a certain financial indicator over another seems to be more or less complicated or fashionable, for example, for finance people the EVA is considered a more accurate measure of control and profitability because it measures an opportunity cost to the money one is investing, and it might yield better returns elsewhere. The calculation allows for an adjustment to compare one project to a hypothetical alternative investment, as well as it is a short term related indicator.(Stewart, 1993). As long as it regards the bonus perspective, in order to be a successful tool, EVA based bonus systems should be long-term, based mainly on changes of EVA and offer considerable bonuses for considerable shareholder value improvements.

For SME's mangers, who most of the time being SME's owners of vartious background, the understanding of financial indicators takes time. However, more important for them nowadays is having to choose certain indicators over others which may make an important difference in the long run, as decision making is time bound. These schools of thought are putting together several controversial issues related to measurements in management and financial strategy, but they do not give a solution to what is best or why it is so.

Theoretically, Norton and Kaplan (1996) have used such methodology for large corporations

Table 2. Shell's business model- (adapted by the author)

Shell's Business Model	
Revenue Growth	Intrinsic Business Value (i.e., Company's Market value)
ROI (return on investment)	EVA (economic value add)

not only for bringing in increased success and lessen managerial burdens, but also to reduce the agency conflict between mangers and shareholders (Jensen, 2001). In addition, this instrument will help more in understanding the strategic need for planning, measuring and having operational objectives for any SME, as well as linking budgets with strategy and HRM and constantly reviewing it, which defines the double loop orientation in risk mitigation and risk management.

Under this study, the aim is not the same, since in SME's managers happen to be the owners most of the time, while at the same time strong–cultural patterns may reveal differences in the way of doing business. Moreover, budget wise, SMEs are not as well equipped as large corporations to invest in all necessary risk packages or sustainable activities. First of all, this is due to environment culture, to lack of dedicated organizations and free funding for SMEs as well as lack of governmental support, in brief due to lack of regulation of this sector. Next by regulation, it is meant not necessary the control aspects, in the sense of fiscal or financial ones, but in the sense of business facilitators.

Therefore, this chapter focuses on channeling the attention of SME's management towards a combined 360degree model that does not have to be a burden in its application nor very time consuming in being utilized by the stakeholders involved. The novelty of this tool is that the BS is useful and interpretable not only by managers of SMEs in context, but also to the other stakeholders of the business for the purpose of transparency (www.bis.org).

Moreover, since risk management models are just like big weather forecasting models that predict financial storms, who may happen or not in the predicted timeframe, obviously, the models aren't always right. Even recognizing that the weatherman got it wrong a lot, it was helpful to know that it might rain tomorrow, because then one could make backup plans. This is not an excuse for any model's inaccuracy, but outlines the idea that some things are utterly immeasurable due to three common misconceptions, also called 'COM' (Hubbard, 2010), namely concept, object and method. A strong misconception about measurement is the objective of measurement itself. If "strategic alignment", or "employee empowerment" seem immeasurable, it is only because they are ambiguous. But if they are important to a business, then they must become measurable aims, and so they must be detectable in some manner, directly or indirectly (Hubbard, 2010).

To these, one may add perhaps the versatility of the risk mitigator and his/her interpretation altogether of the results.

Even by utilizing controlled experiments, variations on random sampling methods, and some very simple but non-obvious methods in many practical business situations, their interpretation is at stake when responding to problem solving. At the same time Hubbard (2009), looks at the way risk is measured by the four horseman, The four classes are actuaries, "war quants" economists, and management consultants. For that reason, it seems important to respond to the exact set audience, or have approached important aspects treasured by all these types of audiences. Hence, last, but not least, one could also add the calibration of subjective human inputs.

The Case Study: The Balanced Scorecard Designed for SME Risk Management

The main concerns of this study were if once utilizing the proposed BS model, the SMEs applying it could obtain any competitive advantage versus the ones not using it ? And, thereafter to see if there are any particular country differences?

The research had indeed used a relatively reduced sample and a relatively short period of time (12/2008-06/2009), considering that a crisis is based on three stages, pre-crisis, during-crisis and post crisis. Because of the pertaining status of the recent financial and economic crisis, it is still hazardous to estimate the completion time for the second and third phases. This is why the study can be expanded for a future larger sample for various other countries or regions, as well as considering a lengthier timeframe.

At the same time, as many of the SMEs are operating cross-industry, the sample chosen includes SMEs categorized based on having over 60% of their activity in either trading, manufacturing or services. In the sample, only about 35% of SMEs were operating solely in one of these industries. This is the reason why the research has concentrated the study's results on the overall response of the SMEs.

Starting from the debate of people vs. numbers the proposed BS has been designed under a five pillars system. The main rule SME's directors were given to follow was that risks can be better comprehended by the people/units most capable of understanding them, and then the SME's strategy and decision making may alter or not the existing risk metrics system.

It is important to mention that the results obtained were after the application of all five pillars at the same time and the results collection and interpretation was performed after the end of June 2010. Also, this means that all SMEs applying the BS have implemented during the analysis' period a risk management structure and have designed their own strategy and governance to follow the BS.

Figure 1, represents the Conceptual Framework of the study under the BS concept, where the BS elements emerged are based on preliminary discussions with SME managing directors, from the literature review, as well as from the author's experience while dealing with SMEs financing over the past 10 years. Five aspects appeared to be essential in the risk assessment and measurement and they are presented under the following cluster:

Next, the study presents and analyses the five pillars characteristics aiming at integrating them under a Risk Management and Control Centre BS.

Pillar I: The Enterprise Responsible Package (ERP)

The ERP measures:

1. The degree of awareness of the need for having socially responsible stakeholders
2. The ability of the SME to connect and network at all levels with its stakeholders (starting first with the government and corporations and then with customers, employees and eventually with non-trading customers, potential future stakeholders (Jensen, 2001))
3. The ethical response of the SME towards the society
4. The protection of the environment, meaning how green the SME can become in the limits of its business and beyond.
5. The relationship management develops with all stakeholders, including customers

This pillar focus is mainly on measuring: customers' satisfaction, asset turnover, number of green activities per entity, market share increase, operational excellence.

The initiatives and spin-offs brought forth by the SMEs applying the offered risk management tool were: cross-industry partnerships, mergers and acquisitions, managerial and staff training, customers' loyalty schemes and development of corporate social responsibility actions, reinforcing SME-lender relationship (Collier, 2009).

Regarding strategy related aspects of this pillar, a broad outline of risk management at this first level has encompassed:

- Organisational structure definition and implementation or redesign;
- Comprehensive risk management approach;

Figure 1. The conceptual framework of the proposed BS for SMEs

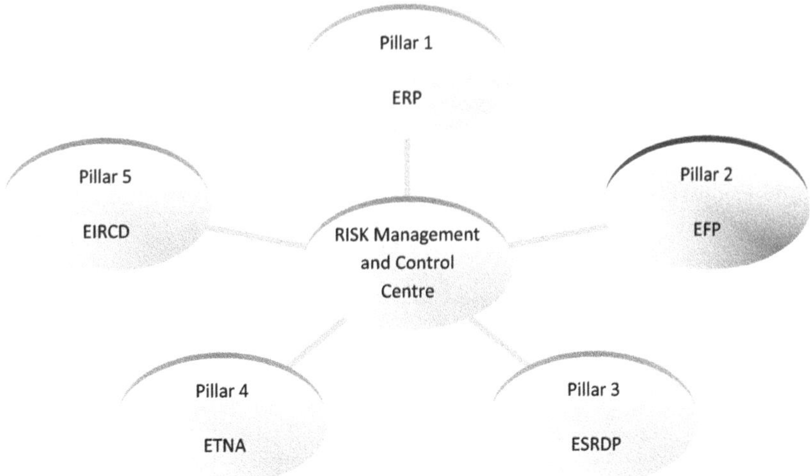

- Risk management policies approved by the Board or the people in charge which should be consistent with the broader business strategies, capital strength, management expertise and the overall willingness to assume risk;
- Guidelines and other parameters used to govern risk taking including detailed structure of prudential limits or thresholds of risk for each identified category (Glantz & Mun, 2008);
- Developing a strong Management Information System (MIS) for reporting, monitoring and controlling risks;
- Well laid down procedures, effective control and comprehensive risk reporting framework (Collier, 2009);
- Separate the risk management framework and team/designated body, independent of any operational department and with clear delineation of levels of responsibility for management of risk;
- Periodical review and evaluation not only for risk levels on short term, but also for policies and procedure on a medium to longer term; and
- Redesign checklists – highlighting updated key industry-specific hazards and control measures (Welham, 2003).

Eventually, the results obtained for the Pillar I are those obtained after the application of all the five Pillars of the balanced-scorecard are presented below.

Table 3 interpretations require attention to the major differences obtained, firstly in the area of increased customer satisfaction, which could translate in a serious image change in the eye of the customer for those SMEs approaching the risk management platform. Secondly, the area of green activities seems to have similar high impact in both countries. The audit evaluations eventually seem to have the strongest impact for both countries and in all sectors, mainly because of the international exposure of all of the SMEs within the sample, but also to the strict correlation to the financial success (that is evaluated next in Pillar II).

Table 3. Pillar I: Results per country

Pillar I Parameters Available	**Equivalent Bilaterally set Benchmarks**	**No application of risk management balanced-scorecard**		**Direct application of risk management balanced-scorecard**	
		Romania SMEs	Cyprus SMEs	Romania SMEs	Cyprus SMEs
a. customers' satisfaction increase	50% for all industries	-10%, -20%, and -22% respectively	-12%, -16%, and -22% respectively	60%, 35%, and 38% respectively	55%, 45%, and 18% respectively
b. asset turnover increase	30% for trading, 30% for manufacturing 40% for services.	0%, 0% -10%	10%, 15% 20%	20%, 25%, 28%	35%, 30%, 45%
c. number of new green activities per SME	10	0	2	5	6
d. raise in the market share	30% for trading, 10% for manufacturing and 20% for services.	Difficult to evaluate per SMEs	Difficult to evaluate per SMEs	Difficult to evaluate per SMEs	Difficult to evaluate per SMEs
e. operational excellence grade (Audit evaluations)	Maximum score 10 out of 10.	-1, 2 3	4, 5, 7	7, 6, 7	8, 8, 9

Pillar II: The Enterprise Financial Package (EFP)

Thanks to the simple financial structure, SMEs can respond faster to changing economic conditions and meet local customers' needs, growing sometimes in their size or becoming bankrupt. Here the study analyzes the set of financial indicators that have to be risk associated and treated for proper adequacy. The reason the research introduced this aspect is also due to finding the response to the Kaplan and Norton (1996) question: How should we appear to our shareholders?

In the original Z-score model of Altman (1968) used five out of 22 ratios as decisive in predicting bankruptcy: Working Capital/Total Assets, Retained Earnings/Total Assets, EBIT/Total Assets, Market Value Equity/Book Value of Total Debt and Sales/Total assets.

Despite having many supporters for the application of Basle II Accord to SMEs (Altman & Sabato, 2005), many criticism have been raised by governments and SME associations that high capital charges for SMEs brings closer credit rationing and could reduce economic growth.

Since banks are the main source of finance for SMEs, and since the Basel II accord allow banks to classify SMEs with turnover under EURO 50 million as either retail or corporate for capital requirements calculation, the banks will tend to treat SMEs under a more automated decision methodology, by scoring their financial ratios.

Several studies show that when banks classify SMEs as retail they obtain considerably lower level of capital adequacy than 8%, (www.bis.org) (i.e 4.76% when utilizing Z-Score models), while if they classify them as corporate, they obtain above 8%, (i.e. 8.6% for Z-score model) (Altman & Sabato, 2005). Why is this important? For this reason, banks who tend to split their SME portfolio in 50/50 retail and corporate, will not necessarily win in terms of risk mitigation and they will tend to develop customized risk prediction models for SME, for which reason, financial ratios of SMEs will gain further importance in acquiring bank funding.

However, the study considers only a small package of financial indicators, which are more appropriate to SMEs: Cash ratio (or the company liquidity position), Return on Investment (ROI) (The higher the ROI, the better the performance, and a higher ratio will also improve the firm's ability to find new investors), Return on equity (ROE) and Economic Value Added (EVA) and a Stress Tested Cash Flow ratio (Stress Tested Cash Flow/ Debt Payment+ Preference Dividend+ Interest).

However the example below may show that ROI and EVA results may be contradictory and more accurate interpretation is required.

Let's suppose of a SBU earning currently a return (ROI, ROIC,ROCE) of 30% and that this firm faces an investment opportunity producing a return of 20%

What happens to the ROI of the SBU if the investment is executed?

Before investment: Capital 100, Operating profit 30 or (NOPAT), Capital cost 10%

ROI = 30/100 = 30%, EVA = 30 - (10% x 100) = 20

Investment's capital requirement 20, return 20%/year: Thus increase in yearly operating profit is 20% x 20 = 4.

After investment: Capital 120, Operating profit 34, Capital cost 10%

ROI = 34/120 = 28%, EVA = 34 - (10% x 120) = 22

In this case decreasing ROI is good for the shareholders, thus ROI should not be maximised and therefore it is problematic controlling tool. Usually large corporations have at least some very profitable units and particularly these units are steered wrongly with ROI.

Due to these contradictions, and for the purpose of having consistent and comparable data and results for interpretation, our recommendation to each SME applying the BS was to stick to its banker's set of financial indicators, as well as to calculate them just like their banker (example: yearend vs. average method of computation, same formulas and periods) (Schoenebeck and Holtzman,2010).

Although ratio analysis is a powerful tool in credit risk evaluation (Bacidore *et al.*, 1997), banks have understood that it is not the only means in this process, as they are other as important parameters for such assessment. Besides, high quality of one ratio may mean a poor quality of another dimension or ratio, as in the box above, also it may not guarantee a sustainable continuous basis of the respective ratio, as they are based on the past.

The results obtained after the application of all the five Pillars of the balanced-scorecard were as presented below and they may represent on average about 20% of the overall credit risk evaluation, where each ratio may have equal strength. The limitation here resides in the calculation of these ratios for the end period and comparison with only the same period a year before not with a trend of at least 5 years.

Last, but not least important, the study employed the stress test of the cash flow from operations, which proved to be also a sensitive issue for the SMEs in Romania, but less sensitive for those in Cyprus. The positive cash flow indicator shows the ability of the SME to satisfy the potential cash flow constraints in case of credit default. At the same time, it is a clear cut indicator compared to meeting just industry adjusted liquidity levels. Although there is no benchmark for the volume, an arbitrary level of 20% was set, which was met by all SMEs applying the risk BS, but not fully by those not applying our risk BS. This may be correlated to the positive results of the other financial indicators, but it proved to be a rather sensitive indicator in this case.

The financial improvement for the SMEs utilizing the risk BS proposed seems overall higher for both countries, which may be or not related to the parallel implementation of the other Pillars' elements, or just as a placebo effect or a noise effect in the market.

The results for these ratios were obtained directly from the SMEs, after being provided with the same formula of calculation. Obviously the element of trust was essential, but also the double calculation based on the companies' key financial statements. However, due to differences in accounting standards and taxation, there are always differences in the input results, hence in the final results. However, since 60% of the firms were medium size companies, they have adopted IAS and therefore the margin of error has significantly diminished, being left only with the input error probabilities. However, to understand risk, one

Table 4. Pillar II: Results per country

Pillar II Parameters Available	**Equivalent Bilaterally set Benchmarks**	**No application of risk management balanced-scorecard**		**Direct application of risk management balanced-scorecard**	
		Romania SMEs	Cyprus SMEs	Romania SMEs	Cyprus SMEs
Cash ratio	0.5 to 1	0.15	0.25	0.35	0.44
ROI	15%	3%	5%	14%	12%
ROE	15%	8%	10%	16%	12%
EVA	15%	2%	4%	12%	10%
Stress Tested Cash Flow	Depends on industry, set up at 20%	No	50% No	Yes	Yes

needs to understand the distribution of potential outcomes and also apply consistently the same calculation/computation formula.

Consequently, it seems there is a definite opportunity to take actuarial type standards to modeling insurance risk for example, and apply them to other operational, financial and strategic risks that organizations face.

As for the stress tests, (Chorafas, 2006) covering stress analysis and the use of scenarios, models, drills, Monte-Carlo simulation, benchmarking, back-testing, and post-mortems, creditworthiness, wrong way risk and statistical inference, probability of default, loss given default and exposure at default, stress testing expected losses, correlation coefficients, and unexpected losses, stress testing related to market discipline and control action, and pillars 2 and 3 of Basel II (www.bis.org)., all these seem to be very complicated to an SMEs, therefore they find their usefulness in banks and the financial sector and rarely in SME sector. For such a reason, the levels were bilaterally settled. This is not only due to their intricate nature and data warehouse needs, but also due to the human factor interpretation or, moreover, response in the business world.

Further stress tests parameters could be added for SMEs performing import-export activities, in order to observe and control associated foreign exchange exposures (Aabo, 2010) and hedging techniques (Woods *et. al.,* 2007).

Pillar III: The Enterprise Supervision and Resource Development Package (ESRDP)

This is a redesigned part of the BS that looks into the development of the learning and growth aspects of the HRM, looking at: employee turnover, investment rate, illness rate, performance pay, gender ratios and frequency of supervision.

Despite arrangements to follow a risk management procedure this part of the BS has encountered difficulties in implementation due to the low level of employment of several firms from our sample, hence results may get a rather poor interpretation.

The only ratios available for Pillar III in this situation are the following:

Before assessing the above indicators the firms applying risk management have acquired several series of knowledge transfer regarding the guidelines for risk management implementation.

There is no clear trend in this series over the analyzed period, except for the decrease of employee turnover for the case of Romania, perhaps due to less people finding opportunities for employment during financial crisis in the SME sector. Interestingly, there is a slight increase of the female employed population versus the male one in the SMEs applying risk BS in both countries, compared to those who do not. But, there is quite a dramatic tendency for unemployment to rise sharply during periods characterized as economic recessions. Both countries have experienced a

Table 5. Pillar III: Results per country

Pillar III. Parameters Available	**Equivalent Bilaterally set Benchmarks**	**No application of risk management balanced-scorecard**		**Direct application of risk management balanced-scorecard**	
		Romania SMEs	Cyprus SMEs	Romania SMEs	Cyprus SMEs
Employee turnover	n/a	20%	5%	5%	4%
Gender ratio	n/a	38% women	30% women	41% women	33% women
Frequency of Supervision	n/a	The same	The same	Increased control and audit	Increased control and audit

superior/effective application of the control and audit function. Indubitably, this last parameter of Pillar III is correlated not only with the other financial and ERP applications of the proposed BS, but also with the way the management empower personnel and allocate the bonus system. Some results also suggest that a firm's incentive or ability to offer fair employee treatment is an important determinant of its financing policy. (Bae, *et al.*,2010).

Also, the relation between incentives and risk is more positive when the manager's effort is more effective in collecting information or in acting upon it. Using data on chief executive officers (CEOs), Shi (2011) found that incentives for CEOs increase with industry-wide risk, a measure of respondable risk. The positive relation diminishes when the CEO is less able to collect information or is less effective in acting upon it. Reason, for which we recommend even more a set up of a tailored scorecard for the Pillar III.

If at bank level, Cash *et. al.* (2010) found only modest evidence that CEO compensation plans promote significant firm-specific heterogeneity in bank risk measures or risky activities, for SME's executives, due to the nature of the activity, this may not be valid.

In this case a SME's general manger or managing director would be in charge to appoint someone to lay down risk parameters and establish risk management control systems. This stage includes the following the below steps:

- Risk Management Architecture / Structure for all risk areas encountered by the industry and the particular firm.
- SME's to get the necessary equipment in getting the ability to identify, measure, monitor and control the various risks.
- Set up an Integrated Risk Management Framework, such as the study's BS or a similar system.
- Move towards Risk based Supervision and Risk focused Internal Audit to enhance risk management functions.
- Set up a Change Management Team or a Crisis Management Team.
- Perform a Risk aggregation and budget allocation for this function and its activities.

Pillar IV: The Enterprise Technological Networking Advance (ETNA)

This particular pillar looks at the technological environment impact on the business and measures indicators such as: level of computerization, number of software packages, number of technological changes in a year time, return on investment in technology.

Regarding ETNA, most SMEs need to build up an Industry Databases which makes available information from an organised centre that offers control for offices and operating levels and help in the risk analysis and in the performance of daily activities.

Table 6. Pillar IV: Results per country

Pillar IV. Parameters available	**Equivalent Bilaterally set Benchmarks**	**No application of risk management balanced-scorecard**		**Direct application of risk management balanced-scorecard**	
		Romania SMEs	Cyprus SMEs	Romania SMEs	Cyprus SMEs
Level of computerization	n/a	The same	Average increase of 2 computers per firm	Average increase of 4 computers per firm	Average increase of 5.5 computers per firm
Software packages	n/a	The same	The same	10% increase	25% increase
ROI in technology	n/a	n/a	n/a	n/a	18% increase

The degree of computerization of the two countries at the sample level have increased, with a larger spur in the Cypriot firms on both hardware and software, and we assume that this was reflected also in the ROI in technology figures. For certain groups of companies operating in manufacturing investment in technology seem to be benefic also in financial terms even on a short term period in Cyprus, while such information is only assumed in Romania, too. In this sense, more explicit details should be obtain as to the purpose of the acquired technology, the business cycle stage of the firm (Hamilton, 2010), if for incremental change, for improvement of daily activity, or for radical change (as in Pillar V), or for facing market demand shifts (despite contraction of demand), or even for future strategic plans.

Pillar V: The Enterprise Innovation and Radical Change Development (EIRCD)

At this stage we assume already that the SME will be exposed to previous learning methods and patterns and we try to identify the possible risk related to implementation of new change processes due to either risks or economic/financial crises. The indicators which may measure this development stage are: employee turnover, investment in innovation, number of radical changes in a year regarding strategy, operational profit margin change post innovation and product leadership per innovation.

This last pillar suggests a high distribution of innovation at the level of SME for both countries, with present both incremental and radical innovations, and with almost double expenditure in Cyprus, but with more results in Romania, although the Cypriot innovations seem to be more profitable. These innovations may account more or less for the employee turnover reduction, where investment in such areas brought a higher stability on freeze of staff on the market, but are certainly provided a cutting edge in the future competitiveness of these firms.

Solutions and Recommendations Based on Discussion of Results

If we consider our model based on a general linear model: Yp=∑ (XB)+εi, where Yp is the study's risk metrics final result or the response variable of each SME applying the suggested model, ∑ represents the sum from 1 to 5 of each pillar's result X, which is the a random discrete variable

Table 7. Pillar V: Results per country

Pillar V. Parameters Available	**Equivalent Bilaterally set Benchmarks**	**No application of risk management balanced-scorecard**		**Direct application of risk management balanced-scorecard**	
		Romania SMEs	Cyprus SMEs	Romania SMEs	Cyprus SMEs
Employee turnover	n/a	20%	5%	5%	4%
Investment in innovation	EUR 50k/ firm	n/a	n/a	EUR20k/firm	EUR40K/firm
Number of radical changes regarding strategy	5/5	0	0	4/5	3/5
Operational profit margin post innovation	20%	n/a	n/a	16% manufacturing 20% services	18% manufacturing 18% services
Product leadership per innovation	Minimum 1 per county	n/a	n/a	2 (service industry)	0

of almost sure equality that belongs to {0,100} or has an alphanumeric equivalent translated into a numerical range belonging to {0,100}, B is a fixed parameter of 0.2 (the equivalent of a equally distributed weight of each pillar), and εi is an unobserved variable that adds noise to the linear relationship between the dependent variable and regressors (X), which for each case we assume it to be close to zero. Eventually the maximum score of the BS will be Yp=100x0.2+100x0.2+100x0.2+100x0.2+100x0.2+0=100.

Where there is no value of the components that make up X, 100 points will be divided for the filled in values only, thus their weight in X become slightly higher.

However, for better interpretation of results, a larger sample together with a more detailed separation of the SMEs according to their representative industries is recommended.

A unique characteristic of this BS is that it can be customized to the specific needs of the company, industry, business cycle, manger and stakeholders, depending also on the emerging risk assessment results and on the capacity to interpret risk of the organisation. This way, the BS will offer the flexibility for the SME's management own judgment and appraisal of the firm's generic and specific risks, as well as offering an adaptable tool to the market change and the technology change in risk management techniques.

A simple Final Rating Model of the BS suggested for SMEs, could look like Table 8.

There is always room for providing a minimum passing grade for each pillar (perhaps 15, depending on industry dynamics) and for the entire scorecard (perhaps 80 instead of 100). However, after getting the score most important becomes the choice of the most appropriate solutions in risk management.

Therefore, in view of the presented results in the above section, one should interpret all these result by correlating both the pillars among them and the indicators within each pillar.

For example on Pillar I, the two countries SMEs obtained mixed results on customer satisfaction on both categories of samples (with no risk and direct risk management application). Although, the results show that Romanian SMEs who have applied risk management as per this study guidelines have obtained higher customer satisfaction on average than the Cypriot SMEs which did so, these results create a paradox interpretation. At the same time we need to understand the results of the following pillars to see if this is a long term or an erratic change.

As for the ratios calculated as per Pillar II, one can notice an improvement of liquidity ratios for the SMEs with risk management in both countries, registration of higher standard deviation for Romania for the SMEs with risk management tools.

Table 8. The Balanced Scorecard final rating and interpretation

Pillars	Pillar I	Pillar II	Pillar III	Pillar IV	Pillar V
Overall Score	100	100	100	100	100
Obtained Score	=90x0.2+0= 18	=80x0.2+0=16	=75x0.2+0=15	=50x0.2+0=10	=75x0.2+0= 15
Areas of risk mitigation (example)	Insufficient green activities	Insufficient ROI and EVA	High employee turnover	Identify benchmarks	Increase investment in innovation
Recommendations	Identify ways of implementing more green activities	Seek alternative sources of capital,	Develop or change the performance pay system	Seek industry Database and build up the MIS, perform data mining.	Supervise innovation output and return on investment in technology.

This may indicate a higher concentration of risk on a particular industry (manufacturing for Romania) or a different application style and management rigor of the scorecard. The results show also a significant change from inefficient investments to efficient ones, especially in the area of manufacturing, and especially for Romania, and at a lesser extend for Cyprus, where investments in trading seem to be more efficient. These results have a standard deviation at a constant average of 8%, while highest average is obtained in manufacturing in Romania and in services and manufacturing in Cyprus.

Again, Romania sees to obtain the highest value of increase in EVA, perhaps due to the ability to pass through the financial crisis.

However, when correlating ROI, EVA and operating profit results for Romania, it appears that the cost of capital is dearer in this country. At the same time the industry with a higher increase is again the manufacturing one, followed by trading one.

When interpreting the results of Table 9 regarding the four financial ratios calculated for the two types of SMEs in the two countries, one can notice the same standard deviation of EVA as for the ROI for SMEs applying our BS risk model in Romania, which is consistent to the two indicators' controversy presented in section 1, Pillar II. Other discrepancies are seen in random higher standard deviations for the SME which applied our BS, particularly in Romania, except for the case of the ROE indicator, where the standard deviation is higher in the case of SMEs applying no-risk model. The same statistical tool shows relatively lower values variation for the SMEs in Cyprus applying the BS, as well as smaller differences in the average values between SMEs applying the BS and those who did not, compared to the Romanian counterparts.

For Cyprus, the highest spur of average values is obtained by SMEs applying risk management tools in the service industry. Also, for Cyprus there is a higher increase in the average ROI vs. the increase of the average EVA for SMEs applying BS. This could be related to the lower ROE and the higher increase in capital for the Cypriot companies for capital expenditure on innovation, radical change and MIS or on computer and technology acquisitions.

The Cash Flow Stress Test shows the ability of SMEs to pass through liquidity problems. Both countries' SMEs which applied risk management scorecard have passed this test. Those SMEs who did not apply the BS failed the test completely in Romania, whilst only 50% of the Cypriot SMEs non BS risk applicant passed the test. This may be correlated to different market mechanisms and different funding facilities available for SMEs in the two analysed countries, or to each company's economic strength, or to the support of the findings of the literature review (Shi, 2011) regarding

Table 9. Pillar II's ratio averages and standard deviations for the cash ratio, ROI, ROE and EVA

	Cash Ratio		ROI		ROE		EVA	
	Average	STDEV	Average	STDEV	Average	STDEV	Average	STDEV
SME (Romania non risk)	0.15	0.07	0.03	0.05	0.08	0.08	0.02	0.04
SME (Romania risk BS)	0.35	0.10	0.14	0.06	0.16	0.06	0.12	0.06
SME (Cyprus non risk)	0.25	0.09	0.05	0.07	0.10	0.08	0.04	0.06
SME (Cyprus risk BS)	0.44	0.05	0.12	0.05	0.12	0.08	0.10	0.06

executive mangers' compensation packages based on their ability to mitigate risks.

In the same manner one can analyse the qualitative indicators, paying attention at meeting the benchmarks, however, there is no exact recipe for getting the best value since one high indicator may mean a possible negative response in another indicator.

Obviously in the end there is a set of the questions that sends us back to Hubbard's literature review of whether we have chosen the right indicators or ratios and if we have calculated them consistently, if the interpretation is the same or if the strategic objectives are met or not eventually. For this issue the fact that the SME develops procedures and involves cross-departmental teams in the area of risk management proves to be an additional risk mitigator, if not even a performance enhancer.

FUTURE RESEARCH DIRECTIONS

It is worth trying to make an analysis as a separate research that crosses the borders of a case study and adds in a large number of SMEs from various countries or regions and upon a larger number of defined indicators to be able to draw a more accurate combination of ratios that are more appropriate in risk evaluation and mitigation for the future generation of SMEs.

Another issue is related to the potential cultural aspects in the way risks are dealt with or risk management is applied in general, the level of uncertainty acceptance, the entrepreneurial spirit in each country, as well as the trends regarding particular reasons for defaulting in one country or another, if any. This may be related to other structures of externalities beside the common affecting ones, such as corruption levels, red-tape level, political implication of actions and others.

From the point of view of comparison against benchmarks or benchmarks translation may be required. It could be mentioned also that a better approach may to set benchmarks not only for periods of boom, normal growth, but also for recession times.

This approach has not been used within this study, the benchmarks being commonly agreed with all the participating subjects (i.e. the managers of the SMEs' sample).

Regarding the fine tuning of the ratios calculation or even choice or the sets of qualitative indicators, the advantage of this balanced-scorecard is that it provides room for adaptation to any industry and it shows how to use and measure or put together under a flexible rating system the obtained results.

CONCLUSION

Upon the analysis of a representative number of SMEs in Cyprus and Romania in the area of trading, manufacturing and services, the findings show the areas where the two countries provide stronger versus weaker appraisal indicators. The strong areas for Romania seem to be manufacturing and trading, while for Cyprus the service industry, all these neglecting the level of the cost of capital in both countries.

The BS utilised indicators have been analyzed from both quantitative and qualitative perspectives and could provide a feedback loop for improvement of SMEs activities as well as for paying attention to other uncovered areas of risk management, such as measuring the impact of a "once in a lifetime" event.

If risk management programs really do work, then it seems logical to assume that companies in a given industry with a (self proclaimed) "highly effective" risk management program would show greater shareholder returns, less earnings volatility, and better safety and regulatory compliance records (Mihai-Yiannaki, 2010) than other companies in their peer group who lack such a program. Yet, there appears to be no valid evidence that current risk management practices, taken as a whole, serve to improve overall corporate performance. The evidence just isn't there, and the research thus has to agree with Hubbard (2010).

The calibration on the rating scale or a balanced-scorecard for SMEs is expected to define also the future deals pricing and other related terms and conditions for the undertaken transaction or operational exposures. Also by tracking down previous transactions' risks, any SME can find resources to discover new business relationships and/or deals, or to improve existing ones, renegotiate or give up certain business relationship or even take legal actions in due time.

Such a prediction or simulation is important not only for forecasting and planning purposes in operational management of the firm, but for accounting and provisioning purposes, as well as for future capital budgeting or even for strategy redesign and resources redeployment.

Moreover, by having such a risk management system in place in parallel with a sound firm strategy, the SME becomes more equipped for the future, its management and human resources more astute on risk mitigation techniques, overall being capable to control better the company and to adapt to crises conditions and adverse risks, altogether creating a cultural-cognitive pressure for environmental sustainability.

Whatever measurement of risk derived out of models especially in operational risks, it is difficult to do back- testing or stress testing. The measurement may be correct (not the absolute figure) in some cases for predicting the trend if other things remain constant. Hence, measurement of tail events does not reflect the true picture of uncertainty in short term (within a year). Therefore, measurement of risks shall provide some strategic inputs towards the direction/movement of risks not in absolute terms.

Thus, it's not just a case of accurately assessing and management individual risks, associated to each SME's activity but also in considering the extent to which there might be a "domino" or "contagion" effect among different risk factors, or among the risk pillars, and how can the SME interpret and integrate the results.

Eventually, this study does show significant improvement in the results obtained by risk management lead SMEs in both countries, compared to those not operating by this norm.

REFERENCES

Aabo, T., Høg, E., & Kuhn, J. (2010). Integrated foreign exchange risk management: The role of import in medium-sized manufacturing firms. *Journal of Multinational Financial Management, 20*(4-5). doi:10.1016/j.mulfin.2010.08.002

Altman, E. I., & Sabato, G. (2005). *Modeling credit risk for SMEs: Evidence from the US market* (December 26, 2005). Retrieved June 2010, from http://ssrn.com/abstract=872336

Altman, E. I., & Sabato, G. (2010). Effects of the new Basel capital accord on bank capital requirement for SMEs. *Journal of Financial Services Research,* Forthcoming. Retrieved June 2010 from http://ssrn.com/abstract=707204

Bacidore, J. M., Boquist, J. A., Milbourn, T. T., & Thakor, A. V. (1997). The search for the best financial performance measure. *Financial Analysts Journal*, (May/June): 11–20. doi:10.2469/faj.v53.n3.2081

Bae, K. H., Kang, J. K., & Wang, J. (2010). Employee treatment and firm leverage: A test of the stakeholders theory of capital structure. *Journal of Financial Economics, 100*(1). Elsevier Inc. doi:10.1016/j.jfineco.2010.10.019

Basel Committee on Banking Supervision. (2003). *The new Basel capital accord.* Retrieved July 2010 from http://www.bis.org

Basel Committee on Banking Supervision. (2004). *International convergence of the capital measurement and capital standards.* Retrieved June 2009 from www.bis.org

Bernstein, P. L. (1996). *Against the gods: The remarkable story of risk. Kindle Edition*. Wiley.

Cash Acrey, J., McCumbera, W. R., & Nguyena, T. H. T. (2010). CEO Incentives and bank risk. Journal of Economics and Business, 63(5). Elsevier. doi:10.1016/j.jeconbus.2010.09.002

Chorafas, D. (2006). *Stress testing for risk control under Basel II*. Butterworth Heinemann.

Collier, P. M. (2009). *Fundamentals of risk management for accountants and managers*. Butterworth Heinemann.

EC. (2003). Commission Recommendation of 6 May 2003 concerning the definition of micro, small and medium-sized enterprises. Retrieved June 2009 from http://www.euresearch.ch/fileadmin/documents/PdfDocuments/Micro_Small_Medium_Enterprise_-_EU_Definition.pdf

Glantz, M., & Mun, J. (2008). *The banker's handbook on credit risk*. Academic Press. Hamilton, J. D. (2010). Calling recession in real time. *International Journal of Forecasting, 27*(4). International Institute of Forecasters. Elsevier B.V. doi:10.1016/j.ijforecast.2010.09.001

Hubbard, D. W. (2009). *The failure of risk management: Why it's broken and how to fix it*. Wiley.

Hubbard, D. W. (2010). *How to measure anything: Finding the value of intangibles in business*. Wiley.

Jensen, M. C. (2001). Value maximization, stakeholder theory, and the corporate objective function. In Andriof, J. (Eds.), *Unfolding stakeholder thinking, Greenleaf Publishing, 2002*.

Kaplan, R. S., & Norton, D. P. (1996). Using the Balanced Scorecard as a strategic management system. *Harvard Business Review*, 76.

Kaplan, R. S., & Norton, D. P. (1996b). *Balanced Scorecard: Translating strategy into action*. Harvard Business School Press.

Kurtzman, J. (1997, February 17). Is your company off course? Now you can find out why. *Fortune*, 128–130.

Lingle, J. H., & Schieman, W. A. (1996). From Balanced Scorecard to strategic gauges: Is measurement worth it. *Management Review*, 85.

Loader, D. (2006). *Operations risk. Managing a key component of operational risk*. Butterworth Heinemann.

Mihai-Yiannaki, S. (2010). Bank shares restore under recent industry bailout. *World Review of Entrepreneurship* [Inderscience Publishers.]. *Management and Sustainable Development, 6*(4), 336–349.

Mooraj, S., Oyon, D., & Hostettler, D. (1999). The Balanced Scorecard: A necessary good or an unnecessary evil? *European Management Journal, 17*(5). doi:10.1016/S0263-2373(99)00034-1

OECD. (2010). *SMEs, entrepreneurship and innovation*. Paris, France: OECD.

Ong, M. (2005). [Academic Press]. *Risk Management, 1*, .

Rohm, H (2004). A balancing act. *Perform Magazine, 2*(2).

Schneiderman, A. M. (1999). Why Balanced Scorecards fail. *Journal of Strategic Performance Measurement, 6.*

Schoenebeck, K., & Holtzman, M. (2010). *Interpreting and analyzing financial statements* (5th ed.). Pearson.

Shi, L. (2011). Respondable risk and incentives for CEOs: The role of information-collection and decision-making. *Journal of Corporate Finance, 17*(1), 189–205. .doi:10.1016/j.jcorpfin.2010.10.003

Spedding, L. S., & Rose, A. (2007). *Business risk management handbook*. CIMA Publishing.

Stewart, G. B. (1993). EVA: Fact and fantasy. *Journal of Applied Corporate Finance*, *7*, 6–19.

Taleb, N. N. (2007). *The black swan: The impact of the highly improbable*. Random House.

Todinov, M. (2006). *Risk-based reliability analysis and generic principles for risk reduction*. Elsevier.

Welham, J. (2003). *Tolley's risk assessment workbook series: Retail*. Butterworth Heinemann.

Wendling, C. (2010). Explaining the emergence of different European Union crisis and emergency management structures. *Journal of Contingencies and Crisis Management*, 18(2), 74-82. Blackwell Publishing Ltd. DOI: 10.1111/j.1468-5973.2010.00606.x

Woods, M., Kajuter, P., & Linsley, P. (2007). *International risk management*. CIMA Publishing.

ADDITIONAL READING

Berge, T. J., & Jordà, O.̀ (2009). The classification of economic activity. *Working paper*, University of California, Davis.

Camacho and Perez-Quiros. (2010). 2010 M. Camacho and G. Perez-Quiros, Introducing the EURO-STING: short term indicator of Euro area growth. *Journal of Applied Econometrics*, *25*, 663–694. doi:10.1002/jae.1174

Chauvet, M. (2010). The beginning and end of the 2007–2009 recession. sites.google.com/site/crefcus/probabilities-of-recession/The-beginning-and-end-of-the-2007-2009-recession1, April 2010.

Chauvet, M., & Potter, S. (2010). Monitoring business cycles with structural changes. *International Journal of Forecasting*, *26*(4), 777–793. doi:10.1016/j.ijforecast.2009.08.003

Croushore and Stark. (2003). 2003 D. Croushore and T. Stark, A real-time data set for macroeconomists: does the data vintage matter? *The Review of Economics and Statistics*, *85*, 605–617.

Dueker, M. (2008). Predicting the trough and a jobless recovery. www.econbrowser.com/archives/2008/12/predicting_the_1.html. December 9, 2008.

Estrella, A., & Mishkin, F. S. (1998). Predicting US recessions: financial variables as leading indicators. *The Review of Economics and Statistics*, *80*, 45–61. doi:10.1162/003465398557320

Gilchrist, (2009). 2009 S. Gilchrist, V. Yankov and E. Zakrajšek, Credit market shocks and economic fluctuations: evidence from corporate bond and stock markets. *Journal of Monetary Economics*, *56*, 471–493. doi:10.1016/j.jmoneco.2009.03.017

Hamilton, J. D., & Owyang, M. T. (2009). The propagation of regional recessions. *Working paper*, UCSD.

Importance of Basel II Regulation/Supervision. (October 17, 2008). *The Icfai University Journal of Banking Law*, Vol. 2, April 2009. Available at SSRN: http://ssrn.com/abstract=1413765

Kash, T. J., & Darling, J. R. (1998). Crisis management: prevention, diagnosis and intervention. *Leadership and Organization Development Journal*, *19*(Issue 4), 179–186. doi:10.1108/01437739810217151

Katayama, M. (2009). Improving recession probability forecasts in the US economy. Working paper, Louisiana State University.

King, T. B., Levin, A. T., & Perli, R. (2007). Financial market perceptions of recession risk. *Working paper, Federal Reserve Board.*

Mihai Yiannaki, S (2008). Regulatory Failure and the Subprime Credit Crunch - The

Nalewaik, J. J. (2007). Estimating probabilities of recession in real time using GDP and GDI. *Working paper, Federal Reserve Board.*

Siomkos, G. J., & Kurzbard, G. (1994). The Hidden Crisis in Product-harm Crisis Management. *European Journal of Marketing, 28*(Issue 2), 30–41. doi:10.1108/03090569410055265

Stock, J. H., & Watson, M. W. (1993). A procedure for predicting recessions with leading indicators: econometric issues and recent experience. In Stock, J. H., & Watson, M. W. (Eds.), *Business cycles, indicators, and forecasting* (pp. 95–153). Chicago: University of Chicago Press.

Wright, J. H. (2006). The yield curve and predicting recessions. *Working paper, Federal Reserve Board.*

Yiang, Q., Yao, D., Garnett, J. & Muller, K (2010). Using a Trust Inference Model for Flexible and Controlled Information Sharing During Crises. *Journal of Contingencies and Crisis Management,* Volume 18, Issue 4, pages 231–241, December 2010. Wiley. DOI: 10.1111/j.1468-5973.2010.00621.x

KEY TERMS AND DEFINITIONS

Balanced Scorecard: The balanced scorecard in general is a strategic planning and management system that is used widely in any business activity worldwide to attain vision and strategy of the respective organization, improve information flows, monitor organization performance against strategic goals. It was originated by Drs. Robert Kaplan (Harvard Business School) and David Norton as a performance measurement framework that added strategic non-financial performance measures to traditional financial metrics to give managers and executives a more 'balanced' view of organizational performance.

Crisis: 'Negative impact turning point', while from a business perspective, it might be a substantial, unforeseen circumstance that has a higher probability than usual to jeopardize a company's stakeholders, products, services, fiscal or legal situation, or reputation. Both definitions contain an element of urgency that requires immediate decisions and actions from management and employees involved and most possibly from other stakeholders.

Crisis Management: The process of preparing for and counteracting an irregular negative event in order to prevent eventually the organization to fall apart. Crisis management resides in a set of well-coordinated actions to implement thorough control of any potential damage and preserve or re-establish public confidence in the system under crisis.

Entrepreneurship: We refer in the chapter text to entrepreneurship regarding SMEs, as its most obvious form of starting a new business, including its recent association to its terms of social and political forms of entrepreneurship.

Operational Risk: The risk of direct or indirect loss resulting from inadequate or failed internal processes, people and systems or from external events. We define SMEs Operational Risk as any risk which is not categorized as the risk of loss arising from various types of human or technical error either internally or externally triggered.

Risk Assessment: As a component of risk analysis, it involves identification, evaluation, and estimation of the levels of risks or hazards that have a certain probability to get involved in a situation, their comparison against benchmarks or standards for a particular industry or category of business, and its determination of an acceptable level of risk for the business.

Risk Management: We considered risk management for SMEs, as that particular combination of strategies, policies, procedures, and practices involved in the identification, analysis, assessment, control, and avoidance, minimization, or elimination of unacceptable risks emerging from future events. A company can use risk assumption, risk avoidance, risk transfer, risk retention or any other strategy in risk management. Risk Management is a planned method of dealing with uncertainties to estimate expected and unexpected losses. Risk Management essentially implies identification, measuring, monitoring and controlling risks. Identification of risk is the first and essential step in risk management to assess the extent to which risk could or should be taken and put in place risk mitigation measures. Risk Management is an on-going process as the dynamics change with time. Guidelines on risk management provides broad parameters and each firm may evolve their own systems compatible to their risk management architecture and expertise.

Risk Measurement: Choosing to ignore risk is the riskiest thing a firm can do. One does not measure risk as a stand-alone metric, instead continually analyses the environment to anticipate and avoid situations that may potentially damage the organisation. Some things are thought to be immeasurable only because it is believed that measurement must be some exact value. But the more pragmatic scientific approach to the term measurement is to treat it as quantified uncertainty reduction based on observation (Hubbard, 2010).

Risk Mitigation: We considered risk mitigation or risk reduction the systematic and managed decline in the level of exposure to a risk and/or to the odds of its occurrence.

Sustainable Innovation: The term of sustainable innovation in connection to the risk management of SMEs relates to that particular management approach that leads any new product or service, companies to analyse the benefits, risks and impacts of not only economic factors, but also social and environmental ones.

Section 4
Creativity, Collaboration and Co-Creation

Chapter 13
Apache Sustained Competitive Advantage in the Web Server Industry

Carlos D. Santos Jr.
University of Nottingham, UK

Márcio A. Gonçalves
Federal University of Minas Gerais, Brazil

Fabio Kon
University of Sao Paulo, Brazil

ABSTRACT

Open source communities such as the ones responsible for Linux and Apache became well known for producing, with volunteer labor innovating over the Internet, high-quality software that has been widely adopted by organizations. In the web server market, Apache has dominated in terms of market share for over 15 years, outperforming corporations and research institutions. The resource-based view (RBV) of firms posits that an organization outperforms its competitors because it has valuable, rare, imperfectly imitable, and non-substitutable resources. Accordingly, one concludes that Apache possesses such resources to sustain its competitive advantage. However, one does not know what those resources are. This chapter is an effort to locate them, answering the question: "What resources enable Apache to outperform its for-profit competitors consistently?" This research draws on the RBV to develop a series of propositions about Apache's internal resources and organizational capabilities. For each proposition developed, methods for their empirical validation are proposed, and future research directions are provided.

DOI: 10.4018/978-1-61350-519-9.ch013

Figure 1. Market share for top servers across all domains (August 1995 - August 2010)

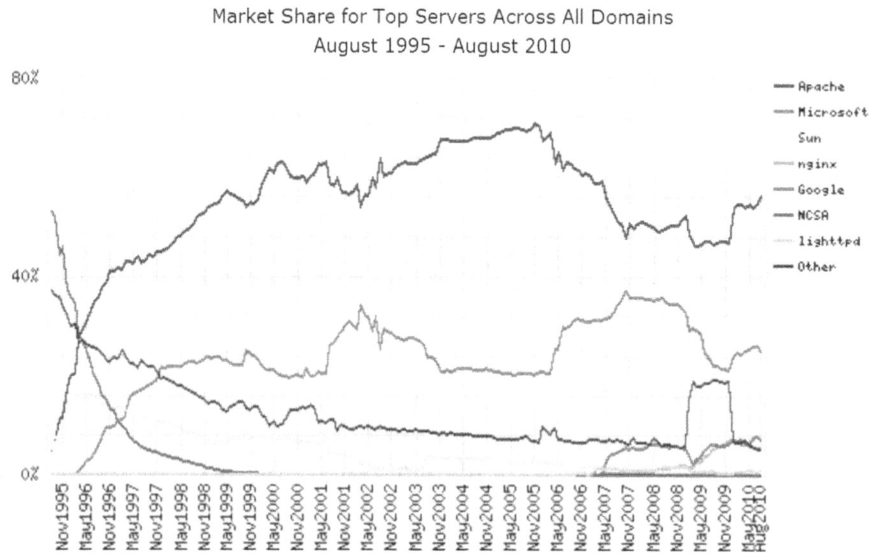

INTRODUCTION

The web server industry has been dominated in terms of market share by an open source product for the last 15 years. A recent survey, which collected data from more than 200 million servers, showed Apache with about 56% of the market, followed by Microsoft with over 25% (Netcraft, 2010). Other participants in the web server market include Sun (Oracle), Google and the National Center for Supercomputing Applications at the University of Illinois, accounting together for the remaining 19% (see Figure 1 for details).

According to these numbers, clearly, Apache is preferred by users and has maintained its competitive advantage in the web server industry for a long time. Therefore, from a resource-based view (RBV) perspective, Apache must possess valuable, inimitable, non-substitutable and rare resources. To point out these resources creates knowledge on how, when and where a volunteer-based organization can sustain competitive advantage against for-profit and research-oriented organizations. That knowledge with due adaptations may apply to a broader class of organizational efforts, covering any attempt to incorporate customers (communities) online input on organizational processes to keep up with stakeholders' changing needs and to communicate back to input generators (Di Gangi, et al., 2010; Gallaugher & Ransbotham, 2010).

This chapter is a model-building attempt to explain Apache's sustained competitive advantage based on its organizational structure, strategies, policies, and, of course, volunteer availability, which represent the resources it possesses. Apache is an illustrious effort in terms of open innovation (Chesbrough, 2003) or even may be termed as crowdsourcing innovation (Howe, 2009). This research identifies Apache's traits that arguably meet all theoretical criteria needed for an organization resource to provide it a sustainable competitive advantage, in a RBV-fashion. Specifically, the study intends to separate Apache from its competitors, especially its closest one, Microsoft (a giant and corporate business entity), through its resources, which would then provide Apache a source of competitive advantage. In specific words, the research has tried to provide an answer for the question: How can an open source software organization outperform its competitors sustainably?

This chapter's goal is to promote awareness of online communities' strengths to outperform corporations of large scope of action, developing theoretical propositions to guide future empirical research, thus assisting its readers to take it as a model of small scale entrepreneurship among the global competition on innovativeness.

Market Scenario

There is an increasing involvement of large scale enterprises or corporations (e.g., IBM and Sun/Oracle) in open source projects (Fitzgerald, 2006). This, along with the assumption that corporate efforts are driven by economic motivation, shows that the corporate world has recognized the potential to reap benefits through the adoption of practices found in open source communities, which were not used by corporations before. Thus, an initial conclusion from this scenario is that open source software communities possess valuable resources that their for-profit competitors do not and want to. The case explored here is intentionally chosen to illustrate the point just made.

Miller's (1996) conclusions that there is a lack of descriptive studies about configuration of organizations, or their structural systems designed toward specific goals, justify studies like this one. It is this configuration of resources and strategies of organizations in the web server market that has been brought here. Similarly, Robinson and Lundstrom (2003) stated that the strategy literature has focused on the adjustment of organizations and their external conditions, neglecting the identification of premier organizational internal resources such as structure, management systems and culture. These authors explained that those external conditions are necessary to properly formulate a plan, but the internal resources are the factors that allow a strategic plan to be implemented and succeed.

The remaining of this chapter is organized as follows. First, a literature review on the resource-based view of the firm is presented. Second, a brief but in-depth discussion of the open source software research appears. Third, a series of propositions is developed in a discussion where parts of Apache's structure and strategies are argued to meet each and every factor the RBV posits to be required for a resource to provide an organization (sustainable) competitive advantage. Fourth, a conclusion section is provided. Finally, suggestions for future research along with the limitations of the model developed are discussed.

LITERATURE REVIEW

The Resource-Based View of the Firm

The RBV of the firm has its roots on the theory of imperfect competition (Melville, Kraemer & Gurbaxani, 2004). Prior to its development, researchers were normally justifying organizations' ability to be profitable according to their positioning in an industry (Porter, 1985). This approach is normally referred to as industrial economics (IO). Although it has provided many insights to explain firm performance, it failed to explain why companies in the same industry experience different amounts of return, and why some go bankrupt, while others do not. Furthermore, the amount of variance explained by environmental (external) variables was not satisfactory, creating the necessity for researchers to look for other factors. That scenario provided the basis for Barney (1991) to introduce the resource-based view of the firm.

In the RBV theory, resources are organizational traits that might be viewed as either strength or weakness. These resources are assumed to be heterogeneous across organizations, which is a necessary condition for the theory to make sense. Otherwise, if resources were homogenous, one would be returning to the condition first described by IO researchers, where the positioning of the organization within an industry would determine its returns, not its internal resources.

Then, assuming that resources are firm-specific, this chapter now turns to the description of what characteristics a resource must possess to be elected as a potential source of (sustainable) competitive advantage. In this chapter, it is acknowledged but put aside the fact that, so far, the RBV literature was not able to define the terms valuable and competitive advantage independently (Priem & Butler, 2001).

Barney (1991) posits that those key resources must be, first of all, (1) valuable in the sense that they should provide the company the opportunity to conceive or implement strategies that enhance its effectiveness and/or efficiency. Second, they have to be (2) rare, meaning that competitors willing to acquire it would have to pay premium prices. Third, these resources must be (3) imperfectly imitable, blocking competitors from developing it, instead of buying. Finally, they have to be (4) non-substitutable, in the sense that competitors cannot use something else to avoid its premium price to be acquired or its difficulty to be developed.

From that, it follows that a firm that does not possess competitive advantage, in order to catch up with its competitor (leader), has to identify what it is that gives the leader an advantage to react properly. That situation is far from trivial, and Barney (1991) discusses some of the issues that make this assessment and reaction a complex task.

Barney (1991) explains further that a resource that meets some or all requirements presented above is also expected to be: (5) causal ambiguous, meaning that most of the times not even the company that possesses and deploys it (let alone an external actor such as a competitor) understands exactly how that resource ended up being what it is; (6) path dependent, in the sense that in order for it to be deployed effectively, a special scenario needs to be first in place for a sequence of events to be triggered, making replications complicated; and (7) socially complex, in the sense that they can be dependent on something almost non-replicable such as an organizational culture or a country specific condition. Therefore, to understand and replicate a key resource, and to develop a strategy using that resource, are complex and costly tasks, allowing the advantage holder to keep its position as leader, where it can reap abnormal returns.

Open Source Software and Communities

The Internet relies on open source software and practices surprisingly more than an inattentive user perceives. According to The Economist magazine (March 16th, 2006), "every time internet users search on Google, shop at Amazon or trade on eBay, they rely on open source software. [...] More than two-thirds of websites are hosted using Apache, an open-source product that trounces commercial rivals."

An open source community is composed of volunteers, in the sense that they are not necessarily paid or employed to improve the software. Those volunteers are dispersed geographically and brought together through IT structure over the Internet (Hertel, Niedner, & Herrmann, 2003). Thus, they are virtual teams too, which are groups of peers who work together on interdependent problems towards a common goal, across space and time, using communication technologies (Lipnack & Stamps, 1997). Virtual teams were also defined by DeSanctis and Monge (1999) as geographically distributed and culturally diverse actors that are connected by electronic forms of communication.

Hertel, Niedner, and Herrmann (2003) described open source communities as organizations with a large number of contributors, being open to anyone, and, thereby, different from regular virtual teams, which normally represent projects of one company or consortium. Open source communities rely on referral and reputation to promote their members and accept their proposed changes in the software. They also have specific procedures for contributing that includes sanctions for violations.

Open source communities are composed of "hobbyists", but the number of paid developers working within them is increasing (Fitzgerald, 2006). Sometimes, these communities have hundreds of collaborators. For instance, Egyed and Joode (2004) stated that the Apache project includes approximately 630 contributors, 90 being from the core developer group. The product (software), as well as its content (source code), produced by those communities are always made available on the Internet free of charge, and the reasons why those developers get involved in such activity have been discussed in the literature.

Studies have found that most of the effort of the volunteers is due to the status attributed to, and the enhanced reputation received by, belonging to the group (O'Mahony, 2003; von Hippel, 2001a). According to Hertel et al. (2003), open source developers are not driven by monetary compensations but by status and reputation. Recognition in these communities appears as a credit file that every software component comes with, attributing task completion-to-person publicly (Lee & Cole, 2003).

Moon and Sproull (2002) pointed out characteristics of open source communities have: (1) a norm that says authority comes from competence; (2) they favor self-selection of tasks and responsibilities are assigned clearly to people after the selection; (3) a software is divided in modules to facilitate independent work and reduce complexity; (4) there is an incentive to release source code often to enable fast development and bug identification; (5) a policy to always acknowledge who contributed what and when; (6) the adoption of an etiquette of online communication; and (7) the use of reliable online communication such as e-mail and forums.

Lee & Cole (2003) stated that open source communities violate the norms of regular communities in at least five dimensions: (1) the unique assignment of intellectual property rights, which advances trust and promotes knowledge sharing; (2) the open membership policy, which leads to a much larger community; (3) non-monetary incentives, differentiating from the employee-type of worker and restricting the means to regulate the behavior of members; (4) members are organizationally and geographically dispersed; and (5) the platform for knowledge creation is based on a many-to-many technology.

The first point identified by Lee & Cole (2003) is because the products, as well as their contents (source code), developed by the volunteers are available publicly. It is possible to use and have access to the software source code and to the e-mails exchanged by community members free of charge via the Internet. In the case of Linux, communication occurs mainly through e-mail lists, where developers send their inputs, discuss technical details, and interact with one another (Lee & Cole, 2003). However, Lee & Cole's (2003) second point can be questioned, as open source projects have leaders who decide what is to be included in the official version of the software (i.e., the core group does not practice an open membership policy). Therefore, the status of a member depends on the recognition of his or her work by senior colleagues (O'Mahony & Ferraro, 2007).

Nevertheless, these open source communities represent viable forms of software development. As discussed, there is at least one case of an open source software leading a market, the Apache web server. Others such as Linux and OpenOffice.org are seen as good substitutes for software developed by corporations like Microsoft (Hert et al., 2003). Lee and Cole (2003) have pointed out that the high-quality of Linux proves the feasibility of a large-scale project that includes online collaboration between developers, users and developers-users, mostly volunteers.

Accordingly, an interesting issue for the IT industry and the academia is of how to identify and make use of open source communities' resources as a building block to deliver business value. To understand what gives an organization the ability to develop competitive software with the help of volunteers with low organizational support in a traditional sense is an important problem both to researchers and practitioners. It demonstrates how an organization can use a social media like a forum or email list to improve its competitive performance, a demonstration so far missing in the literature (Kane, et al., 2010).

Apache Project in a Glimpse

Apache is a software that enables computers to serve web sites, providing then content to the general Internet user through their web browsers. Its initial source code was developed at the University of Illinois and further released to the open source community to foster its development. Apache's version 1.0 was released on Dec. 1st, 1995 (von Hippel, 2001a). The latest version of the system is available on the website for download (http://httpd.apache.org) and is numbered 2.2.16, as of August 29th, 2010. Apache maintains on its website the list of contributors and main contributions received, as a compensation for volunteers' time and effort (http://httpd.apache.org/contributors).

The Apache project is managed by the Apache Group, which puts together a distributed group of volunteers who use the Internet to collaborate, developing source code and creating related documentation. This effort, according to Fielding (1999), is aimed at developing and maintaining a secure, robust, and extensible open source web server. Among Apache's list of contributors, one finds IBM and Collab.Net, but many individuals contribute to Apache as well. Nowadays, Apache Foundation's sponsors include Yahoo!, Microsoft and Google. However, there are also small scale enterprises that have been benefited from using Apache or have contributed to it, such as Day Software and Striker.

MODEL DEVELOPMENT

Organization Configuration: Apache and Its Competitors

This research considers Miller's (1996) definition of organizational configuration as an organizational resource in the RBV terminology. Miller (1996) stated that organizational configuration is what defines the degree to which its elements are orchestrated and connected towards the completion of a single goal. His research had a focus on organization's goals guiding actions towards their achievement. In summary, Miller concluded that configuration is more likely to be the source of competitive advantage than a strategy per se. Furthermore, Miller, Eisenstat and Foote (2002) explained that these characteristics of organizations are not evident, but they can be to some extent observed by their processes and designs. Authors argue that it is in this arena that Apache differentiates from competitors, creating a valuable and unique organizational configuration that allows Apache to hold competitive advantage for such a long period of time. Accordingly, three different organizational areas where Apache's sources of advantage are likely to be are discussed next, (1) human-resource base, (2) organizational structure complexity, and (3) the nature of the relationship between an organization and its customers. In doing so, this study demonstrates how Apache differs from its competitors in each of those areas, and how these differences are likely to be sustained, or not, over time.

Human-Resource Base

Right at first sight, to superficially distinguish the Apache organization from its competitors, especially the Microsoft Corporation, is not very difficult. Apache is mainly composed of volunteers, whereas Microsoft is an employer in the ordinary sense. This leads us to identify a series of differences between them both. For example, the hiring (or attracting in the Apache case) is substantially different in each organization. Also, the types of coordination mechanisms available to the hierarchically higher participants are also different in nature due to the non-employee characteristic of Apache's contributors.

Prior research, such as Hertel et al. (2003), has found that open source volunteers are homogeneous in demographic sense (e.g., age, education, and gender), and in motivational sense (e.g., what they expected to receive in exchange for labor). Open source contributors self-select projects and tasks to get involved with, thereby, naturally forming groups of similar people working on each problem. That suggests a bigger diversity of employees' profiles to be engaged in Microsoft's tasks than in Apache's, creating, consequently, a larger occurrence of conflicts within Microsoft, hindering decision-making processes, task selection and distribution, and, consequently, the achievement of consensus.

The issue of consensus was discussed by Miller (1996). Miller explained that when high levels of it are present, organizational elements, processes and culture become orchestrated, complementing each other and leading to more effective outcomes. Furthermore, this synergy reflects in the structure of power and the shape of the top management team, smoothing decision-making processes. One potential drawback from this homogeneity is the reduction of creativity, but since innovation is not the main focus of an already specified under development software, that should not work against Apache.

Moreover, Apache volunteers have no contractual obligations with the organization, implying that they do not receive direct monetary compensations by simply being members of the organization. Apache participants got engaged in the software development activities already knowing that it is a volunteer work. The simple fact of not being paid for their time makes volunteers more efficiently oriented to results, mainly because they have to take care of their regular jobs as well (Fielding, 1999).

Furthermore, open source developers "compete" with other members in solving problems; and the solution is chosen based on the source code quality. There is no compensation for effort, only for results. These things together provide us ground to argue that Apache contributors tend to be more efficient than Microsoft employees. And without the need for formal control, Apache configuration also reduces managerial costs. By more efficient here, it is meant achievement of the same results in a less costly (time and money) manner. Thus, in a form of proposition, this study proposes:

Proposition 1: Apache's human-resource volunteer base and compensation scheme allows it to accomplish its tasks more efficiently, responding faster to users' demands, in comparison to its for-profit competitors'.

Now, this chapter turns to the discussion of the odds of Microsoft or any other for-profit competitor to be able to catch up with this Apache advantage. First, it is assumed, as expressed in Proposition 1, that an organization with engaged and homogenous volunteers instead of employees is a valuable condition. Second, to possess a reputation powerful enough to attract external people to develop software and contribute with time without expecting to receive any monetary compensation is rare in the software industry. Third, arguably, this Apache advantage could not be easily imitated, for it would be necessary that a corporation of the size of Apache's competitors to back off as a software seller and develop the reputation required to attract volunteers. That attempt would cost a price that none of these corporations, without guarantees of success, would be willing to pay. Fourth, there is no readily substitute to this attraction capability Apache possesses. Apache's attractiveness depends on image and the availability of human resources willing to volunteer. These things cannot be built by imitation (Miller et al., 2002). Given that, Apache's advantage is likely to be maintained over time. Consequently, this configuration is a candidate to be (at least) one of Apache's sources of sustained competitive advantage.

Organizational Structure Complexity

Apache is a case of extreme focus on core competences. Apache is a simplified type of organization. To begin with, it does not have to run payroll functions, manage finance, accounting or human resources department (hiring or firing processes). This oversimplified type of organization facilitates the achievement of synergy, a situation where processes and structures highly complement one another. Consequently, to direct and coordinate people over these simplified processes towards solutions become easier.

Furthermore, there are no stockholders to be factored in during decision-making processes, or other departments to report to. On top of that, demand, competition and regulations are variables that do not matter much, since most developers are users, and an increase or a decrease in "sales" (market share) does not impact organizational plans. Contributors' activities will not change in case of external adversity; the organization does need to collect money to pay them. In other words, Apache's activities are not dependent on economic variables such as sales and competition levels. Apache's operations are affected by the environment only at a minimum. Its for-profit competitors, on the other hand, are affected by environmental uncertainty in the software industry quite differently. A fair amount of environmental uncertainty disappears in the open source organization case.

As said, Apache seems to be a case where an extreme focus on core activity occurs. Every single organizational effort can strictly be related to its final goal, a connection not so easily made in for-profit or highly diversified organizations. These things together make us believe that the management of Apache is much simpler than that of its for-profit competitors (i.e., Microsoft). This is expected to be translated into more complex (i.e., time consuming to be implemented, harder to follow through) and costly coordination mechanisms in for-profit corporations than in Apache. Thus, in the form of proposition:

Proposition 2: Apache's simplified organizational structure reduces significantly the effects of the environment on its operations, allowing a less complex and more focused managerial action than that of its for-profit competitors.

This situation is also likely to be sustained over time. First of all, unquestionably, a situation where an organization is able to make use of less complex and costly policies and procedures is valuable. Second of all, the necessary conditions (i.e., to have a simplified organizational structure without payroll or hiring) for those policies and procedures to be successfully implemented are rare in a competitive market such as the software industry. No other organization has a similar shield against the environment. Third and fourth of all, these procedures and structure cannot simply be copied (developed) or acquired successfully, being then non-substitutable and inimitable. Thus, the exclusive possession of this key resource is expected to be retained over time by Apache.

The Customer-Organization Relationship

As Siggelkow and Levinthal (2006) and Hart (1995) pointed out, organizations have to look for configurations that are consistent with their available resources as well as with their environments, which include their customer-base. It is the fit of resource-environment-strategy that is believed to provide competitive advantage to firms. However, the task of finding this fit is costly and complex. Apache, again, has here an advantage not likely to be copied by any of its for-profit competitors such as Microsoft.

The task of managing the relationship between the organization and its clients is significantly simplified in the Apache way of production. The web server Apache is developed by its own "clients", or users. Open source communities are run fully by and for users (von Hippel, 2001a; von Krogh & Spaeth, 2007). That gives them an enormous edge over ordinary business models, where clients have to be understood by developers, creating opportunities for misunderstandings and delays in detection of changing needs. In practice, there is no need for things such as sales or quality assessment through the users' point of view (i.e., no need for market research).

Apache's developers have at least one "real" job (Fielding, 1999). Thus, if one assumes that Apache's volunteers work as software developers in their real jobs, or at least have a close relationship with the IT department of their companies, it is not very difficult to conclude that they know what a software of that kind must do and how, following any changes closely. Furthermore, they potentially have influence on their companies' web server application decision. Put simply, Apache developers are Apache customers; the customer develops his or her own software, working for the organization (von Hippel, 2001b; Di Gangi, et al., 2010).

This equivalence facilitates the assessment and implementation of users' perceptions in the software, helping Apache to meet client's expectations and needs far easier than its for-profit competitors, which have users and developers as separate entities (von Krogh & Spaeth, 2007). The communication loop between organization-user-organization is neatly closed. That situation does not occur in a general for-profit corporation. In a nutshell, "eliminating" the client from the production process might increase the likelihood of achieving better quality, since quality will be assessed by the same user who developed the software. One assesses to improve its own product. Thus, in the form of proposition:

Proposition 3: User satisfaction towards web server software is higher for Apache in comparison to its for-profit competitors, because users develop Apache themselves, thereby, impacting adoption rates positively in a sustainable way.

For Microsoft or any for-profit corporation to bring its customers to a position as close as the one they occupy in the Apache case would be something not at all easily achievable. Basically, to keep customers to a minimum distance where there is still dependency of the customer towards the corporation is a necessary condition for a company to sell a product. That creates or maintains willingness to pay for the product or service. In the Apache case, this willingness does not exist, since developers are also users and the software can be obtained for free. The same strategy could only be pursued by a for-profit corporation by releasing its software open to the community; a situation that would immediately reduce corporate profits. Thus, this last Apache advantage here discussed is also likely to be sustained over time, especially if its competitors keep the software-for-sale model.

Having developed the propositions, related to three different organizational arenas, this chapter now turns to a brief discussion on how they could be empirically tested. The intention is to facilitate future research endeavors, providing discussion on possible data collection and analysis methods.

Suggested Methods for Empirical Studies

This section has been divided in three sub-sections, separating each theoretical proposition developed. As an overall observation, the studies here proposed would be composed of comparative case studies, given that the focus is on major organizations' characteristics and processes from the web server industry and their respective performance in the marketplace, measured as market-share distribution.

Proposition 1: Apache's human-resource volunteer base and compensation scheme allows it to accomplish its tasks more efficiently, responding faster to users' demands, in comparison to its for-profit competitors'.

For this proposition to be tested, it requires the researcher to collect output information, such as time-to-market performance as both types of organizations have to release software updates frequently. The ideal situation would be to set up an empirical study with two or more organizations, being Apache necessarily one of them. The researcher would have to locate when similar features related to web server software were identified as needed by different organizations (e.g., through press releases), and when they were implemented and released to the public. Efficiency (time to implement) would then be compared between organizations.

Similarly, a researcher could focus on the time to solve bugs (problems in the software) identified by users or by the own researcher. Bugs of similar complexity identified in both the Apache and in the Microsoft software would be reported through the same media (e.g., e-mail), and then the lag from the report time to the implementation time would be stored. Over time, these data could be used as a proxy for efficiency on solving problems in the software, allowing an objective comparison.

Proposition 2: Apache's simplified organizational structure reduces significantly the effects of the environment on its operations, allowing a less complex and more focused managerial action than that of its for-profit competitors.

This proposition can be tested through a first stage of survey research aimed at the decision-makers of the web server industry organizations, intending to describe organizational policies regarding hiring (attracting), promotion, and laying off; as well as procedures regarding reporting bugs and proposing changes in the software or in organizational processes. In a second stage, a focus group with organizational policies' implementation specialists would be performed to evaluate the complexity and diversity of the policies. The opinion of the specialists would be used to compare one organization's policies with the others', contrasting them according to difficulty of implementation and considering whether the environment has influenced the decisions made.

Proposition 3: User satisfaction towards web server software is higher for Apache in comparison to its for-profit competitors, because users develop Apache themselves, thereby, impacting adoption rates positively in a sustainable way.

This proposition could be tested through a survey aimed at a sample of web servers' managers. The intention of this survey would be to evaluate the perception of quality, in its many variant forms (e.g., downtime and ease of use), of the managers regarding the web server currently in use. Having collected the data, an ANOVA-type of analysis could be performed and differences in quality perceptions, according to the software adopted, would then be assessed.

DISCUSSION: APACHE'S ADVANTAGE AS DRAWBACK

Past advantage does not necessarily guarantee a future one. Apache seems to be a case of extreme focus on core competence, which can be seen as a double-edged sword. Miller (1996) made this point, stating that firms too tightly configured with a single dominant perspective that drives their functions may become too simple, leading to failure. Thus, on one hand, this characteristic might be leading Apache to achieve competitive advantage through efficiency. However, on the other hand, prior research on innovation has shown that the lack of diversity found on homogenous open source communities tends to hinder innovation levels, possibly outdating Apache's creations in the future, even if distant (Galunic & Rodan, 1998). This type of strategy, which is very narrow and surrounds very few elements, has also been found to lead organizations to lose their resilience and relevance (Miller, 1996). Consequently, future chances of Apache sustaining its competitive advantage might be diminishing by this same source of advantage. This creates a challenge to open source community managers, who can take advantage of the knowledge developed in this chapter.

As extreme focus on core competence in homogenous teams hinders future innovation, Apache leaders should act to prevent this long-term risk. To recognize that the Apache organization possesses a rapid response capability to community input is the first step in that direction. After that, it is a matter of focusing on effectiveness rather than only on efficiency. To fix bugs reported by the community is important for open source communities to succeed, but it is not enough. Besides that, to develop new functionalities to keep up with users future expectations is also crucial. Thus, to guarantee a constant incoming of ideas for new functionalities, Apache leaders cannot rely solely on the community of developers and should monitor competitors' application, industrial and academic trends, and users' of proprietary solutions behavior to constantly feed the project with forward vision.

LIMITATIONS AND FUTURE RESEARCH

Being a theoretical chapter, the main limitation of the propositions here developed is to draw conclusions based only on prior research and authors' perceptions. Clearly, further empirical research is needed in order to evaluate the accuracy of the propositions developed here, hopefully extending the current understanding of the open source phenomenon. Thus, further research to overcome this limitation and extend the literature understanding on this topic is discussed next.

First, further research is needed to understand why one still find organizations (websites) using other web servers besides Apache. Given that this software is claimed to be of high quality, that it can be obtained free of charge, and that managers tend to be economically rational, how can one explain organizations paying for Apache's competitors products? Is/Are any of the assumptions of this paper wrong? Are they valid only for some cases? What are those cases?

Second, as it was discussed in the end of the literature review, the open source phenomenon is being increasingly influenced by the corporate world. This identification led Fitzgerald (2006) to suggest that one should even change the way to refer to open source. He claimed that OSS2.0 would be a better label to reflect these changes. Thus, has this transformation affected the open source way of production to a point that its differentiation from the ordinary way of software development does not make sense anymore? Or is it just a way of separating open source organizations in two different groups (e.g., traditional non-commercial and OSS2.0 commercial)? What else is in the nature of this transformation?

Thirdly, one cannot avoid asking the question on whether any comparison of open source software with corporate-proprietary ones is legitimate. Since these open source organizations do not compete in the regular sense for sales, how can one reconcile the traditional economic perspectives that permeate the strategic management field with the existent open source phenomenon that affects regular corporations' performance? Surely, the literature understanding of open innovation in software is limited; much research remains to be pursued and should, thus, be encouraged.

Finally, as the case of Apache demonstrates, community efforts to innovate openly over the Internet are viable and may benefit small enterprises, which can adopt and use the high-quality applications and documentation developed by these online communities for free. Also, the Apache case shows that the strategic attempt to create an open innovation community to foster an application development and compete with proprietary solutions maybe interesting, but what it takes to be successful in the endeavor is yet to be fully understood.

ACKNOWLEDGMENT

This research was funded by CAPES (www.capes.gov.br) and the Horizon Digital Economy Research Institute at the University of Nottingham (http://www.horizon.ac.uk/Home). We thank the participants of the 34th ANPAD Meeting in Rio de Janeiro, where a previous version of this chapter was presented.

REFERENCES

Barney, J. (1991). Firm resources and sustained competitive advantage. *Journal of Management*, *17*(1), 99–120. doi:10.1177/014920639101700108

Chesbrough, H. W. (2003). The era of open innovation. *MIT Sloan Management Review*, *44*(3), 35–41.

DeSanctis, G., & Monge, P. (1999). Introduction to the special issue: Communication processes for virtual organizations. *Organization Science*, *10*(6), 693–703. doi:10.1287/orsc.10.6.693

Di Gangi, P., Wasko, M., & Hooker, R. (2010). Getting customers' ideas to work for you: Learning from Dell how to succeed with online user innovation communities. *MIS Quarterly Executive*, *9*(4).

Egyed, T. M., & Joode, R. W. (2004). Standardisation and other coordination mechanisms in open source software. *International Journal of IT Standards and Standardization Research*, *2*(2).

Fielding, R. (1999). Shared leadership in the Apache project. *Communications of the ACM*, *42*(4), 42–43. doi:10.1145/299157.299167

Fitzgerald, B. (2006). The transformation of open source software. *Management Information Systems Quarterly*, *30*(3), 587–598.

Gallaugher, J., & Ransbotham, S. (2010). Social media and customer dialog management at Starbucks. *MIS Quarterly Executive*, *9*(4).

Galunic, D., & Rodan, S. (1998). Resource combinations in the firm: Knowledge structures and the potential for Schumpeterian innovation. *Strategic Management Journal*, *19*(12), 1193–1201. doi:10.1002/(SICI)1097-0266(1998120)19:12<1193::AID-SMJ5>3.0.CO;2-F

Hart, S. (1995). A natural-resource-based view of the firm. *Academy of Management Review*, *20*(4), 986–1014.

Hertel, G., Niedner, S., & Herrmann, S. (2003). Motivation of software developers in open source projects: An internet-based survey of contributors to the Linux kernel. *Research Policy*, *32*, 1159–1177. doi:10.1016/S0048-7333(03)00047-7

Howe, J. (2009). *Crowdsourcing: Why the power of the crowd is driving the future of business*. New York, NY: Crown Publishing Group.

Kane, G., Majchrzak, A., & Ives, B. (2010). Editors' comments – Special issue on enterprise and industry applications of social media. *MIS Quarterly Executive*, *9*(4), iii–iv.

Lee, G. K., & Cole, R. E. (2003). From a firm-based to a community based model of knowledge creation: The case of Linux kernel development. *Organization Science*, *14*(6), 633–649. doi:10.1287/orsc.14.6.633.24866

Lipnack, J., & Stamps, J. (1997). *Virtual teams*. New York, NY: John Wiley and Sons, Inc.

Melville, N., Kraemer, K., & Gurbaxani, V. (2004). Information Technology and organizational performance: An integrative model of IT business value. *Management Information Systems Quarterly*, *28*(2), 283–322.

Melville, N., Kraemer, K., & Gurbaxani, V. (2004). Review: Information Technology and organizational performance: An integrative model of business value. *Management Information Systems Quarterly*, *28*(2), 283–322.

Miller, D. (1996). Configurations revisited. *Strategic Management Journal, 17*(7), 505–512. doi:10.1002/(SICI)1097-0266(199607)17:7<505::AID-SMJ852>3.0.CO;2-I

Miller, D., Eisenstat, R., & Foote, N. (2002). Strategy from the inside out: Building capability-creating organizations. *California Management Review, 44*(3), 37–54.

Moon, J. Y., & Sproull, L. (2002). Essence of distributed work: The case of the Linux Kernel. In Hinds, P., & Kiesler, S. (Eds.), *Distributed work* (pp. 381–404). Cambridge, MA: MIT Press.

NETCRAFT. (2010). Web server survey. Retrieved August 29th, 2010, from http://www.netcraft.com

O'Mahony, S. (2003). Guarding the commons: How community managed software projects protect their work. *Research Policy, 32*(7), 1179–1198. doi:10.1016/S0048-7333(03)00048-9

O'Mahony, S., & Ferraro, F. (2007). The emergence of governance in an open source community. *Academy of Management Journal, 50*(5), 1079–1106. doi:10.5465/AMJ.2007.27169153

Porter, M. E. (1985). *Competitive advantage*. New York, NY: The Free Press.

Priem, R., & Butler, J. (2001). Is the resource-based view a useful perspective for strategic management research? *Academy of Management Review, 26*(1), 22–40.

Robinson, G., & Lundstrom, W. (2003). Market expansion strategy: Development of a conceptual market expansion decision scorecard. *Strategic Change, 12*(5), 259–272. doi:10.1002/jsc.642

Siggelkow, N., & Levinthal, D. (2006). Temporarily divide to conquer: Centralized, decentralized, and reintegrated organizational approaches to exploration and adaptation. *Organization Science, 14*(6), 650–669.

von Hippel, E. (2001a). Innovation by user communities: Learning from open-source software. *MIT Sloan Management Review, 42*(4), 82–86.

von Hippel, E. (2001b). User toolkits for innovation. *Journal of Product Innovation Management, 18*, 247–257. doi:10.1016/S0737-6782(01)00090-X

von Krogh, G., & Spaeth, S. (2007). The open source software phenomenon: Characteristics that promote research. *The Journal of Strategic Information Systems, 16*, 236–253. doi:10.1016/j.jsis.2007.06.001

ADDITIONAL READING

Agerfalk, P. J., & Fitzgerald, B. (2008). Outsourcing to an Unknown Workforce: Exploring Opensourcing as a Global Sourcing Strategy. *Management Information Systems Quarterly, 32*, 385–409.

Allen, R. C. (1983). Collective invention. *Journal of Economic Behavior & Organization, 4*(1), 1–24. doi:10.1016/0167-2681(83)90023-9

Bagozzi, R. P., & Dholakia, U. M. (2006). Open Source Software User Communities: A Study of Participation in Linux User Groups. *Management Science, 52*(7), 1099–1115. doi:10.1287/mnsc.1060.0545

Baldwin, C. Y., & Clark, K. B. (2006). The Architecture of Participation: Does Code Architecture Mitigate Free Riding in the Open Source Development Model? *Management Science, 52*(7), 1116–1127. doi:10.1287/mnsc.1060.0546

Balijepally, V., Mahapatra, R., Nerur, S., & Price, K. H. (2009). Are Two Heads Better than One for Software Development? The Productivity Paradox of Pair Programming. *Management Information Systems Quarterly, 33*(1), 91–118.

Basili, V. R., & Reiter, R. W., & Jr. (1981). A Controlled Experiment Quantitatively Comparing Software Development Approaches. *IEEE Transactions on Software Engineering, 7*(3), 299–320. doi:10.1109/TSE.1981.230841

Benkler, Y. (2002). Coase's Penguin, or, Linux and "The Nature of the Firm". *The Yale Law Journal, 112*(3), 369–446. doi:10.2307/1562247

Bonaccorsi, A., Giannangeli, S., & Rossi, C. (2006). Entry Strategies Under Competing Standards: Hybrid Business Models in the Open Source Software Industry. *Management Science, 52*(7), 1085–1098. doi:10.1287/mnsc.1060.0547

Crowston, K., & Howison, J. (2006). Hierarchy and Centralization in Free and Open Source Software Team Communications. *Knowledge. Technology & Policy, 18*(4), 65–85. doi:10.1007/s12130-006-1004-8

DuBridge, L. A. (1968). Can Success Be the Cause of Failure? *Science, 162*(3859), 1225. doi:10.1126/science.162.3859.1225

Economides, N., & Katsamakas, E. (2006). Two-Sided Competition of Proprietary vs. Open Source Technology Platforms and the Implications for the Software Industry. *Management Science, 52*(7), 1057–1071. doi:10.1287/mnsc.1060.0549

Fang, Y., & Neufeld, D. (2009). Understanding Sustained Participation in Open Source Software Projects. *Journal of Management Information Systems, 25*(4), 9–50. doi:10.2753/MIS0742-1222250401

Fershtman, C., & Gandal, N. (2007). Open source software: Motivation and restrictive licensing. *International Economics and Economic Policy, 4*(2), 209–225. doi:10.1007/s10368-007-0086-4

Herbsleb, J., & Mockus, A. (2003). An Empirical Study of Speed and Communication in Globally Distributed Software Development. *IEEE Transactions on Software Engineering, 29*(6), 481–494. doi:10.1109/TSE.2003.1205177

Johnson, J. P. (2002). Open Source Software: Private Provision of a Public Good. *Journal of Economics & Management Strategy, 11*(4), 637–662. doi:10.1162/105864002320757280

Koch, S. (2004). Profiling an Open Source Project Ecology and Its Programmers. *Electronics Markets. Special Section: Open Source Software, 14*(2).

Krishnamurthy, S. (2002). Cave or Community? An Empirical Examination of 100 Mature Open Source Projects [Electronic Version]. *First Monday (*http://firstmonday.org/htbin/cgiwrap/bin/ojs/index.php/fm/article/view/960/881*), 7.* Retrieved 02/18/08.

Kuk, G. (2006). Strategic Interaction and Knowledge Sharing in the KDE Developer Mailing List. *Management Science, 52*(7), 1031–1042. doi:10.1287/mnsc.1060.0551

Lerner, J., & Tirole, J. (2005). The scope of open source licensing. *Journal of Law Economics and Organization, 21*, 20–56. doi:10.1093/jleo/ewi002

Mockus, A., & Herbsleb, J. D. (2002). *Why not improve coordination in distributed software development by stealing good ideas from open source?* Paper presented at the ICSE '02 Workshop on Open Source Software Engineering.

O'Mahony, S. (2007). The governance of open source initiatives: what does it mean to be community managed? *Journal of Management & Governance, 11*, 139–150. doi:10.1007/s10997-007-9024-7

Osterloh, M., & Rota, S. (2007). Open source software development--Just another case of collective invention? *Research Policy, 36*(2), 157–171. doi:10.1016/j.respol.2006.10.004

Raymond, E. S. (1999). *The Cathedral and the Bazaar: Musings on Linux and open source by an accidental revolutionary.* Sebastopol, CA: O'Reilly.

Santos, C., Jr. (2009). *Open Source Software Projects' Attractiveness, Activeness, and Efficiency as a Path to Software Quality: An Empirical Evaluation of their Relationships and Causes.* SIUC, Carbondale, IL (available at http://opensiuc.lib.siu.edu/dissertations/2/).

Sauer, R. M. (2007). Why develop open-source software? The role of non-pecuniary benefits, monetary rewards, and open-source licence type. *Oxford Review of Economic Policy, 23*(4), 605–619. doi:10.1093/oxrep/grm034

Stewart, D. (2005). Social Status in an Open-Source Community. *American Sociological Review, 70*(5), 823–842. doi:10.1177/000312240507000505

Stewart, K., & Gosain, S. (2006). The impact of ideology on effectiveness in open source software development teams. *Management Information Systems Quarterly, 30*(2), 291–314.

Thomas, D., & Hunt, A. (2004). Open source ecosystems. *IEEE Software, 32*(1), 89–91. doi:10.1109/MS.2004.24

Van Rooij, S. W. (2007). Perceptions of Open Source Versus Commercial Software: Is Higher Education Still on the Fence? *Journal of Research on Technology in Education, 39*(4), 433–453.

von Krogh, G., & von Hippel, E. (2006). The Promise of Research on Open Source Software. *Management Science, 52*(7), 975–983. doi:10.1287/mnsc.1060.0560

Vries, H. d., Vries, H. d., & Oshri, I. (2008). *Standards Battles in Open Source Software: The Case of FireFox*. Palgrave MacMillan. doi:10.1057/9780230595095

Wang, H., & Wang, C. (2001). Open Source Software Adoption: A Status Report. *IEEE Software, 18*(2), 90. doi:10.1109/52.914753

Watson, R. T., Boudreau, M.-C., York, P. T., & Greiner, M. E. W. J., D. (2008). The Business of Open Source. *Communications of the ACM, 51*, 41–46. doi:10.1145/1330311.1330321

KEY TERMS AND DEFINITIONS

Apache: A web server software freely available on the Internet, developed by a community of volunteers.

Competitive Advantage: The organization characteristic or strategy that enables it to operate more effectively than its competitors, sustaining the position of market leader.

Linux: An operating system developed by Linus Torvalds with the help of a community of volunteers spread all over the world. The most prominent and popular example of a software developed using the open source model.

Microsoft: The world's largest software corporation, founded by Bill Gates and Paul Allen in 1975.

Open Source Software: Software with source code available on the Internet for free inspection, modification, use, and distribution. Open source and free software are considered equal in this chapter.

Organizational Structure: Formal and informal elements that constitute organizational processes, defining information flow and hierarchical positions.

Resource-Based View of the Firm: An organizational theory that intends to explain a firm's competitive advantage based on its internal resources characteristics.

Software Source Code: Software commands written by humans that are either compiled or interpreted by a computer to perform a task.

Web Server: Software that delivers web pages using the HTTP protocol, it processes document requests.

Chapter 14
Innovation 2.0:
Creating a Sustainable Business Model and a Win-Win Ecosystem

Luis Angel Galindo Sánchez
Telefonica, Spain

ABSTRACT

This chapter describes a new sustainable business model, Innovation 2.0, led and deployed by the author, with the goal of increasing the innovation level at the telco service portfolio for the end-user, thank to the convergence of Internet and Telco worlds and the 'fresh' and 'effervescent' proposals created by start-ups. This model, created by Win Win Consultants, a Spanish company leader in marketing and strategy issues, under the premises of fairness for the different players, requires an open-minded view of the businesses and, in fact, is a model for open innovation in telcos, which traditionally tend to address innovation from a prism more reactive than proactive. The model proposed has been deployed in Telefonica with very good results, deploying 12 commercial services, which have generated more than 10 Millions € incomes and 8 Millions of savings in investments. Innovation 2.0 is a sustainable model, finding a balance for all parties in a win-win basis and constitutes an appropriate model for the twenty-first century telcos.

INTRODUCTION

The emergence of Web 2.0 (O'Really, 2005) has generated a social change (Fumero & Roca, 2007) that has accelerated, on the one hand, the convergence between the telco and Internet worlds, becoming less differentiated by the user and, on the other hand, the reduction of the margins in telco operators. This has fueled the imperative for them to reposition its business beyond access technologies to an innovative offer in the service portfolio to offer to the end user, first, bringing together aspects of multimedia communications with any other aspects that can range from areas as diverse as health, e-goverment, financial services ... or, now on everyone's lips, SMART (Shapiro, 2006).

DOI: 10.4018/978-1-61350-519-9.ch014

However, operators positioning is more and more under stress. The current economic climate is one more issue to reduce the economic margins in the telco operations; however, it seems the trigger of a situation inherited years ago.

Taking a look to history of any operator over the world, the results are quite similar and are following a Pareto law: 95-5. The 95% of the incomes from services are coming from only a 5% of the services at the service portfolio. This situation states that customer's demands are not aligned with operator's services and the user, who is not aware of the network technology, is not satisfied with the applications offered by the telcos.

Added to the previous scenario, two more issues have accelerated the complex situation faced by many telco companies:

- The increasing entry of ISPs (Google, Yahoo, Microsoft) in the telecommunication arena where they have applied innovation in creating new forms of communication, designing and developing applications where usability is a core parameter, whose responsiveness has been enormous between users; and
- The growing demand for participation by users in different company processes such as innovation, usability, etc (Williams, 2010) that have so far been overlooked by many telcos.

The emergence of Web 2.0 (O'Really, 2005) has generated a social change (Fumero & Roca, 2007) that has accelerated, on the one hand, the convergence between the telco and Internet worlds, becoming less differentiated by the user and, on the other hand, the reduction of the margins in telco operators. This has fueled the imperative for them to reposition its business beyond access technologies to an innovative offer in the service portfolio to offer to the end user, first, bringing together aspects of multimedia communications with any other aspects that can range from areas as diverse as health, e-goverment, financial services ... or, now on everyone's lips, SMART (Shapiro, 2006).

In these contexts, the creation of innovative business models in the ICT sector and, above all, sustainable over time (Svensson, Wood & Callaghan, 2010), is based on a new conception of the environment, which speaks of partners rather than suppliers and is looking for excellence based on two main pillars: a relationship of trust between parties and fair treatment for all the parties.. These two pillars are the foundation on which to develop the Innovation 2.0 model, where each and every one of the players are active in the business development, sharing risks and revenues.

Moreover, the definition of a business model under the above circumstances requires a change in the way of thinking, and even in the way of organizing a telco, and thus enabling it to maximize and capitalize on the value of the model.

This study describes a new proposal for a sustainable business model, with the aim of increasing the level of innovation in the service portfolio of a telco operator to end users. This business model, deployed commercially in mid 2009 uses the available assets in the Telco and Internet convergence and the freshness and effervescence of proposals created by start-ups.

The vision in this chapter is provided by a person working in an operator more than 15 years, who has participated in the definition of the strategy and the innovation. Of course, there are some figures and details that are confidential and cannot be provided here. Nevertheless, the Innovation 2.0 model exposed in this chapter is a brilliant proposal of Win Win Consultores (Win Win Consultants, 2009) deployed commercially at Telefonica, which is starting changing the mix of incomes and that could be useful for other operators over the world.

The structure of this chapter is the following: In section 2, it has described the limitations in the innovation of the operators; Section 3 covers how the convergence of the telco and Internet are increasing the competitiveness in ICT sector causing a downgrading of the economic margins. Section 4 shows the Innovation 2.0 model defined by Win Win Consultores and applied to Telefonica, covering how this model is operated in Telefonica and, at the end, what is quite important, the sustainability of the model, based on the open innovation (Chesbrough, 2006).

THE LIMITED INNOVATION OF TELCOS

Telecoms operators have their core business in providing connectivity services or network access. It is usual for any operator over the world to have a large service portfolio, normally over 300, all of them developed with great rigor and with five nines availability[1], which often cause a high Time To Market (TTM). However, to assume that the large number of services in the operator´s portfolio will place it on the top of innovation might lead to misunderstanding. Analyzing the portfolio of any operator it can observed that the large number of services is for bundles (Williams, 2009), which are mostly related to connectivity and price aspects. So it is usual to see a package of unlimited voice, Asymmetric Digital Subscriber Line (ADSL), and Internet Protocol Television (IPTV) in the fixed world or an offer of a smartphone with a flat rate, of course with a cap, in the mobile.

Another situation that causes innovation to go slowly is that operators telco standards -usually well defined and constructed after intensive discussions hosted by standards organisms as 3GPP, ITU, etc, and committed to set up the ultimate feature, even if it was never used though they take years to develop. To reflect only one example, the standard for the development of IMS subsystem (3GPP TS 23.228, 2009) has taken more than 10 years to have a high stability. This model contrasts with the other widely used models in Internet based on 'de facto' standards by providing an initial set of capabilities, which arise usually gradually as one type of software, data format, or language becomes increasingly popular. Usually, they are not approved by any standards organization, but they are widely used and recognized in industry as a standard. This model is developed and evolved based on the users experience when using services, being a much more dynamic, more open and rich as it is the market itself that evaluates the standard and makes it successful or not, and also giving feasibility to it.

This study argues that this situation of limited innovation occurs because of two fundamental aspects:

1. The huge amount of revenues made, and thanks to commercial offers for access to the network (ADSL, Wireless Broad Band (WBB) and others), which constituted the core of the strategy of many players and have made it less necessary to promote the development and marketing of value-added services around connectivity; and
2. The lack of knowledge about customer needs and the inability of operators to offer more personalized services (Lonergan, 2008) with an appropriate user experience. This point has made this issue more difficult causing commercial releases as failures due to the weakness of the product manager in their approach to the market.

Apart from all of the above, the current economic situation (Krugman, 2009) is hitting the ICT sector, causing operators are, like all businesses, large reductions in *Capital Expeditures (CAPEX)* and pursuing cost efficiency that brings the innovation budget constricts and focuses mainly on process innovation (Polder, van Leeuwen, Mohnen & Raymond, 2010), which constitutes another very interesting area, but far from the visibility that innovation in services provide to the end user and, therefore, in the Society.

THE NEW COMPETITIVE FRAMEWORK IN THE CONVERGENCE OF TELCO AND INTERNET WORLDS

After the 'crash' of the "dot com" in 2001, the pessimistic forecasts predicted an uncertain future for the evolution of the Web and its services. However, in recent years different services have appeared, as shown by their figures, that have been gaining wide acceptance and success among users like YouTube, Twitter with 106 millions (The Huffington Post, 2010) active users, and Facebook with 500 million (Zuckerberg, 2010) active users. This phenomenon known as Web 2.0 has been a revolution in society, practically affecting any field that applies in new technologies. Figure1 shows a possible map of Web 2.0.

To speak about Web 2.0 means to refer to mature technologies, and most of them are being developed by the end of twentieth century, which put its emphasis on the user and his/her experience. But above all, it is about people, the social web, where the user takes control, collaborate, share and express thanks to these technologies, producing and consuming content, so-called "prosumers" (Yelmo, del Alamo, Trapero & Martin, 2010), producers and consumers of content.

However, Web 2.0 is generating changes in the behavior of users. The takeover by the user changes issues so hackneyed such as privacy of the users. Thus, for instance, one can find websites where the person publishes his/her location to anyone in the world automatically and continuously, as is done in Ipoki (Ipoki, 2007).

The key point behind the Web 2.0 Revolution is the change in philosophy in all phases of a service, from concept to market through the design and development. Although currently there are a variety of services, the main guidelines of this new philosophy can be summarized in two fundamental concepts:

Figure 1. Web 2.0 map

1. **The user is the center of this new universe:** A user who has changed his/her position to an active one, participating in various aspects such as co-creation of services (Williams, 2010)., quality, usability, design or marketing of these products (Owyang, 2009).The user behavior is changing dramatically and the immediacy of the things forms a new way of relationship. The 2.0 universe places the user in the center and its main objective is to give him what he/she demands. This ambitious goal, to know what each user wants and its implementation in a service, is complex to perform unless the user is an active part of the service, expressing their preferences, creating content, personalized service functions, etc.
2. **Combination and flexibility:** The worldwide adoption of Internet and IP connection capabilities, along with the intensification and extension of the use of open web Application Programming Interface (API) that allow the remote execution of a particular procedure, have enabled simple mechanisms for composing value-added services, by mixing features of services (mashups) and content (syndication). The possibility of combining, in a simple way, heterogeneous resources from different Internet sites, does that Internet becomes a platform to develop and deliver new profitable services virtually anywhere in the world.

The evolution of this Web 2.0 world has been amazing and has been generated in parallel with the development of the telecommunications world. This development has leveraged in two points. On the one hand the new programming and development technologies (e.g. AJAX, PHP…) that enable easier and faster interaction in the creation of new services. On the other, the emergence of broadband technologies (ADSL, FTTH, HSPA, and others) has served as a catalyst for the development of Internet, but the big lever has been the emergence of affordable flat rates, which has produced the socialization of the Internet (Goleman, 2006). This socialization has spawned a growing number of Internet applications and services that have exceeded the traditional focus of browsing services and have gradually moved towards the world of communications through innovation in new communication's formats that have generated a great response among the user and have encouraged, to give a clear example, the spectacular success of social networks sites.

This situation has caused that convergence, usually considered as the integration between fixed and mobile communications worlds, is not such, but the real convergence is the one taking place between telco and Internet worlds (Galindo, Lozano, Moro, 2008).

This convergence has a great value to end users since where previously there were 'islands', creating a poor and not uniform user experience, now these 'islands' are connected to generate a 'seamless' or uniform user experience (Finstad, 2010). This seamless user experience blends contents with communications with the aim of providing better service to the customer.

However, this convergence means the clash of two worlds with business and organizational models very different. On one hand, the telco world that traditionally has heavy, robust and inflexible structures, with traditional business models (Galindo, Moro, Balakrishna & Al Begain, 2009) focusing on charging for service through subscriptions, fees, etc. On the other hand, the Internet world, with lightweight, flexible and user-centered structures, with free or semi-free business models for the end-user, as a result of this model, who allows loss of service as something inherent to the model (Yarmosh, 2005).

One of the most immediate consequences of this convergence is that the operators' position is becoming more and more stressed. The operators arrive to this merger with a limited innovation, a poor capitalization of their clients' relationship -especially in the knowledge of their needs and their participation in the definition of new products and services- and a not very positive perception from customers of the operators, caused by unfair pricing policies from operators (e.g., margins in voice are higher than 40% in Spain)

For many years, operators considered as an interesting income the so-called 'botonazo', which was achieved by placing the access to the operators' wap site, in a key easily accessible by the client that most of the time was pressed accidentally.

This situation is causing the displacement of the telco business to the business of being providers to the network access, more known as a pipe with possible variants 'golden', 'smart' or 'dump'. This could be interesting for operators, like other 'commodities' such as electricity suppliers, power companies, gas suppliers, etc, have a sustainable business model over time. Electric utilities, gas, etc charge users for consumption- however, the telco model and their way to compete is tremendous, especially in the mobile world, wasting a lot of value in a short time by giving unlimited use with a flat rate.

In this phenomenon, there are several interesting aspects. On one hand, it is interesting to see what has happened globally and that any operator in the world followed these steps. On the other hand, it is interesting to note that this model, introduced initially in the fixed access networks, with the same problem in use, has been transferred to mobile accesses without mediate to modify a pernicious model for the operator. Certainly, it is easier for the user to know in advance how much is going to pay for using their communications, but it remains true that the utilities' client, this is, any person living in a community, has fully internalized the pay-per-use model. Here it can be assumed that the communication actions lead by operators or have not been as effective as they should be or have been misconfigured. Again, the problem of communication and the relationship with customers appears as a barrier, which has not been properly managed by the operator. However, the previous communication problem, added with the high prices resulting from the high margins of the operators, has created a spiral of falling prices, degrading the value of available technology.

In addition to these, operators begun to react in various ways such as building their own social network sites (Casado, 2008) or by purchasing any of the existing ones (Muñoz, 2010), in an attempt to get closer to the end user. But still they have to prove their viability and learn how to add value on complex organizational structures like those of the other operators who have tried to make proposals that blend both worlds (Fogg, Husson, Mulligan & Wiramihardja 2009). In any case, what does seem clear is that telecom operators should improve their traditional communications services to meet the current demand of its customers with advanced multimedia services, leveraging on the competitive position of its core business and create greater value around it.

INNOVATION 2.0: A SUSTAINABLE MODEL

The Environment

Analyzing what has been commented in previous sections and the observation of the current environment, there are three important aspects that make innovation in the ICT sector is not led by telco operators:

- Outside the telco operators' environment, there is a growing innovation in the ICT sector leaded by small companies launching and commercializing their ideas.
- There is a misalignment between the operator's offer and user's needs.
- Operators spend a lot of time in developing and commercializing a service, because of their heavy internal processes and organizational inefficiencies.

Current technologies for software development are focused on developers. Taking a look in recent telecommunications history, to develop services in a telco environment required a thorough knowledge of the intelligent network, its protocols and primitives (Faynberg, Gabuzda, Kaplan & Shah, 1997). These protocols, developed by engineers and focused on traditional telco operations, were complex to understand and to manage and constituted an often insurmountable barrier for those developers who tried to develop in this universe without deep telco knowledge. With this situation, it is easy to understand that the number of developers with skills in intelligent network was limited. However, with the advent of internet protocol (IP) technologies to the telco arena, the picture changes dramatically.

The opening of application programming interfaces (APIs) (Benslimane, Dustdar & Sheth, 2008) that enables to execute operations more easily, without having to know the technology or the underlying protocols, represents a significant step in expanding the number of developers. Furthermore, latest techniques for syndicating contents (Hammersley, 2005) and creating mash-ups (Zang, Rosson & Nasser, 2008) reduce development complexity and allow reusing previously developed code, facilitating the creation of new services that use the various software pieces of existing code. Another important issue is the increasing features of mobile phones and the creation by mobile phones manufacturers, or by the operating systems manufacturers, of Software Development Kits (SDKs) much more powerful and easier to manage. In front of Java ME and Symbian programming, with more tradition in development on mobile phones, the latest technologies such as Android or iPhone (IOS) have covered the market through better user experience (Constantinou, Camilleri & Kapetanakis, 2010).

Talking about misalignment between the operators' offer and the users' needs, the mass market segment is specially facing this situation, where operators try to ride out as best as they can, but in general without success. It seems that users are only expecting price offers and promotions, and, additionally, in the Spanish mobile market, subsidized handsets. This situation is a consequence of the competitive landscape promoted by operators during years, where there have been lacks of value added services with a real value for end user. The voice and basic services offer has represented an environment with serious shortcomings in the user-centered innovation. Innovation was hidden behind a marketing based on price offer and promotion, which ensured an annual growth –at least in mobile networks- up to 50%, focused in getting short-term incomes, but without attention or customer needs and the sustainability of a market with wide margins. With this fact, together with some poor initial experiences in Internet and IP environments, it is understandable that product managers of these companies were focused in creating a better tariff or price plan, more than bringing something innovative for the end user.

For years, operator's product managers felt comfortable competing with other operators in price, so diminishing value more and more. In Spain, for example, with the mobile handset subsidy strategy, the market is still more complex, causing clients to move to other operator based on a better handset and a reduced bill. However, it is also needed to mention the technological limitations of the intelligent network that caused that any service was generally replicated by a competitor in a short time. A comparison of service portfolios between the first and second telco operators in any country would show a similarity near to 95%.

With the advent of the deployments of broadband networks, for both fixed and mobile, with affordable and predictable prices for the user (Galindo, 2010), a fiercer competition among operators was established in the telco market, causing prices to fall even more.

Nevertheless, this has allowed the development of two significant aspects, not considered, at least at the beginning, consciously or unconsciously by telco market players: the first is the Google emergence as a search engine and its strategy to leverage and create new interest assets to the end user based on its proximity with the user; the second is the emergence of a mobile phone, iPhone, designed by and for the user, who breaks the existing hegemony so far and shakes to a company as established as Nokia (Reinhardt, 2010). In both cases, something similar happened: it was felt that these companies, Google and Apple, could never get to compete with operators and device manufacturers, which has led to the strategists of the latter to act in a reactive manner once the core business margins have begun to degrade.

The third point would not necessarily be a negative aspect and the author considers that it is not correct to make a goal of the TTM. A first analysis of this aspect shows that other players, such as Google itself, have a TTM close to two years in new products. A case to be mentioned here was the purchase of Grand Central, an IP telecommunications company, bought by Google in the last quarter of 2007, which has made up, among others, the Google Voice (Diaz, 2010), launched commercially in 2010. The fact is that during that time, Google engineers have worked considering the views of users on a recurrent, continuous basis, and rethinking product with this information. As it can be seen in the above example, Google has a high TTM, however, tends to spend most of this time to market in offering products tailored to the user.

At the same time, operators lose most of the time in such business decisions 'go / no go', which is intended to assess the interest of users in a given product. Typically, operators still continue using assessment mechanisms which are based on unreliable mechanisms such as user panels which test a service or product. Its current use is justified by the inability of operators to engage customers in the various processes of design and marketing services (Owyang, Pflaum, Overby, Bowen Murphy, 2008). In this situation, it is understandable the attempt made by operators to enter the world of social networks as a meeting point with users where the operators obtain information from the users about products and services that are of interest for the operator (McQuivey, 2010).

The Innovation 2.0 Model

The model proposed here, Innovation 2.0, to bring innovation in telecommunications operators, is a sustainable model over time based on the criteria of justice and fairness for the parties. This model has already been successfully implemented by the author in one of four major telecom operators worldwide.

It is an industrialized model, which allows, using these premises, delivering to the market a high number of services and, moreover, focuses on providing services tailored to user demand. The model has great flexibility and, without making it a core strategy, intends to maintain a reference time to market close to four months. The purpose of Innovation 2.0 is to capitalize on existing innovative proposals over the world in the ICT sector and in all sectors that ICT can bring some value, like banking, health, justice, fashion and much more to come in almost any sector of the society.

The model comes from a deep reflection based on recurrent events in the recent history of the operators. It is undeniable that for any innovative start-up offering solutions in the field of ICT, working with operators is very interesting, because it allows access to databases of customers with substantial numbers (e.g. Telefonica has 278 millions of customers). If at this point, it is added the operators' broad brand awareness, the result is very attractive for these small businesses which consider an important opportunity to offer their products and services globally.

However, many attempts to commercialize services from start-ups, within the operators' commercial offer, normally have ended in failure. The clash of two completely different worlds has been a constant recurring. On one hand, large telecom companies with complex organizational structures and almost tight departments, with operation and customer care models based in the experience of traditional telco environments (Andrews, 2008). On the other hand, start-ups have simple and flat structures, focused on efficiency and productivity, but with scarce human resources. The confrontation of two structures are very different, mainly in organization and operation issues, coupled with the inability of the telcos to relax the customer care and operation model, which required the start-ups to have the resources needed to provide a 7x24-365 service availability and this has caused the loss of important opportunities in the development of innovation.

Innovation 2.0 is an open innovation model (Chesbrough 2003) where innovative ideas coming from start-ups are commercialized with the objective of the customer to decide about their interest, marking the success or failure of them. The model is applicable to any segment in the market from large enterprises to mass market passing for the small and medium enterprises segment.

The model defined in Innovation 2.0 includes another new agent in the value chain, which lies between the operator and the third party (the innovative start-up), making of an 'adaptive bridge' between both worlds. This agent, the Adapter, facilitates the relationship between the operator and the start-up, enabling on one hand mechanisms for the adaptation of structures and operating on both sides and, on the other hand, providing a crucial role for the functioning of the model: support customer care and O&M operations that are necessary in order to market the service. However, these tasks of care and O&M are very cost-intensive and for small enterprises account for exponential degradation of margins. But, the introduction of this agent in the value chain enables to create a specialized and sustainable model over time, where the different players are specialized based on their skills. The adapter agent finds its business model in the proliferation of services that allow it to reduce the cost per service of care and O&M teams, and creating economies of scale.

Figure 2 shows the industrialized model of Innovation 2.0. The acronym *Service Delivery Platform (SDP)* has been and is very stale, taking different meanings according to the rapporteur. In this case, SDP is a logical entity (Galindo & Salvachua, 2011) consisting of various service platforms such as an SMSC, MMSC, intelligent networks, voice and video servers, orquestrators and composers of services, etc., thus enabling access to features and capabilities available in the operator's network. Logically, to enable this access it is necessary to integrate any agent who wants to use them. In the industrialized model, all the skills needed by Innovation 2.0 services have been open towards a mirror SDP, outside of the operator's premises. The operator monitors the Service Level Agreement (SLA), previously agreed between operator and adapter, on the border where the operator's domain ends and begins the adapter's domain. This model allows offering of services anywhere in the world. The important thing here is that the SLA (Ludwig, Keller, Dan, King, & Franck, 2003) is faithful to what was previously agreed by the two actors, operator and adapter. Compared to more traditional models, where the SLA becomes a crucial tool in the customer-supplier relationships, here the SLA becomes less important because the relationships between agents of the model is that of partners working towards a common goal: to generate more business with a joint approach to the market, which increases the individual value of each partner.

In the proposed model, each agent focuses on the tasks where they have more experience:

- **Operator.** It focuses on marketing services under their brand or a co-branding, using its various sales channels and customer base that the operator has.
- **Adapter.** This agent handles the more complex task to make by a start-up: second level O&M and customer care. Both tasks, cost-intensive, involve a drag on the margins of the start-up.

Figure 2. Innovation 2.0 model

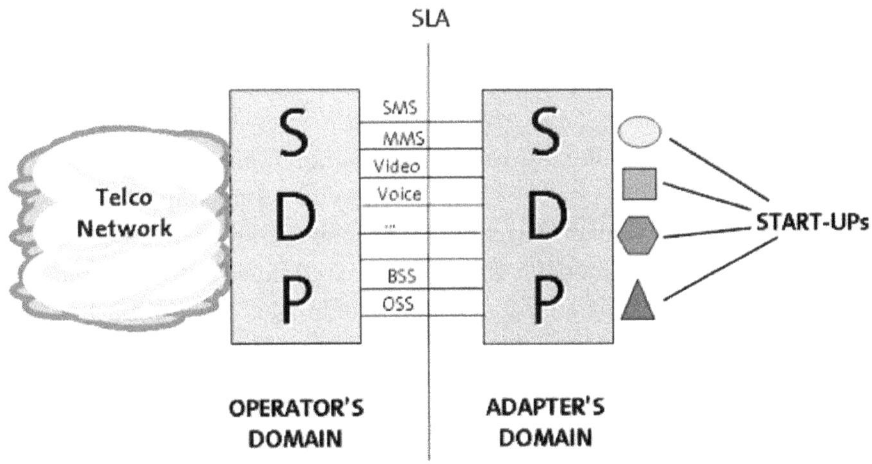

- **Start-ups.** They are the owners of the idea and the service providers. They perform the third level O&M. Also their bright ideas bring freshness to the traditional telco environments.
- **Content Providers.** In addition to those services requiring the use of specialized or differential content, will take care of generating the content and provide it to start-ups to provide the service. This agent can be included, or not, as part of the Start-ups.

To give impetus to the model, there is another very important role. They are the scouts, skilled in finding and representing start-ups worldwide. This agent does a dual role, first knows the strategies of the operator, which serves to find start-ups worldwide related to the strategy of the operator; in the other role, he or she knows the complex organizational structures of the operator, making a work-up and business development for the start-up in the process of marketing services with the operator. All the roles are needed to create a well-oiled machinery "to ensure the success of the model".

ORQUESTRATION AND OPERATION OF THE MODEL

In order to make the proposed model operational and functional, it is necessary to sign a framework agreement defining in advance the main rules of the model. Obviously, start-ups are previously unknown and will vary depending on its value proposition, which is why this framework agreement should be signed with the adapters which are known and which will have a direct relationship with the operator. One aspect to note is that with this framework agreement, the adapter hides the complexity of the business relationship between the start-up and the operator, reducing repetitive tasks in the integration of a new service coming from any start-up. Adapter is a facilitator for the start-ups and can be seen like a concentrator of them.

This model is characterized by a differential trade operation. Compared to traditional models implemented by the operators, customer-supplier models, where the operator was engaged in buying platforms and service development for further marketing, this model arises where the operator establishes a partners' relationship, sharing risk and revenue with them. The proposed model requires a new conception of the other agents (adapter and others) as 'fellow travelers' in seeking to articulate together the necessary mechanisms for the success of each of the services commercialized under this model. As it has been defined, it is a win-win or lose-lose model for all parties in the chain, so, as it is foreseen, all the parties will do the best to generate a successful service.

The framework agreement will form the basis for ensuring the sustainability of the model and allows having a number of critical parameters predefined in order to have a right TTM suitable for the market demands, ensuring service delivery within the window of opportunity The framework agreement should contain, in addition to the various legal provisions, two pillars that form the business aspects. On one side, the minimum SLA required for all services introduced under this model, which must be previously agreed by every unit at the operator involved in the operation of clients (service, operation, monitoring, etc.) and, of course, the adapter, the agent to fulfill it. On the other side, it must include the maximum income to share with the adapter, which, of course, is directly related to the SLA defined, but always keeping in mind the idea of fairness to ensure the compromise of the parties and, at the end, the sustainability. The great advantage of the model is that for the vast majority of services, the SLA defined in the framework agreement is adequate, thus avoiding negotiations with the various technical areas in the definition of it, reducing the negotiations to the business aspects and sharing incomes, which have to be taken with the adapter, on one side, and on the other with the start-up.

As mentioned, the framework agreement reflects the basic rules or rules of departure that both, the operator and the adapter, must fulfill. The important thing is not to reflect contractual legal aspects; typical of any contract, but that the framework agreement reflects the technical and business aspects. The acceptance by both parties (operator and adapter) of a minimum SLA, applicable to most of the services to market and internally agreed in the operator's domain, ensures that the starting point for each new service prevents internal negotiations between the operator areas involved, which saves considerable time and resources (estimations made in Telefonica with this model deployed, show a savings of 2-3 months and 1 *Full Time Equivalent (FTE)* consolidated per service). Something similar happens with the percentages of revenues shared with the adapter. These are defined by their maximum values within the contractual framework that allow reducing negotiations between the operator and adapter.

In the model shown here, the relationship between the operator and the service provider follows two distinct faces:

- **The legal and administrative side:** It is the adapter, which will sign a contract with the start-up that provides the service. This contract will be transparent to the operator, but should include some of the aspects of the SLA included in the framework agreement signed between the operator and the adapter. By delegating this task to the adapter is putting a filter to the inherent complexity of the operator and allows a model of juridical and administrative relation simpler and more flexible; and
- **The profit side:** The side of the business that allows the operator to negotiate with the start-up the percentage of income that the start-up will receive by the marketing of the service.

This peculiarity of the model is very interesting. The operator has the ability to negotiate revenue sharing with the start-up, delegating the complexity of the relationship to the adapter. Once it has been agreed both the percentage that the start-up will receive as the percentage, within the margins defined in the framework agreement, which will receive the adapter, a payment in cascade with the sum of both percentages will be made by the operator. Thus, the adapter will receive its share and that of the start-up, being the adapter, which is who has the legal and administrative links with the service provider, responsible for giving the start-up the percentage previously agreed in the negotiations start-up-operator.

Although from a legal standpoint, the figure of the adapter in this model could be considered as a service provider, but really from the business point of view it is not. In fact as the operator, along with the scouts, looks for start-ups and also the operator negotiates the share of trading income that the start-up will receive.

The contractual model has two distinct parts:

- The framework agreement, which defines the 'rules of the game' and includes, inter alia:
    - The execution procedure
    - Technical Terms
    - Marketing of products and services
    - Term
    - Economic conditions at their peak
    - Confidentiality
    - Privacy
    - Intellectual Property
- The specific agreement to define the specific conditions to be applied directly to a specific service to be traded on the Innovation 2.0 model and are mainly:
    - The percentage of revenue share with the adapter, which will include the previously agreed percentage with the start-up (remember that is a cascade payment); and
    - The modifications, if any, in the basic SLA included in the framework agreement, would be needed for those services to which the basic SLA is not enough. This section is usually not necessary for most services.

The Innovation 2.0 model can reduce time in the marketing of services by defining a framework agreement that includes all the repetitive tasks in the marketing of each service, allowing operator and adapter to focus on what is particular to each service. Although, it could be considered simple to define an agreement between operator and start-up, the situation is far away to be simple. Further to this issue, the integration of the start-up with the operator's service platform is costly in both, time and resources.

The governance model is simple where both the operator, thanks to the scouts, and adapter are seeking new business opportunities to tackle. Once an opportunity is found and feasibility has been discussed, the operator is responsible for defining the service-specific SLA, if necessary, and negotiates with the various agents involved. Previous to the commercial launch of any service, all the parties will agree to the *Key Performance Indicators (KPIs)* of the service that will define if the service is successful or not. Common KPIs, but not the only, used are the incomes reached during a defined period of time. After this period, all the parties will take a decision to continue or not with the service based on the results. In case the decision would be to discontinue with the service, it would be removed from the operator's service portfolio.

SUSTAINABILITY OF THE MODEL AND THE ECOSYSTEM

With the model defined, the value chain is modified as illustrated in Figure 3.

The work undertaken by Win Win Consultants in the field of Innovation 2.0 have permeated the operator with a more just vision to ensure the sustainability of the model. The proposed model is a sustainable model, because it has intrinsic values of equity for each of the agents:

- **Operator:** The operator with this model enhances the service portfolio, becoming an agent of innovation in society and improves their bottom line through revenue earned by each of the services. Do

Figure 3. Value chain in the innovation 2.0 model

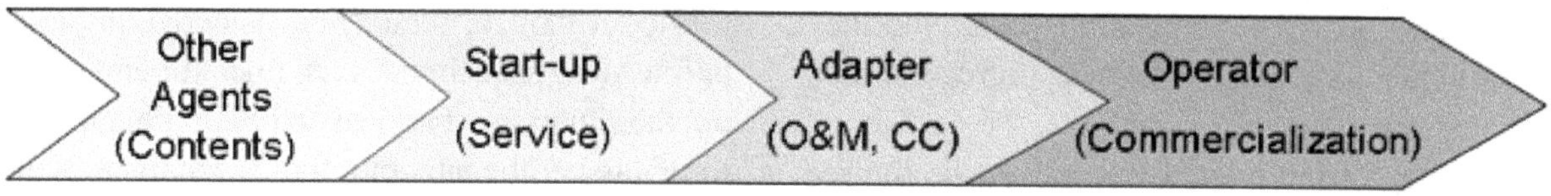

not forget that the cost for the operator is very low, since it does not make investments in infrastructure; CAPEX is taken over by the start-up and the adapter. The operator contributes with the necessary resources to address the marketing of the service, which has an impact on Operational Expenditure (OPEX). Yet this impact is much lower than the market a traditional service because, as mentioned above many of the tasks like SLA definition, definition of contract are already agreed within the defined framework agreement and others like searching for innovative ideas, products and services from start-ups are done mainly by scouters. An additional value for the operator is the integration of a new process of 'try & check' with which it will be more in touch with the reality of the market at a cost of resources needed for the commercial release of each of these services. In any case the cost is much less than the cost of investment in traditional service platforms. The operator provides the more traditional view of the market, which is 'mixed', under the Innovation 2.0 model, with the fresh and groundbreaking vision of the start-up, and creates a mixture of high value for all partners that make up the model.

- **Adapter:** The adapter has a sustainable model based on the wealth of solutions that will enable to have scale economies in the way they have a large number of services that are being hosted, operated and cared in their platforms by the same teams of people. The tasks of the adapter are intensive staffing required to solve problems in the commercial service that the operator is offering to the customers under Innovation 2.0.
- **Start-up:** It is the fundamental agent of innovation, and by using this model it can market its services in a channel, the operator's channel, of high potential and full of customers, through a renowned brand in the market and who also provides a mechanism that avoids complex integrations with the operator and expensive tasks to meet the requirements of the operator. To the extent one has more customers, and thus more revenue in their coffers. They bring fresh ideas to the operator's environment.
- **Other Agents:** Complement the value of the start-up with a differential asset for the service and are successful to the extent that the other agents are successful for that service.

It should be noted that the mixture of freshness, from the start-up, and tradition, from the operator, creates a very suitable environment for innovation and development of new ideas with great value for society.

As mentioned above, the model is sustainable in a win-win basis: it generates more business jointly than the business generated for each of the actors separately. However, it is necessary to integrate the model within the organizational structure of the operator. The Innovation 2.0 model follows the usual procedures for the service development. The difference is that a large part of the tasks to make are already made and collected within the framework agreement, requiring less effort by the project leader (it is estimated for the deployment done in Telefonica that the same project manager can manage 5-6 projects in parallel under the Innovation 2.0 model, compared to 1-2 projects in the traditional model). A barrier that will be presented is that the various technical areas were able to see the benefits of the model arise. A poor communication of the model will cause an adverse reaction in the organization, which could see erroneously this model as a model of disintermediation, particularly in those areas related to O&M and customer care, tasks that are assumed by the adapter. The author, with the help of Win Win Consultants, has addressed this situation in the case of Telefónica and believes it is outside of the scope of this chapter.

To give wealth to the Innovation 2.0 model it is required to analyze a large number of services and business solutions anywhere in the world. This figure of scout associated with this model is very interesting, as it provides services to the operator based on the operator's strategy and of the market demands. The scout will collect information about many and varied products and services, discarding those that are not within the lines of interest, those that are not relevant enough, those that are not evolved, and so on. This pre-orchestrated work brings great value for the operator and for the right development of the model and allows the operator to know many companies with different skills, but all of them aligned with the market and operator' demands.

The model proposed is bidirectional. Thanks to these mechanisms, the operator knows a group of small companies, the start-ups, very interesting and with high potential. The usual way is that the scouts present these pre-selected solutions based on the interest to the operator, which then delivers them to the market in a push mechanism. However, in any country, there are groups not considered by operators like commercial targets due to a not high potential income. Here the study is talking about groups at risk of digital exclusion, such as elderly, blind, deaf, etc. In this line of groups of people, Innovation 2.0 model works in reverse, with a pull mechanism. This allows the operator asked these groups for products or services that meet their needs, transferring after the needs of any of those start-ups in order to, with their own resources, address the necessary developments. Again, this view provides a mechanism for the open innovation, where the user provides his/her real necessities to the service provider.

All the parties compound a sustainable service ecosystem where each of them contributes with its value to create more value at the end. As it has been mentioned above, trust and fairness are key assets for a right development of the model. Although it could be considered idealistic to talk about fairness and trust in a capitalistic world, they are more and more researchers (Debenham, 2009) (Hammadi, Dillon, & Chang, 2009) working on business relationships from a similar perspective that is proposed in the Innovation 2.0 model. As it is demonstrated in "Developing a Conceptual Framework for Business Model Innovation in the Context of Open Innovation" (Wang, Jaring, & Wallin, 2009:page 1), 'a business model in the context of open innovation is not a strategy of a company, but a cognitive tool and a mediating construct between technology innovation and economic value' The Innovation 2.0 model proposed by Win Win Consultants and deployed in Telefonica is a business model for the open innovation in the telecom ecosystem.

FUTURE RESEARCH DIRECTIONS

Any business model can be improved in order to look for an increase in efficiency. In fact, any theoretical model defined requires to be modified once it is applied to a market.

Innovation 2.0 is beginning to show their strengths at the deployment in Telefonica, where the model is taking off. Currently, the author of this chapter is looking for mechanisms to access a greater number of start-ups from anywhere in the world. There are two lines that are being addressed in parallel and will develop in the coming months. However, the negotiations to decide the future of each of them are confidential and are in progress.

Once the success of the model is demonstrated, with important economic figures for all the players in the value chain, Win Win Consultants, the consultancy company that proposed the model to be deployed in Telefonica, is currently extending the model to other operators over the world. In the way that the markets are different from the Spanish market, it is quite probable that the Innovation 2.0 model had to be adapted for each market, creating different flavors of the model, that in the future, could be considered as different business models.

CONCLUSION

Innovation 2.0 is an open innovation model for the ICT sector with the goal to bring more value, practically, to any activity. The model is leveraged in several important aspects. On one hand, it has a key advantage of not weighing down the investment accounts of the operator and allows the operator to be more innovative in its product and services offer, without losing the quality and level of service that serves its customers. Predefined mechanisms in the framework agreement -minimum SLA and maximum income shared- between operator and adapter can reduce dramatically the time of the negotiations, both internal and external to the operator, resulting in an improvement in TTM and giving greater flexibility to the operator environment. The function of the adapter changes the value chain in a very positive way, acting as a catalyst for innovation from the convergence of Telco and Internet worlds, thanks to the adapter who is rushing the tasks towards being more expensive for start-ups: O&M and customer care. On the other hand, the function of the adapter allows a faster deployment of the service, since it is the adapter which makes front-end with the start-up, hiding the complexity, organizational and of integration, of the telco.

The model is sustainable because all agents of the value chain are part of the marketing, sharing the success-or failure-, creating a sustainable service ecosystem. This model allows each of the players is focused on tasks where they add more value, based on their skills and expertise. Innovation 2.0 orchestrates a situation that can generate more business, that the amount of business generated by each of the partners. The role of scouts, looking for start-ups and accompanying them during the commercialization process, is very important for the model to be a success. Scouts are feeding the model with a huge base of innovative start-ups coming from any country over the world, which select following the market demands and operator criteria.

Finally, for the operator it is a great way to increase the service portfolio by improving its limited innovative image. Through Innovation 2.0, the operator may come closer to the customer in the desired satisfaction of their needs.

REFERENCES

Andrews, C. (2008). *The priorities of tech strategists and marketers*. Forrester.

Benslimane, D., Dustdar, S., & Sheth, A. (2008). Services mashups: The new generation of Web applications. [Institute of Electrical and Electronics Engineers.]. *IEEE Internet Computing, 12*(5), 13–15. doi:10.1109/MIC.2008.110

Casado, S. (in press). Telefonica lanza Keteke.com, la primera red social multiplataforma. *ABC*. November 2008.

Chesbrough, H. W. (2003). *Open innovation: The new imperative for creating and profiting from technology*. Boston, MA: Harvard Business School Press.

Chesbrough, H. W. (2006). *Open business models: How to thrive in the new innovation landscape*. Boston, MA: Harvard Business School Press.

Constantinou, A., Camilleri, E., & Kapetanakis, M. (2010). *Mobile developers economics 2010 report*. VisionMobile.

Debenham, J. (2009). *Modelling trust, honour and reliability in business relationships*. Paper presented at Third IEEE International Conference on Digital Ecosystems and Technologies IEEE-DEST 2009

Diaz, S. (2010). *Business week: 1.4 million Google Voice users, global push in the works*. Retrieved December, 30, 2010, from http://www.zdnet.com/blog/btl/business-week-14-million-google-voice-users-global-push-in-the-works/26813

Faynberg, I., Gabuzda, L. R., Kaplan, M. P., & Shah, N. J. (1997). *The intelligent network standards: Their application to services*. McGraw-Hill.

Finstad, R. (2010). The usability metric for user experience. *Interacting with Computers*, *22*(5), 323–327. doi:10.1016/j.intcom.2010.04.004

Fogg, I., Husson, T., Mulligan, M., & Wiramihardja, L. (2009). *To deliver the Vodafone 360 strategy, Vodafone becomes a social network and a handset maker*. Forrester.

Fumero, A., & Roca, G. (2007). *Web 2.0*. Fundacion Orange.

Galindo, L. A. (2010). *2.0. Una oportunidad para las telco. Telos 83: Una sociedad en movilidad: Nuevas fronteras* (pp. 115–119). Mobile: Fundación Telefonica.

Galindo, L. A., Lozano, D., & Moro, D. (2008). *WIMS 2.0: Converging IMS and Web 2.0- New multimedia services for telecom networks*. Paper presented at ICWMC 2008.

Galindo, L. A., Moro, D., Balakrishna, C., & Al Begain, K. (2009). *IMS. A development and deployment perspective*. John Wiley & Sons.

Galindo, L. A., & Salvachua, J. (2011). *Multimedia services and streaming for mobile devices: Challenges and innovations*. Hershsey, PA: IGI Global.

Goleman, D. (2006). *Inteligencia social*. Ed. Kairos.

3GPP. (2009). *TS 23.228 v 10.2.0, IMS: IP multimedia subsystem, stage 2*. Retrieved December 19, 2009, from http://www.3gpp.org

Hammadi, A., Dillon, T., & Chang, E. (2009). Business service composability on the basis of trust. Paper presented at Third IEEE International Conference on Digital Ecosystems and Technologies IEEE-DEST 2009.

Hammersley, B. (2005). *Developing feeds with RSS and Atom*. O'Reilly.

Ipoki. (2007). *Website*. Retrieved August, 20, 2010, from http://www.ipoki.com

Krugman, P. (2009, September 6). How did economist get it so wrong? *New York Times*.

Lonergan, D. (2008). *The ubiquity delusion*. Yankee Group.

Ludwig, H., Keller, A., Dan, A., King, R., & Franck, R. (2003). Electronic commerce research. A service level agreement language for dynamic electronic services. *Springer Management Review*, *48*(2), 21–28.

McQuivey, J. L. (2010). *What people really need*. Forrester.

Muñoz, R. (2010, August 4). Telefonica compra Tuenti por 70 millones. *El Pais*.

O'Reilly, T. (2005). *What is Web 2.0? Design patterns and business models for the next generation of software*. Retrieved from http://www.oreilly.com/pub/a/oreilly/tim/news/2005/09/30/what-is-Web-20.html

Owyang, J. K. (2009). *A consumer product strategy introduction: What works in online company forums?* Forrester.

Owyang, J. K., Pflaum, C., Overby, C. S., Bowen, E., & Murphy, E. (2008). *What works in online company forums?* Forrester.

Polder, M., van Leeuwen, G., Mohnen, P., & Raymond, W. (2010). *Product, process and organizational innovation: Drivers, complementarity and productivity effects*. CIRANO.

Reinhardt, A. (2010). *Nokia's results sag amid faint hope.* Retrieved January, 3, 2010, from http://www.businessweek.com/blogs/europeinsight/archives/2010/07/nokias_results_sag_amid_scant_rays_of_hope.html

Shapiro, J. M. (2006). *Smart cities: Quality of life, productivity, and the growth effects of human capital.* University of Chicago. Retrieved from http://www.mitpressjournals.org/doi/abs/10.1162/rest.88.2.324

Svensson, G., Wood, G., & Callaghan, M. (2010). A corporate model of sustainable business practices: An ethical perspective. *Journal of World Business*, *45*(4), 336–345. doi:10.1016/j.jwb.2009.08.005

The Huffington Post. (2010). *Twitter user statistics revealed.* Retrieved April 30, 2010, from http://www.huffingtonpost.com/2010/04/14/twitter-user-statistics-r_n_537992.html

Wang, L., Jaring, P., & Wallin, A. (2009). Developing a conceptual framework for business model innovation in the context of open innovation. Paper presented at Third IEEE International Conference on Digital Ecosystems and Technologies IEEE-DEST 2009.

Williams, D. (2009). *Building the perfect bundle. Use quantitative methods to pinpoint what consumers want and how much they will pay*. Forrester.

Williams, D. (2010). *Social co- creation. Why product strategy professionals should leverage social technologies to design and improve their product*. Forrester.

Win Win Consultores. (2009). Website. Retrieved January 3, 2010, from http://www.winwinconsultores.com

Yarmosh, K. (2005). *Web 2.0 business models*. Retrieved December 30, 2010, from http://kenyarmosh.com/web-20-business-models/

Yelmo, J. C., del Alamo, J. M., Trapero, R., & Martin, Y. (2010). *A user-centric approach to service creation and delivery over next generation networks*. Computer Communications.

Zang, N., Rosson, M. B., & Nasser, V. (2008). *Mashups: Who? What? Why?* In *CHI '08 Extended Abstracts on Human Factors in Computing Systems* (CHI '08), (pp. 3171-3176). New York, NY: ACM.

Zuckerberg, M. (2010). *Facebook blog*.

ADDITIONAL READING

Chesbrough, H. W., Vanhaverbeke, W., & West, J. (2006). *Open innovation: researching a new paradigm*. Oxford University Press.

Chiaroni, D., Chiesa, V. & Frattini, F. The *Open Innovation Journey: How firms dynamically implement the emerging innovation management paradigm.* Technovation, Volume 31, Issue 1, Open Innovation - ISPIM Selected Papers, January 2011, Pages 34-43.

Christensen, C. M., Anthony, S. D., & Roth, E. A. (2004). *Seeing what's next: using the theories of innovation to predict industry change*. Harvard Business Press.

Dahlander, L., & Gann, D. M. (2010, July). How open is innovation? *Research Policy*, *39*(Issue 6), 699–709. doi:10.1016/j.respol.2010.01.013

Du Chatenier, E., Verstegen, J. A. A. M., Biemans, H. J. A., Mulder, M., & Omta, O. (2009). The Challenges of Collaborative Knowledge Creation in Open Innovation Teams. *Human Resource Development Review*, (September): 350–38. doi:10.1177/1534484309338265

Ferrary, M. (2010). *Specialized organizations and ambidextrous clusters in the open innovation paradigm*. European Management Journal, In Press, Corrected Proof, Available online 18 November 2010.

Harison, E., & Koski, H. (2010, April). Applying open innovation in business strategies: Evidence from Finnish software firms. *Research Policy*, *39*(Issue 3), 351–359. doi:10.1016/j.respol.2010.01.008

Harryson, S. J. (2008). *Know-Who Based Entrepreneurship: From Knowledge Creation to Business Implementation*. Edward Elgar Publishing Limited.

Hippel, E. (2005). *Democratizing Innovation. MIT Press. Carayannis, E. G. & Formica, P. (2008). Knowledge matters: technology, innovation and entrepreneurship in innovation networks and knowledge clusters. Palgrave Macmillan. Backer, K. & Cervantes, M. (2008). Open Innovation in global networks*. OCDE.

Kattel, R., & Karo, E. (2010). *Is 'Open Innovation' Re-Inventing Innovation Policy for Catching-up Economies?* The Other Canon Foundation and Tallinn University of Technology Working Papers in Technology Governance and Economic Dynamics 30, TUT Institute of Public Administration.

Lee, S., Park, G., Yoon, B., & Park, J. (2010). Open innovation in SMEs--An intermediated network model. [March.]. *Research Policy, Elsevier*, *39*(2), 290–300.

Mayle, D. (2006). *Managing Innovation and Change*. SAGE

Mention, A. (2011). *Co-operation and co-opetition as open innovation practices in the service sector: Which influence on innovation novelty?*. Technovation, Volume 31, Issue 1, Open Innovation - ISPIM Selected Papers, January 2011, Pages 44-53.

Rodriguez, J. L., & Garcia, A. (2010). *Open Innovation: Organizational Challenges of a New Paradigm of Innovation Management*. Annals - Economy Series, Constantin Brancusi University [November.]. *Faculty of Economics*, *3*, 99–110.

Schadler, T., & Bernoff, J. (2010). *The HERO Index: Finding Empowered Employees*. Forrester.

Tepic, M., Omta, S. W. F., Trienekens, J. H., & Fortuin, F. T. J. M. (2010). *Governance in open innovation: Exploring the effect of innovation uncertainty and network heterogeneity.* Proceedings of the 5th Research Workshop on Institutions and Organisations, Sao Paulo, Brazil, October 3rd-5th, 2010.

Trott, P., & Hartmann, D. (2009). *Why 'Open Innovation' Is Old Wine In New Bottles*. International Journal of Innovation Management (IJIM). *World Scientific Publishing Co. Pte. Ltd.*, *13*(04), 715–736.

Van de Vrande, V., Vanhaverbeke, W., & Gassmann, O. (2010). Broadening the scope of Open Innovation: Introduction to the special issue. *International Journal of Technology Management*, *52*(3/4), 221–235.

Veugelers, M., Bury, J., & Viaene, S. (2010, February). Linking technology intelligence to open innovation. *Technological Forecasting and Social Change*, *77*(Issue 2), 335–343. doi:10.1016/j.techfore.2009.09.003

Vrande, V., Jong, J. P. J., Vanhaverbeke, W., & Rochemont, M. (2009). *Open innovation in SMEs: Trends, motives and management challenges.* Technovation, Volume 29. *Issues (National Council of State Boards of Nursing (U.S.)), 6-7*(June-July), 423–437.

Warren, A. (2007). *Business Process and Model Innovations Involving "Open Innovation" Threats or Opportunities.* A White Paper in the Corporate Innovation Series from the Farrell Center, The Smeal College of Business at Penn State Huizingh, E. (2011). *Open innovation: State of the art and future perspectives.* Technovation, Volume 31, Issue 1, Open Innovation - ISPIM Selected Papers, January 2011, Pages 2-9.

Watanabe, C., & Shin, J. (2011, January). Heikkinen, J., Zhao, W. & Griffy-Brown, C. (2008). *New functionality development through follower substitution for a leader in open innovation. Technological Forecasting and Social Change, 78*(Issue 1), 116–131. doi:10.1016/j.techfore.2010.04.019

KEY TERMS AND DEFINITIONS

Adapter: It is the figure to act between operator and start-up, hiding the complexity of the operator to the start-up and tackling the customer care and O&M tasks for the service.

Innovation 2.0: The model defined by Luis Angel Galindo to generate more incomes for operators, meanwhile the operator becomes an agent of innovation.

Operator: Commercialize the service under its brand.

Scout: Looks for start-ups over the world and pre-select them.

SLA: Service Level Agreement. The tool for monitoring what is happening in a commercialized service.

Start-Up: The innovator. Provide brilliant ideas to be delivered to the market.

TTM: Time to Market. Time needed to commercialize a service.

Web 2.0: The social web, which is changing the behavior of people.

ENDNOTE

1 Five nines availability means an availability of 99,999%

Chapter 15
Crowdsourcing in Small and Medium Sized Enterprises

Natasha Katuta Mwila
Monash University, Australia

ABSTRACT

The nature of innovation is rapidly changing, and the concepts of open innovation and crowdsourcing need to be embraced. Innovation has quickly become not only the edge on which competition can occur but rather the characteristic of existence-the means to basic survival. Several forces are driving the rapid change in innovation as it is understood today. These forces tend to have a technological impetus: the very impetus that open innovation and crowdsourcing are built on. Small and medium sized enterprises (SMEs) are facing a triple blow in the light of the evolving face of innovation. Survival is the eminent challenge that must be overcome. Firstly, SMEs struggle to keep up with the financial demands of continuous innovation; secondly, SMEs are not in fierce enough competition with one another that the concept of open innovation goes largely dismissed; and thirdly, SMEs appear to a larger extent to severely underestimate the wisdom of the crowd especially in the light of small businesses. In as much as caution must be taken to avoid hyper inflating the advantages of open innovation and crowdsourcing, the relevance of the concepts cannot and must not be dismissed. The purpose of this chapter is to elucidate on the major challenges faced by SMEs in crowdsourcing and the hindrance this presents in open innovation. For this reason, solutions and recommendations are provided to these challenges.

INTRODUCTION

Crowdsourcing in SMEs has been identified as necessary innovation in SME development (Dhillon, Stahl & Baskerville, 2009). As crowdsourcing is an essential component of open innovation, this chapter sheds light on how developing the potential to crowdsource contributes towards open innovation for SME development.

This chapter introduces to the reader the inseparable link that exists between crowdsourcing and open innovation. It presents the relevance of crowdsourcing as a tool in innovation as necessary

DOI: 10.4018/978-1-61350-519-9.ch015

in SMEs. It explores what makes the context of SMEs a unique factor that warrants the devotion of a chapter on crowdsourcing specifically to it, aside from the general literature on crowdsourcing. It identifies what challenges need to be overcome by SMEs and more importantly what can be done to alleviate the challenges.

The background of this chapter briefly elucidates on the emergence of open innovation. It goes on to demonstrate the link between crowdsourcing and open innovation. The background provides an idea of the scope of what is achievable for SMEs through crowdsourcing. The limitations of crowdsourcing are highlighted.

The issues, controversies and problems of crowdsourcing in SMEs are discussed with reference to finance, competition and learning. Financial providers and their stance on SMEs are discussed, particularly how this influences crowdsourcing. Competition among financial providers and SMEs, or the lack thereof, are considered in how they influence the ability of SMEs to crowdsource. Learning is identified as a critical input and output of crowdsourcing. However, the ability and willingness to crowdsource depends on the SME's level of learning just as the SME's level of learning depends on the ability and willingness of the SME to Crowdsource. This bilateral relationship introduces the primary challenge for SMEs.

Solutions and recommendation to the issues, controversies and problems of crowdsourcing in SMEs in the light of finance, competition and learning are provided. The solutions and recommendations presented are based on the experiences of SMEs and the advice of industry experts in innovation and SMEs. The advice presented is not only intended for the managers of the SMEs but rather for the financial provider as well in the light of designing better access to financial products. The relevance of advice provided extends to governmental institutions in enhancing the scope and capacity for learning.

Actual case examples are included in the text to provide the reader a basis on which to understand the applicability of the issues presented. Cases are not sited particularly from any specific context as it has been recognised that although the African context appears as the environment harbouring the greatest challenges to crowdsourcing and open innovation in SMEs, the challenges are not unique to Africa and all experiences in any context can be used to create greater understanding of how to maximise the benefits of crowdsourcing and open innovation.

Further directions in research are indicated and the value they are to add to the topic highlighted. It is envisioned that the detail provided in the chapter will provide sufficient justification for the direction research is proposed to take. Further directions in research are not presented as the only areas that may add value to learning and practice in this area. They are included to provide an insight into what possibilities exist for research. Readers are encouraged to read this chapter with the view of identifying what can be done to improve its usefulness. Such an approach will create in the reader a greater appreciation for the proposed research directions.

The chapter concludes by summarising the main arguments. References and an additional reading section are provided. The key terms and definitions of the chapter are summarised to aid the reader's understanding. Key terms and definitions provided are specific to the context of this chapter- that of crowdsourcing as a vital component of open innovation in SMEs.

BACKGROUND

The commonly known nature of innovation practices, that of being a strategic management tool that needs to be closely guarded, is changing. Being open is no longer an option and crowdsourcing lends itself to the open nature innovation ought to have. Crowdsourcing is an emerging theory of innovation and innovation management (Von Stamm & Trifilova, 2009). Innovation remains the key to the sustainability of an organisation, particularly SMEs, and open innovation and crowdsourcing largely contribute to the extent to which this sustainability is achievable (Von Stamm & Trifilova, 2009). Crowdsourcing is all about collaboration and not working in isolation (Anderson, 2010).

DeCusatis (2008) reports that there is a growing emphasis on collaboration as part of the innovation process owing to sound economic reasons such as the rising cost of technology development, shortening product lifecycles and the difficulty in sustaining closed research and development models. Previous approaches to innovation represent the conventional view of innovation, as driven by large organisations that hold an effective monopoly on their markets. In this context, products of innovation are viewed as highly valued intellectual capital which is protected by patents, as is the case in the technology industry.

The new approach of collaborative innovation has emerged within the past decade and involves the delivery of customer value through the creation of relationships and social networks. Collaborative innovation enables the generation of valuable ideas from essentially anywhere, at any time. With the environment's tremendous emphasis on innovation as a means of firms better differentiating themselves and the products and services they offer, companies are continuously looking for ways to innovate more quickly and more effectively both within and often beyond their core markets and product lines-resulting in the concept of 'open innovation' where firms realise that all components of an innovation do not need to come from within, that they can accelerate their own efforts or perhaps even broaden the scope of their own efforts by sourcing a part of the required technology externally (Witzeman, et al., 2006).

Crowdsourcing is closely linked to open innovation in that it explores the possibilities of organisations in knowledge sharing and in creating value to businesses through profit and quality. Businesses can know what products consumers prefer and, subsequently, what they should further stock or promote by requesting input and reviews (crowdsourcing) (Hamburg, 2007). While crowdsourcing makes it easy to learn from the community about a need for enhancement, it can be similarly beneficial in improving something that is already functioning well but could be superior, as another dimension on innovation.

The term crowdsourcing appears coined from the two phrases 'crowd' and 'outsourcing' or outsourcing to the crowd (Howe, 2009). Howe, who coined the term crowdsourcing in 2006 (Howe, 2009), defines the concept as being the act of taking a job traditionally performed by designated agents, usually employees and outsourcing it to undefined, generally large groups of people in the form of an open call.

The 'crowd' is anyone with access to the internet and in preferably the 'right conditions' which may include a networked virtual environment and motivation and the idea here is that a group with diversity outperforms a group of experts. Diversity is facilitated through the internet by bringing together various problem solving skills, perspectives and abilities, thus enhancing effectiveness as opposed to specialised, homogenous talent (Schee, 2008).

Open innovation and crowdsourcing build on the theories of technology's impact in changing innovation management strategy. Some of the drivers of open innovation and crowdsourcing are global talent mobility, increase of venture capital and government spending enabling small company innovation, more market-focused academic research, push for commercial applications by government labs, emergence of private research institutes, internet and search technology as well as the need to optimise existing supplier networks for strategic value.

The definition of crowdsourcing may vary but the essence of the theory is tapping into the collective intelligence of the public at large, or outsourcing to a third party, to complete the business tasks that would normally be performed by the company itself. Literature uses the synonyms 'fansourcing', 'crowdcasting', 'opensourcing', 'crowdfunding', 'mass collaboration' and 'collective customer commitment' to refer to what is referred to in this chapter as crowdsourcing (Burmann & Arnhold, 2008).

SMEs are lauded for their ability to get close to the customer and address the specific needs of the customer. It goes without saying that employing crowdsourcing would only further enable SMEs to achieve what they are reputed for. One thing is clear, the success of the SME relies on the satisfaction of the crowd and crowdsourcing only gets one closer to this crowd that must be satisfied.

The generation of new ideas, shortening of research and development time, cutting of development costs, and creation of a direct, emotional connection with customers are possible through crowdsourcing. When appropriately integrated, crowdsourcing can be a great way to access new ideas, find solutions to problems or quickly build out that impossible task. Companies can use crowdsourcing to encourage feedback, get others to weigh ideas, have problems addressed and solved early in the process, help market and promote their product, and even reduce risk by giving customers what they want (Hamburg, 2008).

The use of social media and collaborative technology to execute existing programs has led progressive companies to broaden the profile and reach of their innovation, philanthropic, and sustainability programs. Open innovation and crowdsourcing provide a means of sustaining innovation in organisations. This is so because innovative ideas are not limited to an organisation's employees. In addition, customers are set to benefit when they contribute to a product that they too partake in and this would lead to customer satisfaction and sustained interest in products.

Open innovation and crowdsourcing enables companies to be able to respond in a quick and flexible way to changes in the environment and to remain competitive despite the shortening time-to-market and life cycles of products and technologies, the advantages of open innovation for large multinationals are apparent but even smaller companies can benefit from co-operation with other parties (Gawarzynska, 2010).

By far the most favourable advantage open innovation and crowdsourcing carry is that the organisation pays for real outcome not for trial and error. In open innovation one pays only for the favourable outcome and not for the effort which does not yield desired results. The so called huge cost of yielding no result is distributed among many people, making it most of the times insignificant for an individual. The cost of doing collaborative innovation, internal and/or external, is not more compared to strictly in-house innovation. Though a relatively new phenomenon, open innovation and crowdsourcing provide more control and are less risky (Chesbrough, 2011).

Open innovation and crowdsourcing involve such a variable range and complexity of activities that can bring open innovation its own challenges in terms of the effort required to make open innovation happen, the time required of key staff and the extent to which the organisation is stretched in a variety of new directions (Barrett, 2010). Crowds do not always think well and so the context is critical.

The other issue in crowdsourcing is the transfer of control and the reduced ability to control quality and potential exploitation issues (Hamburg, One Degree, 2008). Another challenge in the crowdsourcing arena is how to compensate people fairly for their ideas because by the same token that slack is taken out of the system crowdsourcing could seriously depress wages for many careers in innovation.

Further, a culture based on a need-to-know rather than one based on a need-to-share basis has prevented the transparency necessary to achieve organisational knowledge goals (Girard & Girard, 2009); a failure to bridge open innovation and crowdsourcing. The chapter now discusses three aspects of crowdsourcing namely finance, competition and learning and then provides a case on each of these aspects.

CROWDSOURCING IN SMES: FINANCE, COMPETITION AND LEARNING

The aspect of finance is discussed here because it is identified as the major influence of an SME's ability to develop and in turn crowdsource (Green, Kirkpatrick & Murinde, 2005). Competition is considered in regard to the role it plays in the financial sector and the influence this has on SMEs and their own capacity to be competitive and as a means of doing so crowdsource (Barth, Caprio & Levine, 2008). Learning is presented as the basis on which SMEs embrace crowdsourcing as a necessary concept in their operations (Lloyd-Reason & Sear, 2007).

Finance

International organisations are united on the suggestion that small and medium-sized enterprises do not have as much ease of access to finance as they should that would enable them grow sufficiently let alone innovate for their growth (European Commission, 2004). For most financiers, the very suggestion that SMEs must be financed is a risky proposition that needs more than a sound business proposal as a motivation.

In a survey by the European Central Bank (European Central Bank, 2010), 2009 saw the financial needs of SMEs remaining largely unchanged for half of the SMEs surveyed in the first half of the year compared to the second half. Availability of bank loans deteriorated. While this was true of SMEs, larger firms enjoyed a greater approval rate of loans and stable availability of loans. The irony of this however, lies in the evidence of SMEs remaining largely unaffected by the global financial crisis of 2008 (as they have not been in the past provided with substantial loans that would make them susceptible) in comparison to their larger enterprise counterparts.

Crowdsourcing is a step in innovation that requires technologies to be in place. Employing technology has financial implications and the SMEs have impeded access to finance. The irony of this financial handicap lies in the possibilities that crowdsourcing has for minimising the financial needs of SMEs.

Crowdsourcing would reduce the amount of time and money spent on trials and errors. Efforts would be directed to activities that would be beneficial to both the customer and SME and therefore the extra finance that would have been designated to the trials and errors would largely be saved.

A Case in Finance

Masuke (2007) presents the case of the initiatives taken by the CRDB Bank of Tanzania to support SMEs in Tanzania. The bank has undertaken numerous endeavours such as designing customised products for SMEs, engaging in capacity building, increasing the number of collaterals accepted by the bank and extending the customer relationship concept.

The bank's customised credit products now include working capital loans and investment loans. The flexibility of the products has been increased to range from three to twelve months and from twelve to sixty months respectively. This is a stark contrast to many of the financial products currently on the market to businesses which have rigid repayment structures rather than customised ones.

To build the capacity of the banking staff, CRDB has recruited specialist staff to serve SMEs at both the branch and head office levels. This capacity has been enhanced through class and field training of specialists. It is without doubt that innovation occurs through people. Furthermore, the people centred approach to innovation ensures that ideas put on paper are delivered in an exceptional manner. SMEs are a peculiar case in business, lying on neither extreme of the sole proprietorship- to- large enterprise continuum. This peculiar nature calls for the specialised approach employed by the bank.

In addition to these, training has been provided in how to develop financial statements from the primary data of SMEs. This has been done in order to relax the requirement of audited financial accounts for greater loan applications. With the lack of conventional assets and irregular income structures of SMEs, financial statements need to be developed to suit the circumstances of SMEs.

With regard to the SMEs themselves, free business and loan management training are offered to SME owners. This ensures that the capacity to meet loan obligations is improved. This is particularly necessary in the light of the fact that many owners of SMEs have very basic education backgrounds, especially in the areas of business management.

Many banks are rigid in the type of collateral they consider acceptable. This is consistent with the risk averse nature of many financial institutions and rightfully so. CRDB has gone a step further to define non-traditional collateral that is acceptable from SMEs but has not merely stopped there. Equipping owners of SMEs with the necessary knowledge and information to enable them acquire more traditional collateral means that for any particular SME, the acceptance of non-traditional collateral is not a permanent affair. This extends the scope of financial service providers from which the SME can access finance.

In this aspect, the sustainability of SMEs relies much on the sustainability of finance. On one side, this is the premise on which the bank has extended the customer relationship concept at branch level. On the other side of the tracks, the bank benefits greatly from the continued patronage of SMEs for the bank's financial products.

The initiatives at CRDB were implemented in 2005 and have shown nothing but success evident in the growth of the bank's portfolio and increase in the number of the bank's customers. Evaluation is done regularly by considering the impact of training programs and loans extended. It may be argued by some that a 5 year span over which the project initiatives have been implemented is not sufficient as a testimony of success. However, other financial service providers have nothing to lose in at least a trial and error implementation of initiatives that appear to be working for CRDB and engaging in investigation for maximising the potential of these initiatives. One does not learn from success only though. Any initiatives with regard to extending financial services to SMEs that have failed must equally be studied to identify any loop holes that led to failure and what aspects of the initiatives are salvageable.

Some of the results witnessed by the bank are improved business and loan management from SME owners, an increase in the saving and banking culture among SMEs, an increase in the wealth of SME owners due to business growth, an improvement in the collateral from non-traditional to the more formal collateral and a cleaner portfolio for the bank. These results seem two way benefits satisfying both the banks and SMEs.

The lessons that can be learned from the case of the CRDB are threefold; that SMEs are a profitable segment that SMEs have non-traditional collateral that can be made acceptable and that SME segments require a relationship management concept that can be achievable through crowdsourcing. The biggest lesson perhaps is that the success of the SME greatly influences the success of the bank and, therefore, if banks are more innovative in the products and services they can offer to SMEs, SMEs can in turn have greater capacity to be innovative in the way they do business including engaging in the open innovation activity of crowdsourcing.

Further the financial handicaps experienced by SMEs do not need to exist any longer. Financial service providers can no longer sit idly by and dismiss SMEs as clients by labelling them high risk. SMEs too must take action in increasing their levels of education and knowledge.

Competition

The element of competition affects SMEs in two ways. On one hand the competition within the banking industry influences the accessibility of funds to SMEs and on the other the lack of sufficient competition among SMEs both hinder any strides that can be made by SMEs in open innovation and crowd sourcing.

Banks often do not provide the appropriate financial products for SMEs because taking the additional risk is often considered unnecessary in the light of minimal competition, and rather may seem as collusive behaviour among banks. It has been observed that the behaviour of many banks and similar institutions is collusive in nature. As a consequence the general regard of SMEs as unnecessarily high risk clients tends to permeate and further limit access to finance for SMEs.

In this context, SMEs need to become more competitive and visionary. Taking on this approach will provide a greater motivation to innovate and engage in innovation activities such as crowdsourcing. A lack of competition and vision breeds a relaxed manner of business which not only stagnates growth but profitability as well. Here again a bilateral relationship is observed with competition leading to open innovation and crowdsourcing as well as open innovation and crowdsourcing leading to competition.

A Case in Competition

A 2006 study by Mutula and Brakel revealed that at least 69% of South African SMEs had a minimum of three personal computers yet only 13% of the businesses sampled used the internet to buy products.

Other technologies such as mobile technologies, highly popular in the country are being taken up but at too slow a pace. The reasons for this trend may be attributed to the lack of the sense of competition among SMEs as well as the lack of knowledge, lack of a skills support program and a desire to control risks.

Moreover, the high cost of connectivity and lack of infrastructure have been identified as one of the biggest challenges for South African SMEs. In addition to these, due to both socio-economic and technological reasons information and communication technology for crowdsourcing as a competitive tool may not be easy to adopt.

The issue of competitiveness of SMEs does not rest at strategy, strategy implementation and technology. More importantly is the issue of the personality, capabilities and risk perceptions of the managers of SMEs. If the manager is competitive, this is reflected in the business and its decisions.

Schubert and Leimstoll (2007) agree that the competitiveness of an SME depends on the ways in which information and communication technology (the necessities for crowdsourcing) is used to support business processes. Implementing information and communication technology in a business by itself does not lend itself to competitiveness. Efforts must be made to link information and communication technology to business processes and strategy (as enabled by crowdsourcing).

We are in a knowledge economy and information and communication technology supports it. Staying ahead of the competition is only possible by employing and staying current with technologies. The need for SMEs to adopt information and communication technology is driven by globalisation, and innovation (particularly open innovation), need for flexibility and act as a means to competitive advantage.

Learning

Learning is identified as a critical input and output of crowdsourcing (Lowry & Turner, 2007). The ability and willingness to crowdsource depends on the SME's level of learning just as the SME's level of learning depends on the ability and willingness of the SME to crowdsource- another bilateral relationship. However, the limited budgets for learning and professional development have an influence on the learning and work culture of SMEs. The need to justify every article of spending prior to embarking on any developmental activity means that the benefits of learning need to be clearly and directly identified. The attitude towards learning and education in general are highly influenced by the manager. Surveys reveal that most managers of SMEs do not have a university or college education and this tends to cripple the growth of the entire SME.

Innovative enterprises are often concerned about intellectual property protection. Increased learning increases the means by which intellectual property can be protected. Securing knowledge in intellectual property protection increases the willingness of small firms to engage in crowdsourcing and also adds greater security to the ideas generated.

Traditional organisational learning theory seems to explain very little about learning in SMEs. The learning competencies deserving of attention in the study of SMEs are adaptability, planning, information and knowledge, human resource development and growth. These attributes lie in the paradigm of entrepreneurial learning.

Learning is often triggered by a competition crisis in the external orientation of the SME. The manager's approach and style influence the mode and type of organisation learning process. The mode may be experiential, information based and interaction based. The type of learning process may be adjustive, values based or, otherwise and, continuous. The scope of learning may be individual, group, within the organisation or between organisations. The evidence of learning lies in the actions of the SME, artefacts and discourse. Effective learning translates to a competitive advantage in the SME's external orientation. This learning process of SMEs is based on the interpretation of the conceptualisation by Zhang et al (2006).

Moreover, the external orientation discussed is the exposure of the SME to information and knowledge that resides out of the SME's boundaries, and the study finds that this external orientation may be facilitated by crowdsourcing.

A Case in Learning

The Forest Press Limited is a SME that hosts a learning centre. Established about 36 years ago, the company's attitude towards training and development has been positive.

Similar to other small and medium sized enterprises very few human resource management systems are in place. External training of employees has not been a strong priority but under the influence of the company's managing director collaboration with regard to extending learning has been underway with Nottingham University.

Particularly in the context of information technology learning, the company is benefiting greatly as it is in the printing business. Reliance on information technology is inevitable.

The learning and information centre started off humbly with six personal computers, office software, internet access, video conferencing and rapid prototyping equipment. A tutor was provided for training. The training has since been made accessible to all levels of staff approximately less than 20 in 1984 and 30 presently.

The benefits of the learning centre have been believed to be fourfold; the opportunity itself, the self-paced nature of study, the personal attention and the convenient location. For individual learners without any formal qualifications at the beginning, the learning centre provided a qualification and this greatly boosted their confidence. This confidence translated to less resistance to change.

The centre has enabled the company learn from other organisations as well. This case serves as another prime example of how learning can cause crowdsourcing and how crowdsourcing can lead to learning. Now the study will illustrate a few cases from SMEs who are adopting crowdsourcing strategies in their entrepreneurships.

A FEW SMES MAKING STRIDES IN CROWDSOURCING

The list of firms presented below has been itself crowdsourced from members on the social networking site LinkedIn. They also appear on the 'top crowdsourcing examples' lists compiled on several websites (Ramachandran, 2009).

Goshido

Goshido is an Irish company that has been exemplary in maximising the benefits of crowdsourcing to SMEs. By using only the social networking site LinkedIn, the company managed to raise funds from the selected list of Enterprise 2.0 investors. The fund-raising exercise that could have taken years was only completed in a week. The product offered by the company is interesting in that it promotes crowdsourcing of sorts by providing a project management system that enables members of teams construct, track and complete connected actions.

Goshido's experience shows that crowdsourcing is a feasible solution to the problem of finance that SMEs have to overcome. This further shows that the cost by itself is not a sufficient enough reason to not employ the open innovation approach of crowdsourcing.

Threadless

Canada hosts the Threadless T-shirt Store. The company sources ideas for the designs of the t-shirts it sells from the online community and gives cognition to the idea generators. The rights to the design are owned by the designer and the company, therefore according some intellectual property protection.

Crowdsourcing with the Threadless T-shirt Store occurs in a manner of ways. The crowd may participate by rating the designs they like best, submitting their own designs (with monetary and non-monetary incentives), by scoring and submitting slogans (with a monetary incentive), by participating in refreshing competitions, awarding the best designers and showcasing ordinary individuals modelling the designed t-shirts. The crowd is extended to include children as well, whose creativity truly seems relentless.

The company has a blog spot, embarks on tours and has artist programs to encourage and support designers. The artist programs empower upcoming designers and enable artists to emerge as contributing talents rather than mere one hit wonders. This empowerment has extended to scholarship programs that can be used to finance university studies in any course for as long as the recipient has been awarded for submitting a design voted the best.

The designs presented are provided a unique personality as the designers are able to discuss their thoughts behind designs on the online forum as well as receive critiques. The online community has the opportunity to feel a part of the actual team from miles away benefiting the member with monetary incentives and the store with mass marketing through shared links.

The store has created a great deal of hype for their t-shirts by trend spotting, an activity that is enabled by individuals submitting photos of unaware individuals adorned in a Threadless t-shirt. Interest in the products offered are further generated by collaborations with elements in pop-culture that would draw attention to the t-shirts such as musicians and photographers.

crowdSPRING

This company from the United States of America enables SMEs purchase logo and website designs from designers in the online community. Over 65,000 designers pitch for the project given the SME's time and cost constraints. With concerns over intellectual property protection, the company has specific policies designed to address this concern. The company website indicates that standards of conduct are in place and formal procedures with regards to violations are outlined.

gooseGrade

gooseGrade enables readers and writers to interact with each other. The purpose of the company is to improve not only the quality but also the accuracy of information found online. Content remains controlled by the author but readers put forward correction suggestion that are subject to approval by the author.

gooseGrade demonstrates how quality control can be improved for some products and services through crowdsourcing. Editors are ranked and for some readers, gooseGrade has been a starting point of their careers. The flexibility of the service enables it to be used on any website.

Errors in online publications are often perceived as a mark of poor quality and a discredit to the reputability of the source. The crowdsourcing initiative of gooseGrade therefore seeks to add more trustworthiness and credibility to online resources by distributing the responsibility of editing to an online pool of editors. This also enhances the extent to which readers engage with online material as they feel themselves a contributor to the publications, and rightly so.

Authors can track changes using short message services (SMS) and even twitter! Another avenue through which crowdsourcing has proved its relevance.

Giffgaff

The name of the company means 'mutual giving'- a simple statement of what crowdsourcing ought to be about. The inspiration behind the company was a completion of a study on Web 2.0 businesses (Web 2.0 is a key facilitator of open innovation and crowdsourcing) in San Francisco in 2008 by the company's founder Gav Thompson. This is a mobile virtual network organisation based in the United Kingdom that boasts on its website that it is run by its members (users of the network). Parts of the business are run by the members enabling the business to cut costs and pass the savings back to the members.

Explicitly stated is the company's belief and confidence in the power of the community (or otherwise, the wisdom of the crowd). The community can contribute to the quality of service provided and is rewarded for its contribution. Crowd support is however moderated by Giffgaff staff.

Customer service operations remain a large proportion of the budget of many companies and being able to reduce these costs through crowdsourcing is an opportunity that must not be overlooked especially by SMEs.

CROWDSOURCING IN SMES

Solutions and Recommendations

Crowds should operate within constraints because to harness the collective intelligence of crowds, there need to be rules in place to maintain order. However, not everything can be democratic. Sometimes a decision needs to be made, and having a core team (or single person) make the ultimate decision can provide the guidance necessary to get things done and prevent fanatical ideas and groupthink from wreaking havoc on the organisation's product.

Furthermore, crowds must retain their individuality. Groups must be encouraged to disagree, and efforts must be in place to try not to let any members of the group disproportionately influence the rest. Crowds are often better at vetting content than creating it. SMEs must therefore be realistic about what component of the business they are willing to entrust the crowd with.

More investors and banks need to be willing to take the risk of financing SMEs. Crowdsourcing is necessary at the financial providers' level as well so that good policies are shared and there is possibility to benefit from the experiences of others in financing SMEs. The efforts of the European Union in this regard set precedence for what can be achieved in the financial services community for the benefit of open innovation in SMEs (European Commission, 1995).

Still on innovation in the financial providers' environment this study finds that, the following technologies may need to be considered:

- custom built or commercially available software for the purpose of tracking transactions;
- real time information exchange through network connections;
- better management of SME client records through personal digital assistants;
- computerised credit scoring; and
- improved access to internet and mobile phone banking.

The research argues that these technologies would largely improve the financial provider's accuracy of lending and pricing strategies with regard to SMEs. However, these efforts would be futile in the absence of sound macroeconomic policy to provide a suitable regulatory framework that would protect both the financial provider and the SME.

Moreover, entrepreneurs need to be in possession of a better understanding of the concerns of banks and investors in order to be able to provide the necessary assurances about the soundness of their proposals. This is where learning and good customer relationships have their roles.

Finally, SMEs are increasingly being encouraged to form groups or clusters so that they can share and draw from a common knowledge pool. Such affiliations are likely to ensure SMEs remain abreast of contemporary issues through seminars, conferences and training programs. SME learning can be enhanced through time flexibility, cost efficiency, support mechanisms, accessibility, overall efficiency, quality, self-paced learning, results measurement, content design and learner-centeredness. Crowdsourcing promises to aid the creation of this common knowledge pool and award of the much needed flexibility.

FURTHER RESEARCH DIRECTIONS

The success of open innovation and crowdsourcing depends on a full understanding and appreciation of web 2.0 technology (social networking facilities) which will translate to better decision making and greater value (value 2.0). The employment of these technologies constitutes a paradigm shift in the way that many companies make decisions and has been termed the emerging era of "*Decisions 2.0*" (Bonabeau, 2009).

Moreover, future researches should be able to respond to questions like, how can we advance the use of information and communication technology in SMEs? At the same time there should be investigative researchers trying to find answers in this aspect. Macroeconomic policy has a substantial role here.

Much of the existing literature on SMEs is based on American, European and Asian studies. More research ought to go into the African experiences of SMEs. This is likely to add more relevance to this literature for African SMEs.

CONCLUSION

In terms of cost, quality, delivery, flexibility and human resource development, SMEs are regularly facing new challenges to their survival and growth. Strategies for developing various capabilities and competencies to satisfy their domestic as well as global customers must be formulated in the context of a dynamic market scenario.

Crowdsourcing provides a useful utilisation of globalisation. Globalisation is the reality of business today and crowdsourcing is a way of managing this overwhelming reality.

SMEs need to realise that information and communication technology is an enabler of innovation, open and otherwise. Clients of the knowledge economy demand intelligent products and services. For this reason, knowledge ought to be integrated into products and services.

This research finds that information and communication technology is the vehicle required to pursue the learning agenda.

Crowdsourcing is not met by all without criticism. Some critics argue that it is exploitative (getting something for nothing), that the originality and quality of the opinion and knowledge of the crowd hardly amounts to wisdom and that there is a lack of respect for some professions by bidding them down (again being exploitative). Similarly, critics argue that many of the challenges faced by SMEs simply require a hiring of the right people, appropriate pay and good treatment (VonStamm & Triflova, 2009).

In defence of crowdsourcing for SMEs, the evidence of crowdsourcing success from the examples stated suggest manners of avoiding exploitation. The crowd is the market that purchases the goods and services of SMEs and who better to suggest avenues for improvement than the end user. As for the bidding down of professions, this is arguable as it is still necessary to have 'the expert' on board to 'control' the crowd.

Hiring a few people one can rely on may appeal to many managers as a hassle free way to run their SMEs. However, such an approach greatly limits the extent to which SMEs can be innovative and competitive.

Evident throughout the text is the existence of a co-dependent relationship between crowdsourcing and the success of an SME. This brings us to the question of what should come first, the ability to crowdsource in order to be a successful innovative SME or being a successful SME in order to innovatively crowdsource? It is a well known adage that very few if any businesses start out as being successful. Crowdsourcing is, therefore, one of the tools that may be used to lead a business to success.

No matter how minimally it is employed, a little crowdsourcing has the potential to yield exponential results. It is this exponential reaping of rewards that enables a company to extend its use of crowdsourcing and so the cycle continues (Shirky, 2008).

The challenge is thrown to policy makers and financial service providers to stop resting on their laurels and dismissing the attempts of SMEs at innovation and rather embrace them as partners for development and mutual financial reward. An even greater challenge is given to the managers of SMEs to desist from the tendency to just aspire comfort but rather seek to achieve greater heights. Endless innovation is the limit!

REFERENCES

Anderson, E. (2010). *Social media marketing: Game theory and the emergence of collaboration*. New York, NY: Springer.

Barrett, P. (2010). *Open innovation survey in health care and manufacturing*. United Kingdom: P A Consulting.

Barth, J. R., Caprio, G., & Levine, R. (2008). *Rethinking banking regulation: Till angels govern*. New York, NY: Cambridge University Press.

Bonabeau, E. (2009). Decisions 2.0: The power of collective intelligence. *MIT Sloan Management Review*, *50*(2), 45.

Burmann, C., & Arnhold, U. (2008). *User generated branding: State of the art research*. London, UK: Transaction Publishers.

Chesbrough, H. (2011). *Open services innovation: Rethinking your business to grow and compete in a new era*. California: Jossey-Bass.

DeCusatis, C. (2008). Creating, growing and sustaining efficient innovation teams. *Creativity and Innovation Management*, *17*(2), 155–164. doi:10.1111/j.1467-8691.2008.00478.x

Dhillon, G., Stahl, B. C., & Baskerville, R. (2009). *Information systems, creativity and innovation in small and medium-sized enterprises*. Germany: International Federation for Information Processing.

European Central Bank. (2010, February 16). *Press release*. Retrieved June 15, 2010, from http://www.ecb.europa.eu

European Commission. (1995). *Access to finance*. Retrieved June 15, 2010, from http://ec.europa.eu

European Commission. (2004). *Benchmarking enterprise policy: Results from the scoreboard*. Ann Arbor, MI: University of Michigan.

Gawarzynska, M. (2010). *Open innovation and business success*. Hamburg, Germany: Diplomica Verlag.

Girard, J. P., & Girard, J. L. (2009). The future of management...are you ready? *Canadian Manager, 34*(2).

Green, C. J., Kirkpatrick, C. H., & Murinde, V. (2005). *Finance and development: Surveys of theory, evidence and policy*. Cheltenham, UK: Edward Elgar.

Hamburg, M. (2007). *Crowdsourcing 101: The basics*. Retrieved March 25, 2010, from http://www.onedegree.ca/2007/12/crowdsourcing-1.html

Hamburg, M. (2008). So you think you can crowdsource. *Crowdsourcing 101*. Retrieved March 25, 2010, from http://www.onedegree.ca/2008/02/crowdsourcing-3.html

Howe, J. (2009). *Crowdsourcing: Why the power of the crowd is driving the future of business*. New York, NY: Crown Publishing Group.

Lloyd-Reason, L., & Sear, L. (2007). *Trading places- SMEs in the global economy: A critical research handbook*. Cheltenham, UK: Edward Elgar.

Lowry, G. R., & Turner, R. L. (2007). *Information systems and technology education: From the university to the workplace*. London, UK: Information Science Reference. doi:10.4018/978-1-59904-114-8

Masuke, E. (2007).Initiatives for SMEs. *CRDB Bank*. Retrieved June 18, 2010, from http://www.worldbank.org

Mutula, S., & Van Brakel, P. (2006). E-readiness of SMEs in the ICT sector in Botswana with respect to information access. *The Electronic Library, 24*(3), 402–417. doi:10.1108/02640470610671240

Ramachandran, A. (2009). *About crowdsourcing examples*. Retrieved January 18, 2011, from http://crowdsourcingexamples.pbworks.com/w/page/16668404/FrontPage

Schee, B. A. V. (2008). Review of the book *Crowdsourcing*: Why the power of the crowd is driving the future of business. *Journal of Consumer Marketing, 26*(4), 305–306.

Schubert, P., & Leimstoll, U. (2007). Importance and use of information technology in small and medium sized companies. *Electronic Markets*. Retrieved July 14, 2010, from http://www.electronicmarkets.org

Shirky, C. (2008). *Here comes everybody: The power of organising without organisations*. United States: Penguin Press.

Von Stamm, B., & Trifilova, A. (2009). *The future of innovation*. United Kingdom: MPG Books Group.

Witzeman, S., Slowinski, G., Dirkx, R., Gollob, L., Tao, J., Ward, S., & Miraglia, S. (2006). Sourcing external technology for innovation. *Strategic Alliance*. Retrieved March 23, 2010, from http://www.strategicalliance.com/article/SourcingExternal.htm# Zhang, M., Macpherson, A., & Jones, O. (2006). Conceptualising the learning process in SMEs: Improving innovation through external orientation. *International Small Business Journal, 24*(3), 299-323.

ADDITIONAL READING

Ahari, K. (2007). *Advantages of open innovation.* Retrieved March, 26, 2010, from http://gotastrategy.typepad.com/innovation/2007/10/open-innovation.html

Alsever, J. (2007). *What is Crowdsourcing?* Retrieved March, 25, 2010, from http://www.bnet.com/2403-13241_23-52961.html

Anonymous (17 February 2010). Crowdsourcing, social innovation, and sustainability challenge; Crowdsourcing challenges can bolster reach and ROI. *PR Newswire.* New York.

Badani, J. (2009). *Economics of Open Innovation.* Available from http://blog.ideaken.com/2009/05/economics-of-open-innovation.html

Beltrametti, M. (2009). Conference proceedings of the Grenoble Innovation Fair.

Catone, J. (2007). *Crowdsourcing: A million heads is better than one.* Retrieved March, 25, 2010 from http://www.readwriteweb.com/archives/crowdsourcing_million_heads.php

Chesbrough, H. W. (2005). Open innovation: A new paradigm for understanding industrial innovation. In Chesbrough, H. W., Vanhaverbeke, W., & West, J. (Eds.), *Open innovation: Researching a new paradigm.* Oxford: Oxford University Press.

Chesbrough, H. W. (2006). *Open innovation: The new imperative for creating and profiting from technology.* Massachusetts: Harvard Business School Press.

Hamburg, M. (2008). *Crowdsourcing 101: Episode 2- Are You Afraid of the Crowds?* Retrieved March, 25, 2010, from http://www.onedegree.ca/2008/01/crowdsourcing-1.html

Howe, J. (2008). *Why the power of the crowd is driving the future of business.* New York: Crown Publishing Group.

Jusko, J. (2008). *Kraft crafts and open innovation strategy: Consumer packaged goods manufacturer finds innovation value in collaborating with suppliers.* Retrieved March, 23, 2010, from http://www.industryweek.com/articles/kraft_crafts_an_open_innovation_strategy_17725.aspx

Kraft Food Holdings. (2008). *Innovate with Kraft.* Retrieved March, 23, 2010 from http://brands.kraftfoods.com/innovatewithkraft/default.aspx

Levy, M., & Powell, P. (2005). *Strategies for growth in SMEs: the role of information and information systems.* Oxford: Butterworth Heinemann.

Lloyd-Reason, L., & Sear, L. (2007). *Trading places-SMEs d the global economy: a critical research handbook.* Cheltenham: Edward Elgar Publishing.

McKay, L. (2010). Where does innovation come from? *Customer Relationship Management, 14*(1), 24–30.

Organisation for Economic Co-operation and Development. (1997). *Globalisation and small and medium enterprises (SMEs).* Manila: The OECD.

Organisation for Economic Co-operation and Development. (2001). *Enhancing SME competitiveness: the OECD Bologna Ministerial Conference.* Bologna: The OECD.

Organisation for Economic Co-operation and Development. (2007). *The SME financing gap.* Brasilia: The OECD.

Organisation for Economic Co-operation and Development. (2008). *Removing barriers to SME access to international markets.* Athens: The OECD.

Reynolds, P. D., & Estime, M. F. (1996). *SME: employment, innovation and growth: the Washington Workshop.* Paris: The OECD.

Singh, S. (22 March 2010). Xerox India hub to follow open innovation model. *The Economic Times*. Retrieved March, 23, 2010, from http://economictimes.indiatimes.com/Interviews/Xerox-India-hub-to-follow-open-innovation-model/articleshow/5710009.cms

Sisodiya, A. S., & Rao, N. J. (2008). *SMEs: key drivers for growth*. New York: State University of New York Press.

Webster, L. M., Riopelle, R., & Chidzero, A. M. (1996). *World Bank lending for small enterprises, 1989-1993*. Washington, D.C.: The World Bank. doi:10.1596/0-8213-3518-9

Winsor, J. (16 June 2009). Crowdsourcing: What it means for innovation. *Business Week Online*. Retrieved March, 25, 2010, from http://web.ebscohost.com.ezproxy.lib.monash.edu.au/ ehost/detail?vid=2&hid=6&sid=73b7618c-3db2-493b-9749-2d4246748571 @sessionmgr 4&bdata=JnNpdGU9ZWhvc3QtbGl2ZSZzY29wZT1zaXRl#db=buh&AN=41892898

Xerox Corporation. (2009). *Open Xerox*. Retrieved March, 23, 2010 from http://open.xerox.com/

KEY TERMS AND DEFINITIONS

Innovation: The improvement of products and services offered and manners of delivery.

SMEs: Small and medium sized business with staff numbers of less than 50 often with a single manager, the owner.

Open Innovation: The improvement of products and services using resources that are not restricted to those owned by the business.

Crowdsourcing: Turning to the online community for business solutions.

Web 2.0: Technology that enables social networking such as LinkedIn, Twitter and Facebook.

Decisions 2.0: Technology that enables business sense in decision making via forecasting and deductions.

Value 2.0: Holistic technologies that enable the translation of activities and decisions to value.

Chapter 16
Open Innovation in France: A Case Study of an Emerging Eco-Innovation Cluster

Rachael J. Ritchie
Universud Paris, France & Université de Versailles Saint-Quentin-en-Yvelines, France

Keith C. Culver
Universud Paris, France & Université de Versailles Saint-Quentin-en-Yveline, France

ABSTRACT

France has long been associated with a state-directed "dirigiste" model of linear R&D focussed on large programmes such as development of the TGV high-speed train. Contemporary France has, however, largely left the state-centered model behind, introducing a range of devices since the turn of the 21st century aimed at opening French innovation to international and cross-sectoral collaboration for increased productivity and national competitiveness. This case study traces the opening of the French innovation system and the way one new academic, industrial and government collaboration aims to make use of new features of the system to accelerate development of an eco-innovation cluster focussed on cities of the future.

INTRODUCTION

Throughout developed and developing countries, actors in public, private, academic and NGO sectors are seeking to use R&D to improve productivity and competitiveness. This much is not new: ever since Vannevar Bush announced in 1945 to the US Government that investment in exploration of the 'endless frontier' of science would generate social benefits, science-based R&D has been seen and used as a source of ideas from which desirable innovations can be generated. After a decade of R&D in the twenty-first century, this vision of the social role of R&D remains persuasive to many, while increasingly taking a new form.

DOI: 10.4018/978-1-61350-519-9.ch016

Improvements in productivity and competitiveness are certainly sought in the fruits of researcher-led, publication-oriented discovery research, yet increasingly these improvements are 'by design,' arising from research collaborations designed from the outset to respond to the needs of some specified set of end users, typically in the spirit of 'national systems of innovation' driven by national funders who identify research goals and subsidize research aimed at those goals. This research is increasingly conducted outside the sole control or guidance of the state, as state sponsors are increasingly co-sponsors and partners in international efforts combining groups of diverse actors working at different positions in the value chain or in disparate but related research activities, organized to speed generation of commercially viable and valuable innovations.

This shift in emphasis from individual to team-based research, and discovery to innovation-oriented research, is not, contra some skeptics, simply an elaborate transformation of public goods to the benefit of private interests seeking commercially valuable innovations. While innovation-driven research is still undoubtedly tied to national competitiveness agendas in which private sector success is encouraged by public sector policies and subsidies, the new emphasis on group-based, innovation-oriented research is also driven by gobalisation of markets. As the OECD (2008:p.9) puts it, "Global competition drastically shortens product lifecycles while the growing integration of different technologies makes innovation riskier and more costly."

In response to this situation, a new kind of 'open innovation' method and practice is developing in tandem, and perhaps in tension with development of national systems of innovation seeking to improve national competitiveness. 'Open innovation,' following Chesborough (2003), the OECD (2008), and others, refers generally to a strategy in which an organization seeks to find and use ideas external to the organization, to enhance the effectiveness of the organization's pursuit of its goals – whether cost reduction, technology portfolio diversification, speed or quality of development, or some other aspects of development of a typically technological product, process or method. Often open innovation involves blending collaboration and competition of diverse players, aiming to satisfy the diverse interests of benefit all participants in a collaborative research endeavour. Open innovation as a strategy and social phenomena has arisen and has been characterized and evaluated chiefly in the context of large, multinational firms and increasingly, academic units, in North America (e.g. IBM, P&G, Chevron, Los Alamos National Laboratories, Genome Canada).

Yet the utility of open innovation is clear for SMEs, who typically have fewer resources to commit to innovation R&D, but often have specific niche knowledge ideally suited to systems of open innovation, and other countries seeking to improve competitiveness, and open innovation is now being studied in these contexts. This is not simply, however, a situation of lagging and catching up, or the 'old world' learning lessons from the 'new world.' While the nature and merits of European open innovation structures, and those of SMEs, may bring distinctive lessons to the increasingly global community of innovators, open innovation is not a one-size-fits-all proposition and the supporting infrastructure and policies need to be suited to the appropriate market and cultural contexts.

One new French approach to open innovation can be observed in the context of the newly-created industrial Econoving International Chair in Eco-innovation at the University of Versailles in Paris, France. This Chair is a multidisciplinary partnership between industrial firms and university researchers designed to create and catalyse synergies between players with different backgrounds and competences to generate eco-innovations, and to train the next generation of multidisciplinary eco-innovation managers. Industrial members of the chair include multinational businesses Alstom, GDF-Suez, Italcementi, Saur, SNCF and government partner l'ADEME.

Members of these institutions along with academic researchers of the UniverSud Paris, a consortium of 21 higher education instituted in the southern Paris region, are engaged in applied interdisciplinary research with a central goal: the eco-city of the future. Departing from a scenario-based approach to identifying future needs of cities, the Econoving Chair acts as a catalyst and coordinator proposing and fostering R&D interactions between public, private and academic partners, focusing on the problems and opportunities of the cities of the future, seeding successive cycles of innovation with the outputs of the preceding cycles.

The cyclic nature of the Econoving program, involving elements of reinvestment of intellectual capital, incorporation of new and diverse partners, and expansion into new promising areas of R&D identified each cycle, is an example of a new form of open innovation in France which is sure to help define systems of successful open innovation in Europe and for SMEs. This chapter contains a case study focusing on the model of open innovation deployed by the Econoving Programme in the context of the national innovation system of France, and the growing openness of that system to open innovation in global context.

BACKGROUND

The Rise of Modern National Systems of Innovation

Leaders of states have long understood the importance of technological prowess to national prosperity, from competitive advantage in weaponry to public sanitation protecting the health of citizens. What leaders have lacked, prior to the mid-20th century, is a set of systematic approaches to use of public policy instruments to foster an equally systematic growth in the rate and quality of technological advancement in targeted areas. For many industrialized countries, the 1939-1945 Second World War was a catalyst and galvanizing challenge to previously fragmented national approaches to research, development, production and implementation of technological advances.

The best-known advance of this era, the invention of the atomic bomb, is a standard example of the emergence of innovation as a 'method of invention.' State-organized scientific teams coordinated their efforts in the Manhattan project to understand and resolve the range of questions arising from conception to deployment of the atomic bomb. Similar state-organized teams were responsible for the development of radar, sonar, and the V2 missile.

The success of these projects, and non-military inventions such as penicillin, demonstrated fulsomely the power of directed research to generate innovations for national security and broader prosperity. The success of the new state-coordinated method of invention was not, however, absolute. Japan was unsuccessful in its deployment of fire-balloons to destroy western North American forests, for example, and the British invention of miracle-material pykrete (a mixture of ice and sawdust) did not lead to useful applications. These failures were largely forgotten in the wake of enthusiasm for the productive power of successful coordinated research. No longer, it seemed, was technological progress the result of a frustrating wait for 'genius' to produce an insight ready for further development: a systematic approach could be taken, led by priority-setting states, and managed to completion in an industrial model of production.

In the postwar era nearly all developed countries attempted to convert the useful lessons of wartime research methods for a blend of military and peaceful purposes. The spirit of the time is well expressed in the title of Vannevar Bush's landmark policy advice urging US peacetime investment in scientific research. In *Science, the Endless Frontier*, Bush identified discovery research as the "pacemaker of technological progress" (Bush, 1945) driving military superiority and public welfare – from curing diseases to creation of new goods, services, and associated jobs. Bush's efforts contributed to the 1947 establishment of the National Science Foundation, an independent US federal agency created "to promote the progress of science; to advance the national health, prosperity, and welfare; to secure the national defense..." (National Science Foundation Act of 1950, Public Law 81-507 (Stat. 149)). The subsequent emergence and dominance of the American model of state-driven or 'etatistic' innovation has been thoroughly documented, perhaps to the point where similar developments at smaller scale in other countries have been given too little attention relative to the diverse lessons we might learn from those experiences.[1]

The model visible in these activities is still familiar in outline today: national investment in discovery-oriented research coordinated with downstream development and commercialization efforts in various ways. For several decades the pilots of these approaches saw satisfactory return on investment in this discovery-emphasising model, visible in technological innovations satisfying apparent military needs, and a rise in citizens' longevity and material quality of life (Amable and Hancké, 2001; Nelsen, 1993). However the economic difficulties of the 1970s—increasing inflation and unemployment associated with a drop in productivity—challenged the assumptions underlying post-war national systems of (Laredo & Mustar, 2001; Amable & Hancké, 2001).

In response to these challenges, countries such as France, UK and US shed their Keynesian marcroeconomic policies in favor of more market driven policies (Archibugi & Michie, 1997). Adoption of market driven policies allowed greater influence of market forces on R&D, spurring internationalization and diversification of sources of innovation. While basic R&D receiving national investment was still recognised as a source of radical innovation (Sachwald, 2009; Archibugi & Michie, 1997), incremental innovation was found to arise from interactions with the market or industry, and interactions between companies and between countries increased in importance. By the 1980s leading nations began to shift their innovation-support policies from an emphasis on supplying inputs in the form of research investment, to an emphasis on support for creation of the conditions for successful innovation – from support of state-driven systems of innovation to support of market-driven, feedback-reliant systems of innovation (Kline & Rosenberg, 1987; Nelson, 1993). As the OECD (2008:p. 24) adroitly summarizes this shift from push to pull to mixed innovation models, "Innovation models have evolved from simple linear models like the technology-push model (up to the second half of the 1960s) and the market-pull model (in the 1970s), towards more complex models."

Response to Globalisation: The Rise of Open Innovation

The close of the twentieth century saw a continued decline in the influence and use of the national system of innovation model which had dominated the post-war era. While states still recognize innovation as fundamental to productivity and competitiveness, and continue to compete with one another for power and wealth, the role of state research-led innovation in that competition has changed significantly, due to the cluster of factors often summarized as 'globalisation.' The increasingly unfettered international flow of investment, talent, and products has reduced the scope of meaningful national control of technological innovation through state-based direct investment, coordination of scientific and related research talent, and subsequent control of products, processes and methods as they are introduced to markets. Increased global mobility of talent and investment has been paralleled by blending of markets, and increasing competition amongst globally-active private sector actors. Rising competition has been driven by easier recombination of its elements in regions and states, with concomitant shortening of product lifecycles and an increase in the cost and risk of engaging in innovation targeting those blended global markets.

Globalisation of markets and flows of investment and talents has been matched by globalisation in systems of innovation. Beginning in the 1980s, and increasing to the present day, the United States and other leading economies began to internationalise their innovation systems in an effort to better configure their supply of science relative to market demands. In what amounts in some national situations to 'de-nationalization' of systems of innovation, more typically called 'open innovation,' (Chesbrough, 2003) states and firms have sought collaboration of various kinds to reduce risk and costs associated with innovation-oriented research, development, demonstration and associated activities.

Activities characteristic of open innovation (surveyed extensively in the national and regional innovation systems literature Sachwald, 2009; OECD, 2008), include firms' relocation of R&D operations near market locations or leading international research institutions, regardless of firms' head office or operational location, often accompanied by collaborative activities with research groups or SMEs with knowledge complementary to firms' knowledge and strategic priorities. The benefits of this openness and de-nationalization of efforts are two-fold: nationally-based firms gain nationally unavailable ideas, technologies, and market access, and in some cases increase overall efficiency in their R&D operations by transferring surplus-to-purpose technologies and knowledge to partners, recouping a return on investment which would otherwise have been lost if the firm were operating within the limitations of the 'old' national innovation system and markets.

This new method of innovating, pioneered in the US with notable success by companies such as Xerox and Proctor & Gamble, has been said to represent a paradigm shift (Chesbrough, 2003). 'Closed' innovation conducted by firms working in self-contained silos is increasingly displaced by 'open' innovation characterized by firms' self-understanding as actors in flexible and changing networks, with specialist actors coming together to access markets, use or buy technologies, generate or incorporate innovative start-up firms, diversifying or re-branding as necessary to achieve specific goals. The shift from nationally-rooted silos to internationally-networked firms with a national home-base has been gradual, and has arisen in as many combinations as there are national cultures, market segments technological sectors and state-supported open-innovation initiatives. Throughout this transition from state-led to private sector-led innovation, the state remains present as an enabler of innovation via policy mechanisms, and as a home for the wealth generated by internationally-active firms which nonetheless choose particular states as their homes, for complex reasons of history, culture, and still-present state urges toward national control of technological innovation for the exclusive benefit of state-sponsors of that innovation and associated productivity and competitiveness gains.

INNOVATION IN FRANCE

The practice of innovation in France is of significant intrinsic and comparative interest, in part because France has a lengthy record of successful innovation, and because the historical French preference for state-centered innovation has undergone one of the most radical transformations toward open innovation. As discussed in further detail below, France has moved from an era of 'national champion' firms supported by state-directed research institutions and programmes, toward a strategy and set of policies supporting state-enabled, private sector-led innovation, viewed as a key to increased national competitiveness (Muller et al., 2009; Laredo & Mustar, 2001; OECD, 2008).

State Involvement

French approaches to research and innovation have long been influenced by policy goals of national independence and international leadership, beginning with Jean-Baptiste Colbert's influential development, 1661-1683, of the policies underpinning what was later characterised as the Colbertist state (Chesnais, 1993). This mercantilist approach to economic development identified the wealth of the nation with the stock of cash, typically gold bullion, present in the state, giving rise to emphasis on domestic production and self-sufficiency enabling effective export generating cash. Later approaches varied in specific means and goals yet had effects consistent with the Colbertist approach, as, for example, in the Napoleonic era a form of national independence was fostered by targeted investments in research directly motivated a desire to be independent of reliance on trading lines easily disrupted by warfare, such as British blockades of national ports. It is important to avoid exaggerating the influence of these characteristics of early French economic development on later research and development strategies, yet at the same time it is worth noting these starting points since there are undeniable echoes of early policy visible in later policy and practice (Mustar & Larédo, 2002).

In the thirty years following the Second World War, France made a remarkably rapid return to prosperity and international prominence, significantly aided by state-authored research institutions and 'large programmes' in the larger 'dirigiste' economy so called because of the directional impetus of the state visible throughout French economic activity (Chesnais, 1993). Immediately following the war, the French government, like the governments of other leading states, expanded its engagement in fundamental research. In particular, it expanded the laboratories of the CNRS (Centre National de la Recherche Scientifique) established just prior to the war in 1939.

Applied research was supported through foundation and expansion of several subject-specific public research institutes and combined research-business institutions. Especially prominent examples include: the CEA, the Commissariat à l'énergie atomique et aux énergies alternatives for nuclear, and since 2009 alternative energy research, development and production, (1945); INRA, the Institut National de Recherche Agronomique for agricultural research, (1946); INSERM, the Institute National de la Santé et de la Recherche Médicale, for health and medical research (1961) (reviewed comprehensively in Mustar & Larédo, 2002).

The Colbertist French system incorporated this mix of discovery and applied research institutions into a highly successful national version of the approach to technological invention operated by leading states. The combination of discovery and applied research institutions developed in France was focused on thematic 'large programmes,' explained by Muller (2009:p. 1) as "encompassing civil and defence activities and aiming at technological developments and national independence in specific fields (such as the nuclear, space, aeronautics, telecommunications or defence sectors). Public support strongly focused on large companies active in high- technology fields, and aimed to achieve a leading position for France in research and innovation."

These highly centralized, linear and goal-driven programmes enjoyed considerable success through the 'Glorious Thirty' years of post-war economic expansion, converting increased national R&D effort (Larédo & Mustar, 2001) into noteworthy technological advances such as France's development of a domestic and export-oriented nuclear power industry, and collaborative efforts such as the Concorde supersonic aircraft. By the mid-1960's France had surpassed the UK in per capita GDP and become the fourth largest OECD economy (Amble & Hancké 2001).

In the late 1970s, the limits of the dirigiste economy became evident as productivity and growth decreased in French and European industries (Amable & Hancké, 2006). While France, like many other European countries reacted by shifting away from Keynsian policies, its adoption of a policy of *désinflation compétitive* in the early 1980s set it apart from its competitors. It was hoped that this policy, adopted in anticipation of inflation-limitation policies integral to European Monetary Unification, would force French firms to increase competitiveness through gains in quality, rather than often artificial gains in price competitiveness through currency devaluation. Rather, it had just the opposite effect, and by the 1980s France was not seen as competitive on price or quality (Amable & Hancké, 2001). During this time, there was a steady increase in R&D funding accompanied by liberalization of the French economy (Goyer, 2001).

Despite national austerity and EU competition policies, the state contributed significantly to French firms in the 1980s, reversing the downward trend of the 1970s, with the ratio of GERD/GDP going from 1.9-2.4% (1981-1990) and stimulating private sector R&D efforts (Larédo & Mustar, 2001). The 1980s saw the end of the majority of the large programmes, integration of research conducted by universities and the CNRS was encouraged, regional programmes gained power and financing, and the emergence of European Union project funding provided an incentive to collaborate internationally with diverse public and private actors (Munstar & Larédo, 2002).

These changes, while significant, are nonetheless likely best regarded as variations on a theme: decentralization and international collaboration emerged as new aspects of a system which remained thoroughly state-centered. The source of radical innovations, basic R&D, was still present yet incremental innovation, found to arise from interactions with the market or industry, was generally lacking in France as a result of its historic focus on state-controlled large programmes (Goyer, 2001). Indeed as Goyer (2001) and others have observed, French firms often acquired American subsidiaries as a means of gaining technologies and core competencies, to build their innovative capabilities (Sachwald, 2009). The story of the years since 2000 has been one of rapid government-sponsored change to most parts of the French innovation system, usefully summarized in toto as gradual adoption of the open innovation model.

Innovation System Infrastructure Renewal

The complexity of the changes made to the French innovation system since 2000 defies comprehensive cartography or analysis in the context of a case study. It is possible nonetheless to highlight some defining drivers of change and main loci of open innovation today, together with identification of factors bearing on future evolution of the system. One major external driver of change within France was the Lisbon Strategy's R&D investment objective established in 2000, seeking investment of 3% of GDP in annual research and development, leading Europe to be, in 2010, "The most competitive and the most dynamic knowledge-based economy in the world" (France Diplomatie, 2010).

This objective provoked considerable internal debate and change, driven notably by government commissioning of reports recommending sweeping changes. The so-called Beffa Report of 2005, for example, resulting from the inquiries of a task force of 12 opinion leaders led by Jean-Louis Beffa, chair of Saint-Gobain, recommended a wide range of policies designed to enhance French competitiveness through encouragement of technological innovation.

Among the set of reforms ultimately introduced, those especially relevant to open innovation provide a range of new coordinative and policy mechanisms to encourage cooperation between universities, government and private sector, and in particular, ways to enable a relatively fragmented university-sector to combine institutional expertise in larger, multidisciplinary bodies well-positioned to engage comprehensively with industry.[2] Most recently, President Sarkozy has further driven integration and concentration of research expertise by identifying the Saclay plateau, in suburban southwestern Paris, as a site for a new super-university to be newly constructed on a new institutional model, to assemble French expertise in a globally competitive or globe-leading institution.

Mechanisms connecting university-sector actors include the Réseau Thématique de Recherche Avancée, (RTRA), introduced in 2006 to enable synergies among groups of universities or researchers sharing common research themes or geography. Also introduced in 2006, the 'PRES' mechanism (pôle de recherche et enseignement superieur) has encouraged grandes écoles, universities, and other postsecondary education institutions to coordinate research and teaching agendas and resources, forming in effect university consortia. Technically-oriented institutions in Paris, for example, have joined together to create the PRES ParisTech, while a multidisciplinary combination of grandes écoles and universities in the south of Paris created the PRES UniverSud Paris.

University-sector actors have also been given new tools to engage the private sector, from incubation of start-ups and spin-outs within individual institutions, to participation as individual institutions (or PRES or RTRA members) with commercialization-oriented institutions of various kinds. Among the most important mechanisms are the Pôles de Compétitivé (created in 2002), which bring together companies, research laboratories and training institutions in a given territory to develop synergies and cooperation around innovation projects, and the programme for the Centre Régional d'Innovation et de Transfert de Technologie (CRITT), which creates ties bewteen regional industrial and research actors.

Private sector participation in these collaborations is encouraged by a blend of incentives and direct investment. A new R&D tax credit, the 'crédit d'impôt recherche' has provided incentives for private firms to fund R&D, and funding gaps in innovation cycles have been identified and addressed with direct investment. The Fondation Pierre Gilles de Gennes (2008), for example, links industrialists and academic researchers, providing seed money for research projects and facilitating the transformation of discoveries into innovative applications. L'Agence de l'Environnement et de la Maîtrise de l'Energie (ADEME) provides multifaceted support and funding for later stage research and application through its links with industry, and various public and private associations. L'Agence nationale de la recherche (ANR), created in 2005, also provides project based funding, and funding for the more than 70 pôle de compétitivité across France.

Further SME-specific research support and technology transfer has been encouraged by the Passerelle programme, and since 2005 OSEO (under the auspices of French Agency for Innovation) has provided financial aid, particularly that for SMEs, through partner financing and guarantee of bank loans etc.

Technology transfer in support of innovation has been supported by institutional reforms and introduction of new institutions and tools bridging gaps between existing innovation system members. Many universities and research centres have implemented Cellules de Valorization, accelerating transfer of technology and fundamental research to market by linking industrial partners with relevant technologies. At a larger, inter-institutional level, the ANR has supported the creation of Instituts Carnots, four-year, topically-specified collaborative networks creating knowledge-mobilizing connections among public research organizations and private sector firms.

Results of reforms and evolution have so far been mixed, according to both comparative measures and French expectations (OECD,

2008; OECD, 2010). During 2002-2004 only 13% of French firms undertook new to market innovations and only one in five undertook non-technical innovations (OECD, 2010). While other data from the OECD published in 2008 and 2010 indicate positive signs such as significant numbers of French firms collaborating in innovation activities (12.9% between 2002-2004), judged comparatively this performance is middling. Denmark (22.2%), Sweden (21.4%), and Britain (15.8%) recorded more firm-level collaborative research over the same period, yet France outperformed Germany (10.4%) and Japan (7.4%) (OECD, 2008).

There may, however, be cause for concern to the extent that French, collaboration is primarily with suppliers (65%) and customers (50%), and just over a quarter of companies collaborating on innovation activities do so with University partners (26%), public institutions (18%) and competitors (36%). France's gross expenditures on R&D (GERD) was below the OECD average in 2008 and is declining. Government and business expenditure on R&D is also declining and France now lies below the OECD average in these measures (OECD, 2010). The cooperative practices and associated support for generation and use of R&D, it appears, are not generating desired increases in innovation or competitiveness (OECD, 2009; OECD, 2010). Thus, while the rapidly evolving French system has been highly sensitive to opportunities and criticisms, introducing sweeping reforms consistent with open innovation practices successful elsewhere, the operation and results of the reformed system are not as effective as anticipated.

Challenges to Open Innovation in France

At first examination, the underperformance of the French innovation system relative to competitors and expectations is quite puzzling. The gradual withdrawal of state dominance and introduction of Anglo-Saxon and other elements has produced a system which is structurally comparable to the leaders it aims to match and overtake. It is comparable in terms of an appropriate balance between corporate taxation levels, incentives for private sector participation in research, university autonomy and opportunities for inter-university collaboration and university-business collaboration, patenting rates, incubators and investment for SMEs, and many other familiar Frascati- and Oslo-style indicators. It may well be the case that the evident underperformance is only momentary, and French leaders must simply be patient as they await the positive consequences of their renovation of a system which will, over time, increase innovation, productivity and competitiveness even as they lack direct control of it. After all, France remains widely admired for the quality of life offered to its citizens, routinely ranked as the best country in which to live.

Further performance improvement might yet, nonetheless, be possible, so it is well worth examining even quite speculative alternative explanations for the innovation system's apparent underperformance, both measured and perceived. In particular, it may be worth considering that where innovation system structures have been adjusted, associated indicators indicate that the adjustments are functioning well, yet productivity still lags, other indicators are needed. More specifically, rather than input-based and output-based indicators, innovation system performance may be better measured by indicators capturing *integration* of system elements.

In particular, one might consider whether those indicators ought to address not just the presence and relatively isolated function of structural reforms, but their inter-operation and management by innovation actors bearing what might be usefully called 'cultural competences.' There are certainly data available whose interpretation suggests that this line of analysis may be worthwhile. For example, while French SME five-year survival rates are at first glance impressively high, enabled by various structural reforms including e.g. the creation of OSEO, some French observers lament the fact that compared against British SME performance, relatively few French SMEs become fast-growing 'gazelles' growing revenues and creating jobs while accessing export markets (Cercle d'outre-Manche, 2008). Given that the structures supporting British and French SMEs are broadly comparable, it again seems warranted to investigate the integration of those structures as managed institutions at an operational level by managers and other users.

Investigation of the influence of management factors on inter-institutional operation is complex, and comprehensive understanding often requires expensive and time-consuming multi-method investigations blending, e.g., survey data, opinion leader interviews, and institutional data. Thus, it is not the purpose of this case study to report such data or to propose methods or indicators for such an investigation. Rather, this case study reports the development of one kind of French triple-helix response to a preliminary and synthetic assessment of challenges to French development of an open innovation system delivering desired improvements in innovation, productivity and competitiveness. The Econoving partners described in the case study below, have chosen to take action in a highly adaptable 'learning organization' strategy in which negative feedback regarding operating assumptions regarding challenges can be quickly incorporated into a modified strategy.

Three main perceived shortcomings of the French innovation system motivate the strategy of the Econoving international chair, discussed in the next section. The strategy aims to improve management factors related to: (1) the operational effectiveness of innovation system reforms aimed at increasing system openness; (2) the complexity of the reformed innovation system, and (3) innovation system actors' confidence regarding participation in a rapidly evolving system.

Operational Effectiveness

It is a commonplace observation amongst French innovation actors that the structural reforms implemented in the system have been necessary and well-chosen, yet have not been fully implemented, for reasons to do with both brute financial investment in those reforms, and for reasons associated with management factors. In the case of universities as upstream innovation actors, the vaunted increases to university autonomy have not been advanced with a blend of investment and management reform to enable universities to make use of their structural autonomy to better engage other upstream actors at international level. The PRES system, for example, allows university consortia to pool resources in shared doctoral schools whose concentration of competence will make them more attractive to the best students in turn potentially retained as 'brain gain' for France.

Yet this reform has not been followed up with a shared postgraduate recruiting function aimed at international students. And while the PRES mechanism is intended to allow French universities and grandes écoles to face the world as comprehensive multidisciplinary institutions comparable to competitor universities as judged by the Shanghai rankings and other mechanisms, postgraduate education in France remains largely francophone, even while French institutions and professors are permitted to teach in English, competitors are increasingly oriented toward postgraduate education in English, and top potential postgraduate students are more likely to speak English than French in addition to their mother tongues. Lack of English-language skills has also been pinpointed by observers such as the OECD as a barrier to SMEs initial engagement in international markets (OECD, 2009).

Similarly, while the PRES mechanism allows member institutions' laboratories to collaborate as though they are members of a single university, the incentives to do so are unclear: even while PRES member institutions benefit from PRES membership, they are in competition for resource allocations from central government, so there is a disincentive to thorough collaborations in which the contributions of individual laboratories cannot be easily separated out and used to advance the cause of individual PRES member institutions. And perhaps more troublingly, while laboratories *can* collaborate to produce better quality results in discovery and applied research, with and without collaboration with private sector or other actors, there is a lack of clear incentives for professors to change their cultural competences to include new forms of collaboration. As civil servants whose initial appointment at the rank of Professeur des Universités is the terminal rank, and is normally a career-length appointment with limited further performance review, professors have little incentive beyond curiosity and patriotism to engage in new forms of collaboration.

Recent reforms permitting professors to participate in start-ups while retaining their professorial posts seem to many to be beside the point: from an innovation system growth perspective, what is needed from professors is not a willingness to be entrepreneurs, but to be professors who are more engaged in the best possible discovery research and applied research in global context. In short, then, and using the upstream actor role of universities as just one example, there are reasons to believe that large-scale structural forms have yet to be thoroughly implemented via appropriate and effective operational mechanisms.

Complexity of the Reformed System

The rapidity and complexity of French innovation system reform may have inadvertently and paradoxically created an obstacle to innovation. At the level of the university sector as an upstream actor, universities and grands écoles are faced with a striking diversity of choices as they attempt to choose the collaborative relations which will bring them opportunity and external recognition. The demands of membership in a PRES, for example, must be managed along with the choice to engage in an Institut Carnot or a Pôle de Compétitivité, while at the same time internal incorporation of a Cellule de Valorization must be considered, all while European-level collaborative opportunities are monitored and selectively taken up, and while eyes are kept on the horizon for the next generation of changes to the innovation system.

The sheer complexity of the system is a substantial barrier to both improvement of its operational effectiveness, and to the university sector's preparation for engagement with partnership opportunities at both national and international level. It is noteworthy, for example, that most French university-sector and industry websites remain either French-only, or present non-comprehensive versions in English or other languages, in contrast to the well-developed bi-lingual or multi-lingual websites of other European and international competitors. Potential international partners are possibly deterred by the relatively inaccessibility of information regarding their French counterparts, accentuated by the fact that major electronic search tools for expertise and partnership, e.g., Le Moteur de la Recherche, operate in French, and on a members-only basis.

The effects of this lack of university-sector openness to international partnering are compounded by the culture of French private sector actors, who tend to have international operations only once they are relatively large, and often have international operations only for the purpose of access to technologies as opposed to operating national subsidiaries of France-based firms (Sachwald, 2009). Private sector actors tend instead to rely on French research infrastructure to generate innovation, omitting engagement with European-level research networks, expending energy instead on use of the many complex French mechanisms for exchange of ideas between researchers and industry.

In sum, the complexity of the innovation system likely inhibits improvement of its operations, and needlessly burdens its internal actors with difficult, time-consuming choices amongst options for internal French coordination and its integration with international partners. These burdens are experienced by both university-sector and private sector actors, and may tend to increase the extent to which they focus on French activities at the expense of potentially useful international participation, which is additionally hampered by the fact that France-focussed innovation tends to be presented to potential partners in inaccessible ways, i.e., in French-language materials or tools, sometimes available by subscription only.

Actor Confidence

The operational inefficiency and sheer complexity of the evolving system poses a further challenge whose effects on collaboration are no doubt substantial even while difficult to gauge. It is widely observed that French efforts to develop effective SMEs have lagged behind British and German efforts. It should of course also be observed that development and support of SMEs, and especially desirable gazelles, is a difficult task in any country, but nonetheless French performance lags French ambitions. There may yet be, nonetheless, specific French weaknesses to be addressed. It may be, in French society, that change is too often associated with unnecessary risk.

The relative lack of gazelles may be evidence that American-style tolerance for multiple entrepreneurial failures prior to success has not taken hold in France, despite structurally similar supports for entrepreneurship, from the presence of angel investors through business incubators, gap financing facilities, and other elements. Or perhaps this increasing tolerance of failure has simply not yet diffused to all potential actors in the French system: SMEs, markets, suppliers, industry, researchers, government, local authorities, etc.

Thus some of the potential benefits of the open innovation model are not being realized. In particular, France appears to lack the range of incremental innovations typically brought by SMEs to counterbalance transformative innovations arriving from discovery research, conveyed to market as push innovations. The lack of comprehensive diffusion of an innovation culture is associated with factors which suggest a lack of actor confidence in the benefits to be gained by participation in the system, and perhaps in evolutionary pathway of the system. French firms outsource less than 10% of their R&D (Sachwald, 2009), and while French firms' co-operation with public sector research actors has increased, it remains relatively low in comparison with other European countries: only 25% of French firms (compared with 50% in Germany) collaborate with academic researchers. As fundamental research is widely regarded as a key source of radical or transformative innovations, this is a significant disadvantage. Effective, not merely structurally permitted means must be found and trialed to bring together private sector actors and public sector research bodies.

Econoving Case Study: A Solution in the Making

As the preceding discussion tends to show, the strategic 'opening' of the French innovation system is significantly incomplete, and operational and cultural challenges remain to be overcome. The complexity the French infrastructure in support of open innovation does bring, however, myriad of opportunities for determined actors to make gains by piloting and incorporating the lessons of particular collaborative configurations within this system. The Econoving programme, operated under the auspices of a collaborative international chair in eco-innovation, has been designed to pilot one such collaborative configuration. In addition to fostering knowledge generation and mobilization amongst its partners, the programme has been specifically designed to target and respond to the three challenges identified in the preceding section.

The Econoving chair programme was created in 2008 by a consortium of the laboratories of the PRES UniverSudParis together with five industrial and one government partner. Lead by the University of Versailles Saint-Quentin-en-Yvelines, UniverSud Paris members Supelec, Ecole Centrale de Paris, the Ecole Centrale Superieur de Cachan, and the University of Paris XI agreed to contribute academic support to a chair-holder to be recruited on an international basis to coordinate mutually reinforcing aspects of a comprehensive programme of post-graduate level teaching and research in the area of eco-innovation. These university-sector contributors were joined by industrial partners Alstom Power (electricity), GDF SUEZ (gas), Italcementi (cement), Saur (water and waste), SNCF (rail transport), and later the French ministry of environment, ADEME, to create a private-public partnership united and focussed on a shared, market-driven set of opportunities selected in consultation with the private sector partners.

In order to reflect and advance the programme's commitment to open innovation, the partners chose additionally to use English as the working language of the programme in research, teaching and administration, and to identify the programme's international outlook explicitly via its title, identifying the programme as the Econoving International Chair in Generating Eco-innovation. Following the recruitment of the chair-holder, anglophone Professor Keith Culver, a Canadian and British national, the new chair-holder began in mid-2009 to guide partners toward a focusing of their priorities. Professor Culver analysed partners' self-identified priorities in energy, water and waste, construction, and transport, and soon found that those priorities were expressed in a single unifying context: the need to eco-innovate in our cities. From this analysis Professor Culver derived the mission of the chair, expressed on the chair website and annual report as follows: To develop a cluster of eco-innovation actors creating green technologies and social innovations to renew our cities as vital contributors to the reconciliation of sustainability with economic competitiveness and growth.

The partners' emphasis on 'doing things differently' in the French context gave rise to commitments which enabled the chair-holder and partners to develop a specific kind of strategy to achieve the agreed mission. In particular, operation under the direction of a non-French chair-holder was intended to bring fresh experience, perspectives and methods to operation of the programme in the French context. The highly collaborative and open approach to initiation and development of the Econoving Chair and associated program has resulted in adoption and development of a very collaborative model of governance. The Chair answers to a board made up of representatives from the chair partners who meet quarterly. The Chair benefits from the infrastructure, an in kind commitment, of its host institution but is in effect the CEO and COO of the research and teaching programme. Challenges regarding the complexity of the French system and confidence of the Chair partners were also recognized and addressed by the structure of the programme, in which a limited number of partners contributed substantial cash and in-kind resources for deployment within a collaborative programme of defined scope and shared resources, united by shared focal concern with eco-innovation in cities of the future.

These initial commitments enabled the chair-holder and partners to avoid repetition of some practices previously found to be unhelpful to generation of sustained partnership-driven innovation. By committing to development of an enduring eco-innovation cluster, a long-term view was taken, avoiding the trap of treating partners and their resource-commitments as investments in a time-limited project whose success would be judged solely in terms of research results enabling problem-solution.

Taking partners' expressed research priorities as both near-term needs and indicators of future shared research areas, the cluster-building strategy develops the Econoving programme not as a stand-alone concentration of expertise relevant to partners' near-term priorities, but as a catalyst and coordinator of collaborations designed to use existing partner capacity and competence to address near-term priorities while creating a network of contacts and partners enabling flexible adaptation to emerging opportunities. Those emerging opportunities are identified by consultation of European-level foresight exercises which inform European and member state R&D investment strategies.

By aiming to both address near-term problems and to create a cluster which functions as a learning organization ready to take up future opportunities, learning lessons from each collaborative research action advance subsequent collaborative research, the Econoving programme aims strategically to become self-sustaining as a cluster of actors sharing a common purpose, and united by a shared set of research practices and a shared history – a shared culture of innovation. That shared culture of innovation is intended to drive increasingly rapid cycles of lab-to-market innovation supporting productivity and competitiveness.

Rapid Cycles of Innovation and Three Levels of Partnership

Within the context of this mission, academic and industrial partners collaborate in a cyclical process of capacity and competence mobilization, analysis of priorities and projects in the context of shared interests and available opportunities, followed by R&D leading to pilot and transfer of results (Figure 1). Following and accompanying collaborative R&D efforts, partners assess the merits of their collaborative work, and embed and disseminate lessons learned, and R&D results. The strategy is self-consciously designed as a repeating cycle in which the entire organization of public, private, and university sector partners learn from one another and implement lessons learned in subsequent rotations of the R&D and learning cycle, to generate a genuine learning organization.

As the cycle turns, the Econoving research programme operates at three mutually reinforcing levels of partnership. At the first level, the programme coordinates problem-solving applied research, matching industry and government problems with university-sector professors and students, generating incremental innovations. In this problem-solving activity the programme promotes translation and transfer of problems and solutions while building working relationships between university and industry. At the second level of interaction the programme coordinates and catalyzes collaborative research in France, bringing together interdisciplinary, university and industry teams to bear on eco-innovation problems and opportunities in the city. At the third level of interaction the programme strives to bring a distinctive French contribution to European and international partnerships, sharing experience, skills, risk, and reward in technical and social research into cities.

Throughout this research, true transformative innovations may arise as results are mobilized to benefit partners, and others are published in peer-reviewed international journals, contributing to the pool of knowledge from which yet other innovations may arise. Knowledge[3] is also translated by eco-innovators incubated in close proximity to Econoving researchers, and connected by the programme to finance actors and larger eco-innovative firms to improve the quantity and quality of green technologies and practices available to city-focused actors in France.

Development and catalysis of synergies between Chair partners on a central research focus is only part of the Chair programme. The Econoving strategy aims to create and accelerate cycles of learning in a cluster of actors from public, private, and university sectors. (In this strategy students are far from being passive absorbers of well-honed knowledge handed down from professors. Rather, students are living bridges between theory and practice, and when working in group projects together, are a kind of 'innovation pump' surfacing new ideas for consideration as research priorities for the Econoving chair R&D group and cluster partners.)

The Econoving knowledge community's core capacity is in the programme's appointees and founding partners, including the chair-holder, a manager of partner relations, a manager of financial sector relations, and project managers for both French and European-level projects. This team is supplemented by an Entrepreneur in Residence who counsels students on development of entrepreneurial opportunities, former students developing eco-innovation startups in an incubation environment, and PhD and post-doctoral fellows fostering exchanges of problem definition, knowledge creation, and problem solving amongst founding partners and UniverSud labs. Further international connections are maintained by Global Fellows, scholars from around the world who visit the Master's and PhD programme to lecture, give talks to UniverSud Paris labs, and engage collaboratively in research projects. From this core competence as a coordinative and catalysing body, the Econoving programme's network seeks competence adapted to the size of the problems it engages, associating researchers in France and across Europe in research projects and relations with our industrial partners and other labs.

In seeking to develop both research activities and an innovative learning organization the Econoving strategy recognizes that innovation is not just about new products and processes. Innovation is very much about new technologies, but it is just as much about a new way of generating new technologies and linking these methods with fundamental research, the source of transformative innovation. Understood fully, including new products and technologies and the production system which generated them, the programme is best understood as a new way of inventing, driven by alignment of interests in closely related actors. Put more formally, the Econoving strategy understands innovation as the systematic organization of competence and capacity to generate new ways of understanding problems and opportunities, and new ways to develop products, processes and policies in response.

In this context, an innovative learning organization is a body of people and institutions interacting productively together according to a shared set of commitments and goals, constructing habits and practices of interaction enabling improvements to high-quality productive through synergies, efficiencies and insights gained from interaction over time. The Econoving learning organization engages in self-assessment and learning of lessons to improve the quality of interactions between researchers, industry, government and end-users as we work to develop new technologies and social practices for sustainable cities. At each step of Econoving research, development and applied innovation cycle, the Econoving programme is interdisciplinary, future-focussed, and always concerned to integrate student learning so every eco-innovation activities is accompanied by development of the next generation of responsible users of technical and social innovation.

Success of the Econoving model of innovation is measured against several key performance indicators. These indicators are a combination of recognized indicators of innovation (and open innovation) such as the number of collaborations with research and industry groups, the number of SMEs included in collaborations, and the level of participation in international, European, national and regional collaborations. The key performance indicators also include indicators developed to measure progress towards other learning and cluster building outcomes. These include the number of programme graduates translating internship work into longer-term contracts, the number of SME's incubated within the programme, and the number of doctoral and post-doctoral 'bridges' between industry and academia.

Early Successes

Mid-way through the second year of the program, progress has been rapid, in developing the capacity, competence and linkages characteristic of an emerging cluster, and in identifying relatively internal challenges in organization of core partners, and external challenges in the operational environment of the Econoving programme. Project managers at the heart of the programme are negotiating problem definition, selection and use of methods, in addition to performing a traditional process-management function uniting diverse partners operating from geographically proximate yet separate bases. Project managers and partners are aided in their coordination efforts by an e-tool developed by an Econoving Master student: a web-based search engine which associated search key words with particular laboratories and actors in UniverSud Paris and specific French databases, enabling rapid identification of potential R&D partners.

The first class of Master students has graduated, having spent 50% of their time in eco-innovation management project internships in the laboratories and research departments of Econoving partners, bringing skill and imagination to projects on the leading edge of partners' commitment to innovation. The first year's graduates are now pursuing a range of different opportunities. Some are furthering their education at the PhD or Masters level, others have obtained employment with their internship host and yet others are bone fide eco-entrepreneurs, pursing eco-innovations developed during year.

Incoming PhD students and postdoctoral fellows are amplifying the human connection between academic partners and industrial partners, as these students will be physically resident in partner labs while receiving academic supervision from academic partners, in effect serving as 'living bridges' between academic research and industrial research and application. These students and collaborating academic researchers are engaged in specific research projects addressing near-term industrial partners' needs, and beginning collaboration with European-level partners to create a network of city-focused innovation clusters.

By 2010, the Econoving was incubating two eco-innovation start-ups, and was in discussion with a third SME. One of these SMEs is headed by a recent graduate of the program. Given the importance of SMEs within an open innovation system and in cluster building this is a very positive and tangible outcome of the Econoving programme.

Success Drivers

The programme has been challenged by operational and cultural factors in its immediate environment. The program is also faced with institutional challenges external to the programme. However as the programme has developed, two factors have emerged as being important for its early successes. First the programme's autonomy as a research and innovation cluster has enabled an agile approach to problem solving. This approach has allowed the programme to take advantage of opportunities as they arise, with little gap between opportunity and action, awating referees reports, completion of funding agreements, or the like. The freedom to act conferred by the programme's autonomy has benefitted design and operation of the programme's cycle of problem co-definition, triage of approaches, taking action and disseminating results. Second, while remaining open to collaboration and cognizant of the benefits of open innovation, we only work with those actors who make sense for us, and we are not at the same time forced to pursue other goals such as support of construction of a European or other particular identity. In other words, while the Econoving programme certainly aims to foster development of an innovation culture, it does so within and among willing partners and is not slowed by obligations to work with partners who reject or are otherwise at odds with the goal of developing an innovation culture.

Challenges Ahead

In the course of the Econoving programme's rapid early progress, a number of future challenges have been identified – challenges both inherent to the complex business of operation of an innovation system as a learning organization, and challenges specific to the French context. It must nonetheless be emphasised that these challenges arrive as an interconnected whole best regarded in summary as a challenge of sustaining confidence on the part of all actors. While partners are naturally very keen to recoup a timely return on what is perceived to be relatively large-scale investment of cash or labour, speedy return on investment in R&D is difficult to achieve in a young partnership intentionally organized to attempt new ways of working together, all in a system undergoing rapid change.

More specifically, investment in R&D processes is notoriously risky even while it is in the longer term essential to innovation, productivity, and sustainable prosperity. Current investors in the chair programme perceive themselves to have invested a great deal, in cash and in-kind contributions, and anticipate a relatively rapid return on investment. Policy incentives for future, greater investment in high-risk, high-reward R&D are either so poorly disseminated or so ill-suited as to be of limited attractiveness to many of these actors, with the effect that their focus on near-term results is naturally reinforced since incentives for tolerance of high-risk are weak.

On a related front, even willing actors in the Econoving programme's eco-innovation cycle may lack both internal innovation experience, and collaborative innovation experience. In bringing together academic, industrial and government partners, the Econoving programme is encouraging interaction between actors who until recently have been highly incentivized to operate in silos. Those silos have had what can only be fairly assessed as minimal interaction when considered in light of the depth of collaboration needed to co-define problems, identify and develop methods, and carry out research, development and demonstration resulting in technical and social innovations improving the environmental performance of cities. Experience in this sort of interaction is critical to effective management of what are in effect 'mergers' of institutional cultures, yet there are relatively few actors with experience of this sort of 'translation' activity - actors who have been active in two or three silos and may serve as ambassadors between silos helping identification of common and complementary interests, methods, skills, etc.

Even as new mechanisms and incentives, are used to assist partners in undertaking high-risk innovation activities, e.g., sharing of risks and benefits in multi-partner European-level projects, the pace of change in the French innovation system remains an asset and an obstacle. The rapid pace of development of the Saclay plateau super-university project, for example, opens new collaborative possibilities which are attractive yet insufficiently developed to provide meaningfully actionable opportunities, so the nascent but still largely promissory prospect of massive investment into a new super-university likely distracts participants in pre-cursor institutions such as PRES from continued development of those institutions. It is difficult, in a rapidly changing situation, for potential partners to gain and maintain confidence that each partner's evolution over the coming months and years will maintain the features which make that partner an attractive collaborator.

FUTURE RESEARCH DIRECTIONS

In a case study examining a rapidly evolving governance context for open innovation, and a new activity in that context, the results that can be reported are less research results than practical lessons learned regarding cluster-building, perhaps usefully regarded as a collegial sharing of experience. In the context of the Econoving programme's mission to develop a city-focussed eco-innovation cluster, the goal of the programme remains as it was initially set: to pilot technological and social innovations in our home territory in Versailles and Saint-Quentin-en-Yvelines, then to transfer lessons learned to our public and private sector partners for application in other contexts.

More specifically, this research recognizes that in Europe an important gap and opportunity exists: implementation of eco-innovations in mid-sized European cities of 250,000-500,000, similar to the French city of Saint-Quentin-en-Yvelines. While the centres of Paris, London and Berlin require and can generate *sui generis* solutions to their environmental challenges, the next 'level' of cities lacks the concentrated innovation power of great capitals, and needs to operate at a European partnership level sharing problem-definition, R&D, risk-taking, and rewards. The study continues to target this opportunity. As the research does so, it draws from the experience to date three pieces of collegial advice to similarly situated innovators.

First, the research experience confirms the widespread view that construction of an open innovation cluster requires adoption of a long-term view. That much is scarcely new. The study would add from this experience, however, that the activity of taking the long view in a collaborative situation is substantially easier if appropriate measures of progress are identified, used, and communicated. In particular, since cluster-construction involves a great deal building of institutional capacity (from non-disclosure agreements through intellectual property-sharing agreements, and doctoral student exchange agreements to equipment-pooling agreement), it is important to measure and communicate to partners the growth of that capacity.

By the same token, it is important to measure and communicate to partners the growth of cluster competence through even small R&D collaborations and commercialization victories. And, further, the research results strongly advocate communication of growth of linkages to other organizations, to permit and foster whole-of-cluster participation in growth of multiple dimensions of linkages – from researcher to researcher, to exchange of lessons on purchasing, recruitment, or other items of shared concern.

Second, the research draws from an experience in a complex and shifting environment the lesson and advice that open innovation cluster-builders are best off when they allow the organization of their cluster to be 'always beta', following the lead of software developers who never try to finalize a piece of software, but let it always be 'beta-ware' or usable but incomplete, as a consequence of incorporating leading-edge developments. A commitment to continuous development and adaptation helps to avoid becoming ossified and outdated by surrounding developments. In practice this means being open to reconfiguring partnerships as needed. Measures taken may include, for example resisting the temptation to insist on the letter of agreements being followed where respect for the spirit of the agreement will advance a common cause more efficiently, and at the same time, remaining open to the possibility that partner involvement in cluster-building will evolve over time, with some partners especially present at particular phases of cluster development, less so in others.

Third, the study advocates a resolutely international perspective and openness as a key means to gaining and contributing new insights and methods both technical and cultural, sharing risk and reward, and providing a comprehensive 'reality check' with respect to the quality of R&D and innovation activities. It is all too easy to regard leadership at home as leadership everywhere, and a thoroughgoing commitment to international participation, and international recruitment, tends to alleviate this tendency while nonetheless remaining rooted in the local cultural context and its requirements.

CONCLUSION

In France and in the specific context of the Econoving programme, elements of open innovation have gradually displaced all but a few vestiges of the state-led dirigiste approach to research and development. The transition to the open innovation model remains, however, incomplete. The transition is incomplete in both the sense that the gains anticipated from open innovation and productivity have yet to be realized relative to competitors, and, as we have illustrated via our discussion of the Econoving programme, creation of the innovation networks and clusters capable of taking advantage of an open innovation model is a lengthy and complex process.

However, motivation and technical capacity to take up new collaborative opportunities is not enough, even when supplemented by skilful use of new devices together with familiar tools such as non-disclosure agreements, material transfer agreements, personnel secondment agreements, and so on. Multi-partner innovation-oriented collaborations require trust-building in collaborative problem-definition, development of research methods, conduct of research and absorption of results by end-users. Successful collaborations further demand that partners accept the need to adapt continuously, facing open innovation not as a model to be implemented as the 'new normal' but as yet another step in evolution of innovation models on something very much like Vannevar Bush's "endless frontier."

REFERENCES

Amble, B., & Hancké, B. (2001). Innovation and industrial renewal in France in comparative perspective. *Industry and Innovation*, *8*(2), 113–133. doi:10.1080/13662710120072930

Archibugi, D., & Michie, J. (1997). Technological globalization and national systems of innovation: An introduction. In Archibugi, D., & Michie, J. (Eds.), *Technology, globalization and economic performance* (pp. 1–23). Cambridge, UK: Cambridge University Press.

Boris, P., & Vaissié, A. (2009). *L'université et la recherche: Moteurs de la création d'enterprise*. Studyrama and Le Cercle d'outre Manche.

Bush, V. (1945). *Science, the endless frontier: A report to the President by Vannevar Bush, Director of the Office of Scientific Research and Development, July 1945*. Washington, DC: United States Government Printing Office. Retrieved from http://www.nsf.gov/od/lpa/nsf50/vbush1945.htm

Cercle d'outre-Manche. (2008). *Du brain drain au brain gain*. Le triangle d'or Université, Recherche, Entreprise. Retrieved from http://www.cercledoutremanche.com/files/pdf/Rapport4.pdf

Chesbrough, H. (2003). *Open innovation: The new imperative for creating and profiting from technology*. Boston, MA: Harvard Business Press.

Chesnais, F. (1993). The French national system of innovation. In R. Nelson (Ed.) *National innovation systems: A comparative analysis*. Oxford, UK: Oxford University Press.

France Diplomatie, Ministere des Affaires Etrangeres et Europeens. (2010). *The Lisbon strategy: Partnership for growth and employment*. Retrieved July 30, 2010, from http://www.diplomatie.gouv.fr/en/european-union_157/european-policies_2190/growth-and-employment_5476/the-lisbon-strategy-partnership-for-growth-and-employment_8764.html

Goyer, M. (2001). Corporate governance and the innovation system in France 1985-2000. *Industry and Innovation*, *8*(2), 135–158.

Griffith, R., Heurgo, E., Mairesse, J., & Peters, B. (2006). Innovation and productivity across four European countries. *Oxford Review of Economic Policy*, *22*(4), 483–498. doi:10.1093/oxrep/grj028

Kline, J., & Rosenberg, N. (1986). An overview of innovation. In Landau, R., & Rosenberg, N. (Eds.), *The positive sum strategies: Harnessing technology for economic growth*. Washington, DC: National Academy Press.

Larédo, P., & Mustar, P. (2001). French research and innovation policy: Two decades of transformation. In Larédo, P., & Mustar, P. (Eds.), *Research and innovation policies in the new global economy: An international comparative analysis*. Cheltenham, UK: Edward Elgar Publishing.

Muller, E., Zenker, A., & Héraud, J.-A. (2009). France: Innovation system and innovation policy. Fraunhofer ISI Discussion Papers. *Innovation Systems and Policy Analysis*, *18*, 35.

Mustar, P., & Larédo, P. (2002). Innovation and research policy in France (1980-2000) or the disappearance of the Colbertist state. *Research Policy*, *31*, 55–72. doi:10.1016/S0048-7333(01)00107-X

National Science Foundation Act of 1950, Public Law 81-507 (Stat. 149)

Nelson, R., & Rosenberg, N. (1993). *National innovation systems: A comparative analysis*. Oxford, UK: Oxford University Press.

OECD. (1997). *National systems of innovation*. Paris, France: OECD Publishing.

OECD. (2008). *Open innovation in global networks*. Paris, France: OECD Publishing.

OECD. (2009). *The challenge of restoring French competitiveness*. Paris, France: OECD Publishing.

OECD. (2010). *Science, technology and industry outlook 2010*. Paris, France: OECD Publishing.

Sachwald, F. (2009). *Global networks of open innovation, nation systems and public policies*. Ministère de l'enseignement supérieur et de la Recherche, France.

West, J., Vanhaverbeke, W., & Chesbrough, H. (Eds.). (2006). *Open innovation: Researching a new paradigm* (pp. 285-307). Oxford, UK: Oxford University Press.

ADDITIONAL READING

Burns, P. (2007). *Entrepreneurship and small business*. Hampshire, UK: Palgrave Macmillan.

Cercle d'outre-Manche (2006). 5 fiches prêtes à l'emploi. http://www.cercledoutremanche.com/rapports.shtml

Cercle d'outre-Manche (2006). My neighbour is rich: Ou comment la France peut s'inspirer du réformisme permanent d'Albion pour accélérer son retour à l'emploi et à la croissance. http://www.cercledoutremanche.com/rapports.shtml

Cercle d'outre-Manche (2007). 6 mois, 6 mesures pour que la France repasse devant le Royaume-Uni. http://www.cercledoutremanche.com/rapports.shtml

Chesbrough, H. (2003). Open Platform Innovation: Creating Value from Internal and External Innovation. *Intel Technology Journal*, 7, 3 (August): 5-9.

Chesbrough, H. W. (2003). The Era of Open Innovation. *Sloan Management Review*, 44, 3 (Spring): 35-41.

Christensen, J., F., Olesen, M., H., & Kjær, J., S. (2005). The Industrial Dynamics of Open Innovation — Evidence from the transformation of consumer electronics. *Research Policy*, 34, 10 (December): 1533-1549.

Cooke, P. (2005). Regionally asymmetric knowledge capabilities and open innovation: Exploring 'Globalisation 2'—A new model of industry organisation. *Research Policy*, 34, 8 (October): 1128-1149.

Cooke, P. (2006). Regional Knowledge Capabilities and Open Innovation: Regional Innovation Systems and Clusters in the Asymmetric Knowledge Economy. In Breschi, S., & Malerba, F. (Eds.), *Clusters, Networks & Innovation*. Oxford, UK: Oxford University Press.

Dittrich, K., & Duysters, D. (2007). Networking as a Means to Strategy Change: The Case of Open Innovation in Mobile Telephony. *Journal of Product Innovation Management*, 24, 6 (November), 510-521.

Dodgson, M., Gann, D., & Salter, A. (2006). The role of technology in the shift towards open innovation: the case of Procter & Gamble. *R&D Management*, 36, 3 (June): 333-346.

Gemünden, H. G., Salomo, S., & Hölzle, K. (2007). Role Models for Radical Innovations in Times of Open Innovation. *Creativity and Innovation Management*, 16, 4 (December), 408-421.

Héraud, J.-A., & Lévy, R. (2005). University-industry relationships and regional innovation systems. In Llerena, P., & Matt, M. (Eds.), *Innovation policy in a knowledge-based economy* (pp. 193–219). Berlin, Heidelberg: Springer-Verlag. doi:10.1007/3-540-26452-3_8

Kirschbaum, R., (2005). Open Innovation in Practice. *Research-Technology Management*, 48, 4 (July-August): 24-28.

Maula, M., Keil, T., & Salmenkaita, J.-P. (2006). Open innovation in systemic innovation contexts. In H. Chesbrough, W. Vanhaverbeke & J. West (Eds.), *Open Innovation: Researching a New Paradigm* (pp. 241-257) Oxford, UK: Oxford University Press.

OECD. (1999). *Boosting Innovation: The Cluster Approach*. Paris, France: OECD publishing.

OECD. (2001). *Innovative Clusters: Drivers of National Innovation Systems*. Paris, France: OECD publishing.

OECD. (2005). *Building Competitive Regions*. Paris, France: OECD publishing.

OECD. (2005). *Business Clusters: Promoting Enterprise in Central and Eastern Europe*. Paris, France: OECD publishing.

OECD. (2007). *Competitive Regional Clusters: National Policy Approaches*. Paris, France: OECD publishing.

OECD. (2007). *Competitive Cities in the Global Economy*. Paris, France: OECD publishing.

Perkmann, M., & Walsh. K. (2007). University–industry relationships and open innovation: Towards a research agenda. *International Journal of Management Reviews*, 9, 4, (December), 259-280.

Simard, C., & West, J. (2006). Knowledge networks and the geographic locus of innovation. In H. Chesbrough, W. Vanhaverbeke & J. West (Eds.), *Open Innovation: Researching a New Paradigm* (pp. 220-140) Oxford, UK: Oxford University Press.

Van Pottelsberghe, B. (2008). Europe's R&D: Missing the Wrong Targets? *Intereconomics: Review of European Economic Policy, 43*(4), 220–225. doi:10.1007/s10272-008-0254-y

Vanhaverbeke, W. (2006). The Inter-organizational Context of Open Innovation, In H. Chesbrough, W. Vanhaverbeke & J. West (Eds.), *Open Innovation: Researching a New Paradigm* (pp. 205-219) Oxford, UK: Oxford University Press.

West, J. (2006). Does Appropriability Enable or Retard Open Innovation? In H. Chesbrough, W. Vanhaverbeke & J. West (Eds.), *Open Innovation: Researching a New Paradigm* (pp. 109-133) Oxford, UK: Oxford University Press.

West, J., & Lakhani, K. (2008). Getting Clear About the Role of Communities in Open Innovation. *Industry and Innovation, 15*, 3. doi:10.1080/13662710802033734

KEY TERMS AND DEFINITIONS

OECD: Organisation for Economic Co-orporation and Development.

Colbertist: Espousing qualities of economic policies of Jean Baptiste Colbert (1619-1683), a French Politician and Controller General of Finances of France (1665-1683) under King Louis XIV. In particular, economic policies or programmes that support creation of commerce for national wealth. Sometimes associated with protectionism.

Dirigiste: An approach to management of economic and social affairs which emphasises direction by highly-centralized state organs.

Etatistic: Relating to political movements and trends which support development of state-led and directed growth.

Globalisation: The process by which regional and national economies, societies and cultures have become integrated through a global network of communication, transportation and trade.

Protectionism: Government-led protection of domestic industry from external competition by placing of duties or quotas or similar means on imports.

ENDNOTES

1 In Canada, for example, the National Research Council (http://www.nrc-cnrc.gc.ca/eng/about/index.html) was established in 1916 and in the post-war era the Council's technology-develop activities were significantly expanded with the aid of national investment. Similarly, in Britain, Industrial Research Associations were formed and supported, encouraging exchanges and collaborations, and in Germany, "basic research of a high international standard" was carried out at the Max Planck Gesellschaft, created in 1948 as a successor to the Kaiser Wilhelm Society for the Advancement of Science (http://www.mpg.de/english/aboutTheSociety/history/index.html). In France, state-funded research was concentrated in the CNRS, the Centre National de la Recherche Scientifique, an organization founded in 1939 (http://www.cnrs.fr/en/aboutCNRS/overview.htm) to evaluate and carry out all research capable of advancing knowledge and bringing social, cultural, and economic benefits for society.

2 Members of the Cercle d'outre-Manche, for example, an association of UK-resident French business leaders, have drawn attention to the fact that with roughly the same population as the UK, France has many more degree-granting postsecondary institutions, often monodisciplinary: "Pour le même nombre d'étudiants, il existe 4 305 structures d'enseignement supérieur en France, pour beaucoup monodisciplinaires, contre 169 structures pluridisciplinaires au Royaume-Uni." [For the same number of students, there are 4,305 institutions of higher education in France, largely monodisciplinary, compared against 169 multidisciplinary institutions in the UK] (Boris and Vaissié, 2009).

3 Here we follow standard usage in the innovation literature by using a multi-faceted understanding of knowledge, including dimensions of tacit knowledge, or 'know how" which involves learning and skill but which cannot easily be transferred to others, and explicit knowledge, that which can be articulated or 'codified' and shared easily with others. We recognize also the existence of embodied knowledge held and expressed by agents as expertise or skills operationally embedded in an institution or multi-institution system.

About the Contributors

Hakikur Rahman is an academic over 25 years, and has served leading education institutes and established various ICTD projects funded by ADB, UNDP, and World Bank in Bangladesh. He is currently serving as a Post Doctoral Researcher at the University of Minho, Portugal. He has written and edited over 15 books on computer education, ICTs, knowledge management and research, and contributed over 50 book chapters, journals, and conference proceedings. Graduating from the Bangladesh University of Engineering and Technology in 1981, he has done his Master's of Engineering from the American University of Beirut in 1986, and completed his PhD in Computer Engineering from the Ansted University, BVI, UK in 2001.

Isabel Ramos has a Doctorate degree in Information Technologies and Systems, specialization in Information Systems Engineering and Management (2001) and a Master's degree in Informatics for Management. She is an Assistant Professor in the Information Systems Department of the Minho University, Portugal and Chair of the Information Systems Master Programs of the University. Dr. Ramos is Associate Editor of the International Journal of Technology and Human Interaction and member of the editorial board of Enterprise Information Systems. She is Secretary of the Technical Committee 8 (Information Systems) of IFIP – International Federation for Information Systems and awarded with the IFIP Outstanding Service Award in 2009. She is author and co-author of two books and more than 4 dozen scientific and technical papers.

* * *

Keith Culver is Professor and Econoving International Chair in Generating Eco-Innovation in the UniverSud Paris, France, since July 2009, following twelve years in the University of New Brunswick, Fredericton. His research brings his expertise in legal theory to team-based engagement of interdisciplinary questions regarding innovation, with particular emphasis on the social acceptability of technological innovations. His broad background as a researcher and research team-leader includes several projects funded by Genome Canada, Industry Canada, and the Canadian Network of Centres of Excellence program. Culver has been an invited academic visitor at the universities of Oxford and Stirling, and in 2008 he was McCain Visiting Professor in the University of Edinburgh's SCRIPT law and technology centre. His most recent book, *Legality's Borders,* co-written with Michael Giudice, was published in 2010 by Oxford University Press. He describes himself as a collaborative, problem-focussed researcher who likes best the kinds of problems which cannot be resolved from any one point of view.

Carlos Denner dos Santos Jr. is a Research Fellow in the Horizon Digital Economy Research Institute at the University of Nottingham, UK. He earned his Master's in Strategic Management from UFMG,

Brazil; PhD as a Fulbright Scholar in Management Information Systems from Southern Illinois University at Carbondale, USA; and postdoctorate in Computer Science from the University of São Paulo (USP), Brazil. His research interests include open innovation, free and open source software, adoption of information technology, technology diffusion, systems development, statistical methods, empirical studies, and organizational theory.

Catarina Figueira is Director of Executive MBA Programme and Senior Lecturer in Applied Economics at Cranfield School of Management. She was previously a Robert Schuman Scholar at the Scientific and Technological Options Assessment Division at the European Parliament. Her research interests lie in the areas of financial economics and management, particularly related to financial services' performance, regulatory reforms, and housing markets. Catarina has published in leading international journals as well as produced chapters for several international volumes and is a frequent contributor to a wide range of international conferences. She has advised the OECD, the Asian Development Bank, and the European Commission on regulatory issues and a number of financial institutions on the UK housing market.

Luis Angel Galindo Sánchez has executive MBA from IESE Business School, Master's in Services and Security in IP networks from Technical University of Madrid, Master's in Production Management, and MSc degree in telecommunications from the UPM. From his position, has managed various processes in the area of M&A. In the last, from the acquisition of Tuenti, has led the development of activities necessary to provide MVNO capabilities. Luis is a 'creovator' (creative+innovator) bringing innovation to the user. Luis is the founder and leader of various initiatives (WIMS 2.0, Innovation 2.0, etc). A regular writer in journals on business and strategy issues, he is also the author of several books. He conducts research to define a new sustainable business model in today's society. Luis performs in-company courses to outside companies addressing issues from organizational behavior to socio-organizational aspects in the transformation of a traditional company to a '2.0' company. He is also a freelance consultant of business and strategy.

Robert Girling is a member of the faculty at Sonoma State University's School of Business and Economics. He received his PhD from Stanford University and has taught and consulted in 20 countries. His previous professional experience includes service on the faculty at the Federal University of Bahia, and the Federal University of Minas Gerais in Brazil, the University of the West Indies, and American University in Washington, DC. He has consulted with the World Bank, USAID, the United Nations, the Caribbean Development Bank, and the International Center for Research on Women. He previously served as Director of Regional Development for the Government of Jamaica, president of the California Faculty Association, Sonoma, and on the boards of non-governmental organizations. Dr. Girling is the author of over 50 articles and books including Structures of Dependency (1973); Multinational Institutions and the Third World (1985); Education; Management and Participation(1990); A Escola Participativa (2006). He was the International Project Director at the Federal University of Bahia, Brazil and a faculty member on the University at Sea. He has served as a Fulbright Senior Scholar.

Marcio Augusto Gonçalves is Senior Lecturer in the Business Department of the Universidade Federal de Minas Gerais – UFMG, Brazil. He earned his PhD in Business from Aston University –Aston Business School, UK in 2002. His research interests include finance, the impact of accounting information systems in business management, methods of performance measurement, the use of cost information as meta-information in the health sector, and organizational theory applied to IT.

Matthew Heim is the President of NineSigma – the world's leading open innovation firm. Matthew has over 25 years of global strategy and innovation management experience. Prior to joining NineSigma, Matthew served as Managing Partner at Clarkston Consulting, Managing Director at Price Waterhouse Coopers, and Principal at KPMG Management Consulting in Frankfurt Germany. Matthew served on the faculty of several major universities, and has written and co-authored many articles, books, and conference materials. Matthew has earned a Bachelor of Science degree in Computer Science from the University of Maryland, a Master of Science degree in Systems Management from the University of Southern California, a Master of Arts degree in Human Development, and PhD in Human and Organizational Systems from the Fielding Graduate University.

A.T. Juntunen is a Senior Officer at the Finland's National Police Board. She has graduated from the Marketing and Management Department at the Helsinki School of Economics, and is a post-graduate student at the Department of Political and Economic Studies at the University of Helsinki, and her research interests include innovation and business networks, SMEs and new business development, E-Government, and e-services.

Andrew Kakabadse is Professor of International Management Development at Cranfield School of Management. He is Visiting Professor at the University of Ulster, at Macquarie Graduate School of Management, and at Swinburne University of Technology, Melbourne, Australia. He is also Visiting Scholar in Residence at Thunderbird, School of Global Management, USA. Andrew has consulted and lectured in the UK, Europe, USA, SE Asia, China, Japan, Russia, Georgia, the Gulf States, India, and Australia. His areas of interest focus on improving the performance of top executives and top executive teams, excellence in consultancy practice, leadership, corporate governance, conflict resolution, and international relations. He has been awarded a £2 million research grant to examine Governance and Leadership in the private sector and with governments. Andrew has published 32 books, over 200 journal articles, and 18 monographs. Best-selling books include: Essence of Leadership, Politics of Management, and Working in Organisations. Andrew is co-editor of the *Journal of Management Development* and *Corporate Governance: The International Journal of Business in Society*.

Nada Kakabadse is Professor in Management and Business Research at the University of Northampton Business School. She has co-authored seventeen books, including *Rice Wine With The Minister: Distilled Wisdom to Manage, Lead and Succeed on the Global Stage*, (2010); *Global Boards: One Desire, Many Realities* (2009); *Citizenship: A Reality Far From Ideal* (2009); *Leadership Teams: Developing and Sustaining High Performance* (2009); *Leading the Board: The Six Disciplines of World-Class Chairman* (2008) and *Leading for Success: Seven Sides to Leadership* (2008). She has also contributed with over 60 chapters to international volumes and has published over 150 scholarly articles. Nada has worked for international organisations in Scandinavia, the Middle East and North Africa, as well as the Canadian Federal Government. In a consulting capacity, her clients in the private sector have included Alliance & Leicester, Citigroup, Microsoft, Motorola, Vodafone Australia, and in the public sector, a number of UK government departments.

Fabio Kon is an Associate Professor of the Department of Computer Science at the University of São Paulo. In 2000, he received a PhD degree in Computer Science from the University of Illinois at Urbana-Champaign in the area of Automatic Configuration of Component-Based Distributed Systems. His research interests include distributed systems, agile software development, grid and cloud computing, health informatics, and computer music. He is, currently, one of the Directors of the Open Source Initiative (OSI) and the Director of the USP FLOSS Competence Center.

Elham Metwally is an Assistant Professor of Strategic Management and Entrepreneurship at Misr International University. She is also an Adjunct Professor of Human Resources Management for Government and Non-profit Organizations at the American University in Cairo, Egypt. She earned both her Doctorate of Business Administration (2009) and Master of Philosophy (2006) with distinction, from Maastricht School of Management, the Netherlands. She earned her Master of Business Administration (2002) and Bachelor of Arts in Economics (1982) with honors, from the American University in Cairo. Dr. Metwally is a member of the Academy of Management in USA, and on the Executive Committee of Africa Academy of Management (AFAM), an affiliate of the Academy of Management. She is also a member of the Middle East Council for Small Business & Entrepreneurship (MCSBE), Holland Alumni Network in the Netherlands Organization for International Cooperation in Higher Education (NUFFIC), and the National Association of Student Financial Aid Administrators in USA. She has several publications, and her research and teaching interests include strategic management, organizational behavior, human resources, organizational development, banking, and entrepreneurship and small businesses. Currently, Dr. Metwally is also the Associate Director of Student Financial Affairs for Scholarship Management and Budgeting at the American University in Cairo. Earlier, she made more than a dozen years of experience in the realm of banking, namely in the Hong Kong and Shanghai Banking Corporation (HSBC).

Natasha Katuta Mwila is a graduand in an honours degree in business science management, expected in February 2011 from Monash University. She is a first class honours graduand, having completed her course in the upper second percentile of her class and receiving recognition on the dean's honours list. She graduated in a Bachelor of Business and Commerce degree majoring in the disciplines of Management and Economics in 2010 from Monash University. She completed this course in the upper third to fifth percentile of her class and received a dean's commendation. Her research interests lie in innovation management, the central topic of her honours thesis and forthcoming doctoral thesis. Hailing from the developing African nation of Zambia, she has particular interest in small and medium sized enterprises as they provide promise for the much sought after economic emancipation of less developed countries.

Amiram Porath is an expert on Innovation focusing on Collaborative Research and cooperation between Research and Society. Following a career in the Israeli high-tech industry, Dr. Porath founded the EU office at Tel Aviv University (TAU) combining this work with the function of a Technology Transfer Officer for TAU. Dr. Porath consulted the EU Commission by serving on several relevant expert committees, performing among other task the appraisal of EU member states research systems. He contributed to several research agreements, and was recognized for this by the Italian Republic in 2008 by a knighthood (Cavaliere of the OSSI). He published several topic related articles (www.amiramporath.co.il), relating to knowledge management, IPR, economics, and policy. Last year his first book, *Collaborative Research in Management Inside Out*, came out in print. The book has been described in the press as: "a ready reference for understanding the concept of collaborative research and its practical applications" (Deccan Herald, January 2011).

Denis Remon has taught at the university level for a period of 10 years, following his PhD in Education. In 2004, he began business administration studies, completing first a MBA, then a DBA (Doctor in Business Administration). Through his doctoral thesis in Business Administration, the author has studied Open Innovation in Small and Medium size Enterprises (SMEs), using organizational capacities to understand how SMEs succeed when implementing open innovation principles, practices, and policies. The author does consultancy with SMEs since 2005.

Rachael Ritchie is the Project Manager (Europe), Econoving International Chair in Generating Eco-Innovation in the UniverSud Paris, France, since February 2010, following eleven years at a Provincial Research Organization in New Brunswick, Canada. Ritchie is a life sciences Researcher and Manager (D.Phil, Oxford and postdoctoral training) working at the border between science and its industrial application to improve the productivity and competitiveness of individuals, businesses, and communities. Ritchie has worked with SMEs, midsized companies, and international corporations to help them identify, develop, apply, and implement technical solutions to a range of business and economic development challenges. In addition to her formal education in molecular biology in the field of human health, she has considerable experience in the aquaculture, fisheries, and food sectors, and has recently expanded to the energy and water sectors.

Nicholas Theodorakopoulos lectures at Aston Business School and is Director of both the MSc HRM and Business (PT) and the MSc Strategic HRM for Professionals programmes. His award winning PhD is in the field of Strategic Entrepreneurship. His specialist areas of research, academic publications, and teaching include strategic management, entrepreneurship, and innovation and diverse entrepreneurship. Apart from contributing to a number of books, his papers have been published in reputable academic journals, including *Work, Employment and Society*, *International Journal of Urban and Regional Research*, *Public Administration*, *International Journal of Entrepreneurial Behaviour and Research*, and *International Journal of Entrepreneurship and Management*. He has over 10 years of business experience in the private and public sectors, in the UK and abroad, prior to his academic career. Furthermore, he has acted as a consultant in the areas of new venture strategic planning and SME development for public and private sector organisations.

Kam Hou Vat is currently a Senior Instructor in the Department of Computer and Information Science, under the Faculty of Science and Technology, at the University of Macau, Macau SAR, China. His current research interests include learner-centered design with constructivism in Software Engineering education, architected applications developments for Internet software systems, information systems for learning organization, information technology for knowledge synthesis, and collaborative technologies for electronic organizations and virtual communities.

Simona Mihai Yiannaki, is Assistant Professor and Programme Coordinator for Banking and Finance at the European University Cyprus. Dr. Mihai-Yiannaki is a former International Banker, C.F.O., and Economist. She is also an Executive Banking Consultant and Trainer. Her research interests are in the area of risk management, M&As, corporate and SMEs banking, bank bailouts, creative education, interdisciplinary knowledge and cross-cultural management. She has led and participated in EU funding research projects (TEMPUS IV, ERASMUSIP) and bottom up M&A transactions in Europe. She is also Country Director for Cyprus and track chair in the EuroMed Business Research Institute and EMAB, Conference session organiser on Risk Management, and editor in various international Journals, like *GBER,* and *Megabyte*.

Index

S

T

U

V

W

X

Z

www.ingramcontent.com/pod-product-compliance
Lightning Source LLC
LaVergne TN
LVHW062139300926
849161LV00010B/2578